Handbook of Genitourinary Medicine

EDITED BY

S.E. Barton

Consultant and Clinical Director
HIV/Genitourinary Medicine Services
Chelsea and Westminster Hospital
Fulham, London

AND

P. Hay

Senior Lecturer and Honorary Consultant
Department of Genitourinary Medicine
St George's Hospital Medical School
London

A member of the Hodder Headline Group
LONDON • SYDNEY • AUCKLAND
Co-published in the USA by
Oxford University Press Inc., New York

First published in Great Britain in 1999 by
Arnold, a member of the Hodder Headline Group,
338 Euston Road, London NW1 3BH

http://www.arnoldpublishers.com

Co-published in the United States of America by
Oxford University Press Inc.,
198 Madison Avenue, New York, NY10016
Oxford is a registered trademark of Oxford University Press

©1999 Arnold

All rights reserved. No part of this publication may be reproduced or transmitted in any form or by any means, electronically or mechanically, including photocopying, recording or any information storage or retrieval system, without either prior permission in writing from the publisher or a licence permitting restricted copying. In the United Kingdom such licences are issued by the Copyright Licensing Agency: 90 Tottenham Court Road, London W1P 9HE.

Whilst the advice and information in this book are believed to be true and accurate at the date of going to press, neither the authors nor the publisher can accept any legal responsibility or liability for any errors or omissions that may be made. In particular (but without limiting the generality of the preceding disclaimer) every effort has been made to check drug dosages; however, it is still possible that errors have been missed. Furthermore, dosage schedules are constantly being revised and new side-effects recognized. For these reasons the reader is strongly urged to consult the drug companies' printed instructions before administering any of the drugs recommended in this book.

British Library Cataloguing in Publication Data
A catalogue record for this book is available from the British Library

Library of Congress Cataloging-in-Publication Data
A catalog record for this book is available from the Library of Congress

ISBN 0 340 74084 1

1 2 3 4 5 6 7 8 9 10

Composition by Genesis Typesetting, Rochester, Kent
Printed and bound in Great Britain by The Bath Press, Bath

What do you think about this book? Or any other Arnold title?
Please send your comments to feedback.arnold@hodder.co.uk

My dedication is to the late Dr Jimmy Fluker who inspired his students to try to emulate the knowledge and compassion which he used in his practice of medicine.

Simon E. Barton

My dedication is to Professor David Taylor-Robinson with whom it was a privilege to work. He combined scientific rigour with humanity and made research stimulating and enjoyable.

Phillip E. Hay

Contents

Colour plates appear between pages 86 and 87.

List of contributors	vii
Preface	ix

1. Introduction to history taking and principles of sexual health — S.E. Barton — 1

2. The clinical examination and how to obtain specimens — A.J. Robinson — 5

3. Diagnostic procedures in genitourinary medicine: practical laboratory aspects — D. Taylor-Robinson, B. Thomas and C. Ison — 19

4. Principles and problems of organizing clinical trials in genitourinary/HIV medicine — V.S. Kitchen and G.P. Taylor — 49

5. Establishing a new genitourinary medicine clinic service — J.S. Bingham — 59

6. Setting up genitourinary medicine clinic services in a developing country — Y. Adu-Sarkodie — 69

7. The diagnosis and management of genital dermatological conditions — B.T. Goh — 73

8. The diagnosis and management of vaginal discharge — P.E. Hay — 83

9. The diagnosis and management of genital ulceration — S. McCormack — 97

10. The diagnosis and management of urethral discharge in males — P.J. Horner and R.J. Coker — 123

11. Lumps and bumps of the external genitalia: diagnosis and management — M. Murphy and C.J.N. Lacey — 139

12. The diagnosis and management of prostatitis — C. O'Mahony — 149

13. The diagnosis and management of pelvic inflammatory disease — S.N. Mann, J.R. Smith and S.E. Barton — 161

14. The diagnosis and management of psychosexual problems in genitourinary medicine — P. Woolley — 171

15. Management of asymptomatic and early symptomatic patients with HIV infection (with special reference to antiretroviral therapy and viral resistance patterns) — D.A. Hawkins, G. Moyle and M.S. Youle — 185

16. The management of opportunistic infections in AIDS — A. Pozniak — 211

17. The diagnosis and management of tumours in patients with AIDS — F.C. Boag and E.M. Carlin — 255

Index — 269

Contributors

Y. Adu-Sarkodie
  Genitourinary Physician and Lecturer in Clinical Microbiology, School of Medical Sciences, University of Science and Technology, Kumasi, Ghana, and Head, Public Health Reference Laboratory, Ministry of Health, Kumasi, Ghana, West Africa

S.E. Barton
  Clinical Director and Consultant Physician, Directorate of HIV/Genitourinary Medicine, Chelsea and Westminster Hospital, London, UK

J.S. Bingham
  Consultant in Genitourinary Medicine, Department of Genitourinary Medicine, Guy's and St. Thomas's Hospitals, London, UK

F.C. Boag
  Consultant HIV Genitourinary Medicine Physician, Chelsea and Westminster Hospital, London, UK

E.M. Carlin
  Consultant Physician, Department of Genitourinary Medicine, Nottingham City Hospital, UK

R.J. Coker
  Consultant in Genitourinary Medicine, Jefferiss Wing, St Mary's Hospital, London, UK

B.T. Goh
  Consultant Physician, Department of Genitourinary Medicine, Ambrose King Centre, Royal London Hospital, London, UK

D.A. Hawkins
  Consultant Physician, Directorate of HIV/Genitourinary Medicine, Chelsea and Westminster Hospital, London, UK

P.E. Hay
  Senior Lecturer and Honorary Consultant, Department of Genitourinary Medicine, St George's Hospital Medical School, London, UK

P.J. Horner
  Consultant in Genitourinary Medicine, The Milne Centre, Bristol Royal Infirmary, Bristol, UK

C. Ison
  Department of Medical Microbiology, Imperial College School of Medicine at St Mary's Paddington, London, UK

V.S. Kitchen
  Senior Lecturer and Honorary Consultant in Genitourinary Medicine and Communicable Diseases, Imperial College School of Medicine, London, UK

C.J.N. Lacey
  Senior Lecturer, Genitourinary Medicine and Communicable Diseases, Imperial College School of Medicine, Jefferiss Research Trust Laboratories, London, UK

S.N. Mann
  Research Fellow, Directorate of HIV/Genitourinary Medicine, Chelsea and Westminster Hospital, London, UK

S. McCormack
Consultant Physician in HIV/Genitourinary Medicine, Chelsea and Westminster Hospital, London, UK

G. Moyle
Associate Specialist, Directorate of HIV/GU Medicine, Chelsea and Westminster Hospital, London, UK

M. Murphy
Infection and Immunity Group, St Bartholomew's Hospital, London, UK

C. O'Mahony
Consultant Physician in Genitourinary Medicine, Countess of Chester Hospital, Chester, UK

A. Pozniak
Consultant in HIV/Genitourinary Medicine, Chelsea and Westminster Hospital, London, UK

A.J. Robinson
Consultant and Senior Lecturer, Department of Genitourinary Medicine, University College, London, UK

J.R. Smith
Consultant Gynaecologist, Chelsea and Westminster Hospital, London, UK

G.P. Taylor
Senior Lecturer in Genitourinary Medicine and Communicable Diseases, Imperial College School of Medicine, London, UK

D. Taylor-Robinson
Department of Genitourinary Medicine, Winston Churchill Wing, Imperial College School of Medicine at St Mary's Paddington, London, UK

B. Thomas
Department of Genitourinary Medicine, Winston Churchill Wing, Imperial College School of Medicine at St Mary's Paddington, London, UK

P. Woolley
Department of Genitourinary Medicine, Withington Hospital, Manchester, UK

M.S. Youle
Director HIV Clinical Research, Royal Free Hospital, London, UK

Preface

The specialty of genitourinary medicine is developing rapidly. Specialists are confronting an explosion of viral infections, including the AIDS epidemic, and considerable increases in the numbers of patients with genital herpes, genital warts and hepatitis virus infections. Increasing ease of travel produces a variety of hitherto exotic imported infections. The availability of new molecular and immunologically based techniques for the diagnosis and management of viral infections has revolutionized clinical practice. Such techniques even offer the prospect of improved sensitivity and specificity in the diagnosis of chlamydial and other bacterial infections.

In this book we have aimed to bring together contemporary aspects of genitourinary medicine from the field of sexual health through to the specialized management of HIV infection. Topics have been chosen specifically to reflect areas of practice which are changing, to produce an up-to-date and clinically based guide for the practitioner.

The contributors have been chosen for their expertise both in basic and clinical research, as well as their recognized skill in communicating their ideas. All are experienced researchers and teachers who have been encouraged to develop their own style of communication. This should make the book varied and stimulating for the reader.

We have not set out to produce a comprehensive reference tome, but to stimulate the reader. Therefore we have not included photographs of genital infections which are readily available in colour atlases. We hope that our readers, whether trainees or experienced specialists, will be stimulated to review their practice in the management of sexually transmitted infections and that they will develop new ideas and concepts for further research.

We would ask you to let us know directly of any errors, or even passages which you disagree with, but above all we hope the book will be informative as well as challenging.

Simon E. Barton
Phillip E. Hay

1 Introduction to history taking and principles of sexual health

S.E. Barton

SEXUAL HEALTH CARE

The global ramifications of the HIV pandemic have, over the past decade, focused much attention on the promotion of sexual health and improvements in services for the treatment of sexually transmitted diseases. The recognition of the synergistic effect which ulcerative and inflammatory sexually transmitted infections exert on the transmission of HIV have, in the absence of an effective anti-HIV vaccine, led to public health-care strategies focusing on the prevention of other sexually transmitted diseases (STDs). These infections have constituted a global health problem for considerably longer than the AIDS epidemic.

The majority of previous efforts targeted against STDs have involved attempts to improve the treatment, follow-up and, in particular, contact tracing of those diagnosed as having a venereal infection. But, as with HIV infection (Table 1.1), there are many other conditions that are transmitted sexually which can remain asymptomatic yet be infectious, either continuously or intermittently throughout their natural history. The morbidity and mortality associated with HIV/AIDS exemplify the severity of the sequelae of such an infection.

A hundred years ago, before the discoveries of salvarsan and then penicillin, syphilis was an untreatable condition producing many visible manifestations leading to many of the same social phenomena HIV does today. It is self-evident that HIV is just one of many conditions which may lead to silent pathology followed by visible clinical symptoms and signs. Furthermore, ectopic pregnancy, hepatoma, invasive cervical carcinoma and pelvic sepsis following failed attempts at unwanted pregnancy terminations cause mortality and morbidity worldwide, yet often fail to get as much media attention or become the subject of public health initiatives.

Table 1.1 Examples of sexually acquired conditions which may be asymptomatic or subclinical during their natural history, but cause significant morbidity and/or mortality

HIV infection > immunosuppression > AIDS
Chlamydia trachomatis > PID > ectopic pregnancy/infertility
Hepatitis B infection > chronic hepatitis > hepatoma
Sperm > unwanted pregnancy > abortion > postabortal sepsis
Oncogenic HPV cervical infection > CIN > cervical cancer
Treponema pallidum > latency > tertiary syphilis manifestations

CONTROL OF STD

In the light of the above, a more integrated multidisciplinary and preventative approach to the control of sexually transmitted diseases is clearly indicated. This has been embodied within the concept of 'sexual health'. In conjunction with HIV/AIDS, sexual health was identified as a UK *Health of the Nation* initiative, as one of the major five focus areas for action to improve health. This approach advocates enhancing the historical approach of treating symptomatic conditions more effectively, along with improvement in awareness and knowledge about STDs, amongst both doctors and patients. It emphasizes the need to improve the multidisciplinary and multispecialty collaboration in detection and management of sexually transmitted diseases – for instance, screening for sexually transmittable pathogens amongst women attending for termination of pregnancy, as well as offering improved family planning advice to women attending with STDs. A study in our own unit found that up to 70% of women attending our genitourinary (GU) medicine clinic were not using a reliable method of contraception. Unwanted pregnancy is certainly a sexually transmitted condition and, as such, must not be ignored in any health-care setting. Improving access to services by tailoring clinic hours, facilities and approaches to cater for consumer needs will increase the number of asymptomatic individuals who feel able to use the service for a check-up. Such a check-up needs to include, with consent, screening for a whole range of local and systemic infections as well as providing advice, education and information, with emphasis on promoting health. Where conditions, symptomatic or asymptomatic, are diagnosed, it is vital that adequate follow-up and especially contact tracing are carried out.

The five principles of improvements in sexual health care can be summarized as follows:

1. improved access and awareness;
2. improved diagnostic techniques;
3. improved therapies;
4. improved follow-up and contact tracing;
5. availability of the above to wider section of the population.

Finally, sexual health involves more than just patients being educated and informed to change their ideas about the prevention and treatment of STDs. It is vital that the medical, nursing and paramedical professions do not see sexual matters as being just the province of GU medicine clinics or dermatovenereologists and, sometimes, gynaecologists and urologists. How best to achieve this improved coordination and collaboration has been the subject of considerable discussion, as exemplified in a *British Medical Journal* editorial by Stedman and Elstin.[1] The range of correspondence generated from doctors in different specialties[2] illustrates the diversity of thinking about the best approach to this. It is most likely that a range of approaches with a common goal will be individually tailored to the needs of different local populations and the resources available. Primary care physicians, as well as those working in the above specialties and accident and emergency medicine and student health, need to work together to utilize all their expertise to improve the sexual health of the nation.

SEXUAL HISTORY TAKING

The most important aspect of the taking of the sexual history is that the patient needs to be reassured that privacy and confidentiality are paramount. It is essential to ensure that the consultation cannot be overheard or disturbed by others. Only experience and skill in taking a sexual history can sympathetically encourage the patient to reveal details about their private life which will be of use in the subsequent choice of examination, laboratory investigations, diagnosis, follow-up and, in particular, contact tracing. All information sought should be justifiable on one or other of these grounds. In some cases, it will often help to be explicit about the reasons, to allay the patient's fear or embarrassment about answering a specific question. The standard pro-forma for sexual history taking is shown in Table 1.2.

Sexual history taking holds the key to the proper practice of GU medicine for several reasons. It provides the firm base for gaining information about the patient's risks and routes of acquiring an infection as well as setting the agenda for issues concerning specific risks to be further explored. It does this against the background of reassuring the patient concerning your professionalism and the maintenance of confidentiality and privacy within the clinic situation. It also raises questions for the patient, who will often interject questions such as 'What if I do/did have that

Table 1.2 Sexual history taking: an example of a structured approach

1. When did you last have sexual intercourse?
2. With a man or a woman?
3. Was he/she a new partner?
4. What types of intercourse did you engage in?
5. For each type, e.g. oral/vaginal/anal, were barriers/condoms used? (For heterosexual sex, was any contraception used? Relate to risk of pregnancy by asking LMP)
6. Does your partner have any symptoms?
7. Have you had any other partners in the past six weeks? If so, return to Q2. Offer questions which may lead into other problem identification.
8. Did you have pain during or after intercourse?
9. Have you ever had any previous STD or other infection? (May require information or counselling.)
10. Have you ever had a sexual health check-up before? (If not, explain and offer.)
11. Have you had an HIV/hepatitis B/syphilis test before? (If not, offer.)
12. Have you been vaccinated against hepatitis B? (If not, assess risk and offer.)

infection?', which leads on to discussion of what effects it might have on them and others. As such, the sexual history taking, although taking only a few minutes, can often be the most potent stimulus for the transmission of information, counselling and support to the patient in front of you.

It would be wrong to say that any individual physician will ever be completely skilled in sexual history taking. Within the clinical situation, there will always be an element of risk in raising subjects and topics which the patient may find difficult or painful to discuss. The well-prepared and trained physician will be sympathetic and understand the sensitivity with which questions must be asked. Occasionally, it may be necessary to terminate the history taking until a second visit or even later during follow-up, when the patient is more fully ready and able to give details of their story. In some cases patients will never find this possible. Even in such cases, it is of paramount importance that as much information as possible is gained, that as many different infections as possible are screened for and that as much follow-up and contact tracing is done as possible. Only by these methods can the best attempts be made at improving the sexual health of our patients, their sexual partners and their partners' past, present and future sexual partners.

In different settings, such as primary care, sexual history taking may be much more difficult to achieve. The different skills of the doctors, the different time constraints and differences in perceived confidentiality may all influence both general practitioners and their patients. This has been clearly analysed in an excellent study by Curtis et al.[3] Despite these drawbacks, until improved awareness of sexual problems, sexually transmitted diseases and the importance of people's sex lives impacts on more elements of practice in hospitals, primary care and other settings, attempts to improve the nation's sexual health will be fragmented.

One of the most important issues which can underpin this will be sex and HIV education in schools. Despite occasional episodes of media outrage complaining of inappropriate sex education 'encouraging' teenagers to experiment in sexual behaviour, great improvements have been made for the provision of sex and HIV/AIDS education in schools and these have been shown not to hasten the onset of sexual experience.[4] Clearly, thorough education about sexual health can only be achieved if this starts before the majority of individuals commence their sexual experiences; 50% of schoolchildren do so by the age of 16. It is vital that health-care professionals recognize this most important part of the drive towards improved sexual health.

CONCLUSION

The concept of sexual health emphasizes the need to promote health by education and information. It recognizes the fact that many sexually transmitted infections (including HIV) are asymptomatic for a considerable part of their natural history and encourages a further step in the public and medical acceptance that sexual activity is and should be a healthy and fulfilling behaviour. Furthermore, with clear evidence[5] that the treatment of STDs in a developing country can significantly reduce the new acquisitions of HIV infection, it is vital that financial resources designed to reduce the spread of HIV infection should be targeted at improving sexual health and reducing the number of sexually transmitted diseases in general.

REFERENCES

1. Stedman, Y. and Elstin, M. (1995) Rethinking sexual health clinics. *Br. Med. J.*, **310**, 342–343.
2. See eight letters in *British Medical Journal* (1995), **310**, 1193–1195.
3. Curtis, H., Hoolaghan, T. and Jewitt, C. (eds) (1995) *Sexual Health Promotion in General Practice*. Radcliffe Medical Press, Oxford.
4. Wellings, K., Wadsworth, J., Johnson, A.M. et al. (1995) Provision of sex education and early sexual experience: the relation examined. *Br. Med. J.*, **311**, 417–420.
5. Grosskurth, H., Mosha, F., Todd, J. et al. (1995) Impact of improved treatment of sexually transmitted diseases of HIV infection in rural Tanzania; randomised controlled trial. *Lancet*, **II**, 530–536.

2 The clinical examination and how to obtain specimens

A.J. Robinson

THE IMPORTANCE OF CLINICAL EXAMINATION

An adequate clinical examination is central to the practice of genitourinary (GU) medicine. The history, essential in any medical consultation, may not be as accurate at predicting the most likely diagnosis as in other medicine specialties. This is because it is often difficult for the patient to look at the area involved, symptoms are rarely specific for one particular condition and many sexually transmitted diseases (STDs) can remain latent or asymptomatic.

A complete physical examination is not required routinely for all patients. Some sexually transmitted diseases such as syphilis and the acquired immunodeficiency syndrome (AIDS) have systemic manifestations. Sometimes, either the initial history or the findings of the genital examination will indicate the need for a more extensive assessment. For example, the combination of polyuria and vulvovaginitis may indicate a search for manifestations of diabetes mellitus. It is essential that the genital examination is thorough and systematic to avoid missing subtle signs. With experience, this can still be performed quickly.

Privacy during the consultation is essential, particularly as the examination involves the exposure of intimate areas of the body. Some patients like to see a doctor of the same sex and where possible this should be accommodated. A skilled chaperone helps the patient to relax and can assist in specimen collection and correct labelling. Warm hands and a confident manner will help your patients to relax. Avoid sudden noises or intrusions and do not leave them waiting on the couch. For the examiner, a good source of light, a comfortable seating arrangement and well-fitting disposable gloves are essential.

A good laboratory facility for immediate diagnosis ('hot lab') must be an integral part of the first evaluation of the patient to provide a good quality service. Laboratory facilities for confirmation of initial diagnoses and more extensive investigations are also necessary. Remember that the results from the laboratory can only reflect the quality of the specimens received.

EXAMINATION OF WOMEN

Menstruation is not a contraindication to examination and swab taking, although some of the tests may not be as useful. Blood makes the microscopy of slides of vaginal and cervical material more difficult. It makes vaginal fluid more alkaline, rendering the pH test invalid. It does not usually interfere with specimens for culture, but the performance of some antigen detection tests for *Chlamydia trachomatis* can be reduced. With direct immunofluorescence false-positive results may occur. Urethral tests should not be affected by menstruation. Nevertheless, some women may find it too embarrassing to be examined at this time and the examination may have to be deferred.

Many women find the genital examination an ordeal. The development of good rapport between doctor and patient during the history taking and reassurance at the start of the examination may help the patient to relax, making the experience more comfortable. Time spent waiting for the doctor in a half-undressed state only serves to increase anxiety and is counterproductive.

Nurses play an important role. A female chaperone must be present with a male doctor to avoid any misunderstandings which could result in legal action. For such an intimate examination, a chaperone should be present irrespective of the gender of the doctor. Nurses can amplify information already given to the patient and help with education, especially in relation to safer sex and health promotion. They usually help patients into the correct position for the examination.

Genital examination

The genital examination is performed most easily with the patient in the lithotomy position. The knees are supported on knee rests or with feet in stirrups, depending on the type of examination couch. This may leave the woman feeling helpless and vulnerable. A degree of modesty can be ensured by the use of a blanket or drapes. A full explanation of the examination procedure is also helpful.

First, the inguinal region and surrounding area should be inspected. Palpate for evidence of lymphadenopathy, noting the distribution, consistency and tenderness of any nodes. Inspect the pubic area and surrounding skin for evidence of pubic lice, scabetic nodules, other infestations, folliculitis or other skin lesions.

Next, focus on the external genitalia and introitus. Gently separate the labia majora and labia minora, wiping off any discharge with cotton wool or a gauze swab. Look for labial erythema, fissuring, scars, oedema, scratch marks, ulceration and other vulval lesions, e.g. warts. Observe the urethral meatus, particularly the openings of the paraurethral glands on either side (Skene's glands). The orifices of the greater vestibular glands (Bartholin's glands) open on the inner side of the labia minora external to the hymen. By inserting a finger into the vagina and moving it laterally behind and internal to the labium majus, the gland may be palpated if it is fibrotic or enlarged. Massaging the area gently may express any secretion into the duct orifice, which can then be collected with a plastic sampling loop. A smear and culture may then be prepared.

Finally, the perineum and perianal area should be inspected. This necessitates gentle separation of the buttocks. A better view may be achieved by asking the patient to place her feet on the knee rests to raise her buttocks.

Examination of the vagina and cervix

After explaining the procedure to the patient, introduce a warmed Cusco bivalve speculum into the vagina. Whether the speculum is inserted with the handle up or down depends on the type of couch and personal preference of the doctor. However, with the handle up, remember to avoid undue pressure on the *mons veneris* or nipping any pubic hairs. Before insertion the speculum must be fully closed. The vaginal introitus has a larger diameter vertically than horizontally in the resting state. Introduce the speculum at an oblique angle, turning gently into the horizontal as it is inserted. Warn the patient before opening the speculum. If the cervix is difficult to visualize, ask the patient to cough. This increases intraabdominal pressure and often pushes the cervix into view. Avoid undue pressure and stretching caused by opening the speculum too widely. The patient then becomes more tense, which results in the cervix becoming even less accessible. If the cervix cannot be seen, take the speculum out, insert a finger gently into the vagina and feel for the position of the cervix. Sometimes the cervix may be very posterior or deviated laterally. Ascertaining its position enables the speculum to be directed more accurately on reinsertion. Asking the patient to lift her buttocks slightly and resettle can also tilt the pelvis to a better angle for visualizing the cervix.

The speculum can be secured, to leave both hands free, which is necessary if the doctor has to take swabs without an assistant, a circumstance which is not recommended. With practice, controlled tension can be exerted with one hand continually holding the speculum, leaving the other free for swab taking. This lessens the sound of clanking metal which can often perturb the patient. It also makes the whole procedure more speedy. Only open the speculum sufficiently wide to see the cervix clearly. Opening to a greater angle only causes more discomfort for the patient. Whilst taking specimens from the vagina, note the following.

- *Vaginal discharge* – consistency, colour, smell
- *Mucosa of the vaginal walls* – erythema, ulceration, adherence of discharge
- *Cervical discharge* – clear, mucopurulent or purulent
- *Cervix* – the position of the squamocolumnar junction (ectopy) (see Plate 1)
- *Cervical os* – previous parity, tears, previous instrumentation, polyps, IUCD(s), threads
- Cervical mucosa – glazed (e.g. after previous laser treatment), inflamed (cervicitis), Nabothian follicles, warts
- The presence of foreign bodies.

The application of 5% acetic acid to suspicious lesions on the genital skin, vagina and cervix is not recommended as routine practice because of the nonspecific nature of aceto-whitening. In some circumstances it may be helpful but should be performed by an experienced doctor.

Sampling from the vagina and cervix

Each doctor will develop their own system for taking swabs depending on personal preference as well as the microbiology services available. It is important to discuss with the microbiology department the tests to be taken and find out which are the most appropriate transport media and swabs to obtain the best results. The following is a logical, quick and efficient regime which a doctor and nurse working together can accomplish in less than a minute.

Vaginal samples

Plastic sampling loops are most appropriate for taking samples of vaginal discharge. Sweep the loop from the posterior fornix along the vaginal wall. Smear it on to a microscope slide for Gram staining. Then, mix the remaining contents of the loop with a drop of normal saline on another slide to prepare the 'wet mount'. Place a cover slip over the drop of saline. Take a further sample from the vaginal wall with the same loop. Inoculate the culture medium by agitating the swab in the liquid. The culture medium should allow for identification of *Candida albicans* and *Trichomonas vaginalis*.

At this stage a high vaginal sample taken with a cotton-tipped swab from the posterior fornix can be placed into semisolid transport medium (e.g. Amies' or Stuart's) for identification of other organisms if required, e.g. ß-haemolytic streptococci.

pH/KOH test

The normal pH of the vagina is 3.5–4.5. A pH above 4.5 is one of Amsel's[1] criteria for diagnosis of bacterial vaginosis (see Chapter 8). The pH can be measured by collecting vaginal fluid on a sampling loop and applying to narrow range pH paper placed on a piece of gauze. The same loop can be placed on to a microscope slide with a drop of 10% potassium hydroxide solution (see Plate 2). Mix well. In the presence of bacterial vaginosis, a characteristic fishy smell is released. If bloody secretions are present, the pH will already be alkaline but the KOH test should still be valid.

Cervical cytology

A cervical cytology sample must be taken before the cervix is swabbed. Wooden spatulae are cheaper than plastic, but either is satisfactory.[2] The position of the squamocolumnar junction (SCJ) dictates the type of spatula to be used. When there is a large ectropion present, an Ayre's spatula is more appropriate. An Aylesbury spatula is better to ensure that an adequate sample is obtained when the SCJ is in the cervical canal. A cytobrush should be used to take an additional sample if the SCJ is not visible and the os is small.[3] This is often the case when women have had previous excision and/or ablative therapy.

Take the sample by inserting the fulcrum of the spatula into the os. Sweep the spatula around a full 360° to obtain scrapings of the superficial cells. Spread thinly on to a microscope slide. Fix immediately with 95% industrial methylated spirits. The cytobrush specimen can be placed on the same or a separate microscope slide. The cervical cytology laboratory may have particular requirements, so check the policy before undertaking cervical sampling.

Cleaning the cervix

Remove vaginal and excess cervical secretions with a large cotton-tipped swab or gauze held in sponge forceps to avoid contamination of the cervical specimens by vaginal material. This makes the interpretation of direct microscopy of cervical samples much clearer.

Cervical samples

Insert a plastic sampling loop into the endocervix to obtain a specimen of cervical secretion. Avoid contamination with vaginal secretion. Place the loop onto a microscope slide and spread thinly for Gram staining. Then smear the loop across a plate of selective culture medium for identification of *Neisseria gonorrhoeae*. A second sample may be used for the culture although this is not necessary if enough material is obtained initially.

Identification of Chlamydia trachomatis

There are many methods of detection for this organism and it is important to read the manufacturer's instruction for each kit and check with the laboratory. *Chlamydia trachomatis* is an intracellular organism, infecting columnar epithelium. Therefore endocervical cells are required. The appropriate swab should be rotated in the cervical canal for at least 10 s. If there is cervical ectopy, swab around the SCJ also. If culture is the technique for identification, remember that certain materials may inhibit the growth of *C. trachomatis* (e.g. wood).[4] Use a cotton-tipped swab to obtain the specimen. Mix the swab in the correct transport medium, squeeze it out and discard the swab. The swab should not be left in the medium.[4]

Herpes simplex virus (HSV)

At this stage, if indicated, a swab can be taken from the cervical canal for culture of HSV. Take as for *C. trachomatis*, but remember to use the appropriate viral transport medium.

Human papillomavirus

If there is suspicion of wart virus infection of the cervix or upper vagina, washing this area with 5% acetic acid will enable any warts to be seen more clearly as acetowhite areas (see Plate 3). Patients with cervical warts or acetowhite areas should be referred for formal colposcopy. If suspicious vaginal or cervical lesions are seen, with or without the aid of acetic acid, colposcopic examination is recommended.

Removing the Cuscoe speculum

Withdraw the speculum carefully. Keep enough tension to allow the curved ends of the speculum to be clear of the cervix before allowing them to close naturally. The cervix is often sensitive and catching it as the speculum snaps shut can cause considerable pain. Also, beware of trapping pubic hairs or nipping the vaginal walls as the speculum is removed.

Sampling from the urethra

The urethral smear and culture are best taken last because some patients find this test most uncomfortable and become tense, making subsequent speculum insertion more difficult. Removal of the speculum also milks the urethra.

Place a small plastic loop into the urethral orifice (twisting or turning the loop causes severe discomfort and should be avoided). Spread on to a microscope slide for Gram stain and on to selective medium for culture of *N. gonorrhoeae*. Urethral and cervical specimens may be put on different halves of the same slide to economize on cost. The position of the sample should be consistent to avoid misinterpretation of the two specimens.

The yield for identification of *C. trachomatis* may be increased by taking a urethral swab in addition to a cervical swab.[5] A cotton-wool tipped wire swab (ENT swab) can be used. The specimen can either be placed in separate transport medium or in the same transport medium as the cervical swab. For antigen detection methods, the male urethral sampling system can be used to sample the urethra in women. The swab should be inserted far enough to get an adequate urethral sample without causing excessive pain.

Sampling from the rectum

This may be appropriate in women who have had anal intercourse, who have gonorrhoea themselves or are contacts of partners infected with *N. gonorrhoeae*. Omission of rectal sampling can leave 3–5% of infected women undiagnosed.[6]

Sampling for culture of *N. gonorrhoeae* can be done blind or after insertion of a proctoscope. Use of a proctoscope is essential if rectal symptoms are present, microscopy is required and/or culture for other organisms such as herpes simplex virus or *Chlamydia trachomatis*. The small paediatric proctoscope is less traumatic and is useful if there are skin tags, haemorrhoids or fissuring which make insertion more difficult. Lubricate the proctoscope with saline or warm water. The proctoscope can be inserted with the patient in the lithotomy position, but remember to direct the proctoscope slightly anteriorly and towards the umbilicus. Pass a swab beyond the tip of the proctoscope and sample from the rectal mucosa. Note any rectal mucosal lesions or inflammation. Place the swab directly on to selective culture medium for *N. gonorrhoeae*. Direct examination of a Gram-stained microscope slide is not essential and may be misleading,[6] because of the risk of false-positive results. The exact significance of polymorphs on a rectal slide has not been significantly evaluated but some clinicians may decide to treat as 'non-specific proctitis.'

Bimanual pelvic examination

This examination reveals further information about the vagina and cervix. It also allows the assessment of the size and position of the uterus and any abnormalities in the adnexae. It is easier to perform and more reliable if the bladder is empty. Insert the gloved index finger of the right hand into the vagina, closely followed by the middle finger, if possible. Lubricating jelly (e.g. KY water soluble) may be necessary. Locate the cervix. Place the left hand on the abdomen suprapubically. This examination is always uncomfortable. 'Cervical excitation' is present if movement of the cervix causes pain or extreme discomfort (see Chapter 13). The position of the uterus is identified by palpating the fundus of the uterus with the left hand whilst pressure is applied from the fingers inside the vagina. The fundus of the retroverted uterus can be felt through the posterior fornix. The adnexae can be assessed by placing the fingers to the left and right of the cervix. Any mass can be outlined between the fingers of the right hand and the left hand on the abdomen. Swellings may be felt posteriorly in the rectovaginal pouch (pouch of Douglas). A rectal examination may be necessary to evaluate fully any pelvic swellings. An abdominal examination may also be required at this stage.

Urine tests

Simple analysis

At the first visit to the clinic, it is good practice to examine the patient's urine for the presence of glucose, protein, blood, etc. Simple diagnostic methods with dipsticks are available widely (Table 2.1). If any of the tests are abnormal, the urine should be rechecked at a follow-up visit. Should the abnormality persist, the patient requires further investigation.

Table 2.1 Relevance of abnormalities of dipstick analysis in genitourinary medical practice

Test abnormality	Relevance	Action
Glucosuria	Pregnancy Low renal threshold Diabetes mellitus	Exclude Check fasting blood sugar
Proteinuria	False-positive with very alkaline urine Infection – NGU, UTI Renal disease, e.g. nephrotic syndrome	If persists, refer for 24 h urinary protein collection
Haematuria	Menstruation Infection – UTI Traumatic – post swabs Renal tract disease	Urine microscopy if persistent abnormality
Bilirubin	Hepatitis Obstruction of biliary system	Check hepatitis A,B,C Drug history Travel history
Urobilinogen	Hepatitis Haemolysis	

Pregnancy test

A pregnancy test can be performed on a urine sample. Although an early morning specimen (EMS) has the strongest concentration of human chorionic gonadotrophin (HCG), side room pregnancy tests are usually sensitive enough to be positive in any urine specimen within one week of a missed period.[7]

Urine culture

A midstream sample is required to check for urinary tract infection. Simple instructions should be given to the patient. Ideally, the vulva should be cleaned with water. Avoid disinfectants because traces can get into the specimen and destroy any bacteria. Hold the labia apart while passing urine. The first void of urine should be passed into the toilet and then the next part of the stream caught in a suitable sterile container. The final part of the stream can be passed into the toilet. The urine specimen should be cultured within one hour or stored at +4°C immediately.[8]

Pharyngeal sampling

This can be done conveniently either at the time of venepuncture or after the pelvic examination. Throat swabbing is indicated in patients who have oropharyngeal symptoms or a history of passive oral intercourse or when the patient has been in contact with gonorrhoea. A few clinics do throat swabs routinely on all patients. This decision depends on facilities available and on local gonorrhoea prevalence and sexual behaviour patterns. Take a swab from the posterior fauces and plate directly on to selective medium for culture of *N. gonorrhoeae*.

For serological tests and other specific examinations and sampling techniques, see below.

EXAMINATION OF THE MALE

Physical examination should take place with the patient lying flat on a comfortable couch. The genitalia should be exposed fully, with both trousers and underpants lowered to beneath the knees with the lower abdomen also visible. Merely exposing the penis for examination is not satisfactory.

Genital examination

This should include inspection and palpation, followed by collection of microbiological specimens. Firstly, inspect the inguinal and suprapubic regions. Some conditions may be diagnosed clinically, such as pubic lice infestation, genital warts, molluscum contagiosum and other skin dermatoses (see Chapter 7).

Scrotum and contents

The scrotum should be inspected for asymmetry, erythema and superficial lesions such as sebaceous cysts or telangiectasia. Palpate the scrotal contents carefully. If there is a unilateral problem, feel the normal testicle first. A slight variation in testicular size is common, but if there is marked asymmetry, enquire about past trauma, torsion, surgery or infection (e.g. mumps, orchitis). If the scrotal contents seem enlarged, attempt to differentiate the epididymis from the testis. If only the testicle is enlarged, consider orchitis or neoplastic conditions. If the differentiation is impossible, epididymoorchitis is a possibility. Transillumination of the swelling may support the diagnosis of a hydrocele. A varicocele is more commonly present on the left side, because of the higher pressure venous drainage system of the left renal vein. However, a newly apparent varicocele can occasionally be a marker of a renal tumour. A renal ultrasound is useful to look for associated renal pathology.

If a scrotal abnormality is found, examine the patient standing up, when small varicoceles and herniae become more apparent. The hernial orifices and vas deferens can also be palpated readily. An ultrasound scan of the scrotum often allows a definitive diagnosis to be reached.

Penis

Inspect the shaft of the penis from the base to the tip, noting any abnormalities. Genital warts at the base of the penis can be missed easily, as can plaques of psoriasis. Examine the prepuce (if present) and retract as far back as it will go. The subprepucial area, glans and coronal sulcus can then be inspected. Look for balanitis (inflammation of the glans) and posthitis (inflammation of the prepuce). Take care to look at the entire circumference of the coronal sulcus, as genital warts can be overlooked. Note that some coronal papillae can be quite large and mistaken for warts. Finally, inspect the urethral meatus for evidence of meatitis, urethral lesions or discharge. Note congenital abnormalities such as hypospadias. If no discharge is apparent immediately, the urethra should be milked gently by massaging from the base to the tip. Look for other clues such as discharge on the inside of the front of the underpants.

Perianal region

The perianal region should be examined for genital fungal infection, warts or skin abnormalities. The patient lies on his left side with knees bent up and buttocks over the edge of the couch. The buttocks should be separated either by the patient or the doctor to allow inspection of the natal cleft, perineal and perianal areas.

Collection of specimens

Urethral tests

Urethral smear
If a urethral discharge is present, wipe the glans with a gauze swab. Sample the urethra with a disposable plastic inoculating loop. Do this by everting the lips of the meatus and passing the swab down the urethra approximately 2 cm. This is less painful if gentle traction is applied to prevent the swab from sticking in the fossa navicularis. If no discharge is present, milk the urethra first and consider using a cotton-wool tipped swab, although this does tend to be more traumatic than a plastic loop. Ensure that the foreskin is retracted to avoid contamination of the prepuce. Place the loop onto a microscope slide and spread thinly for Gram staining. Then smear the loop across a plate with selective culture medium for identification of *N. gonorrhoeae*.

Wet smears can be taken if *Trichomonas vaginalis* is suspected – for example, if the partner has *T. vaginalis* or when urethritis does not respond to first-line treatment. Put a drop of normal saline on to a clear slide, mix the sample loop in the saline and cover with a cover slip.

Urethral culture
A urethral culture for *N. gonorrhoeae* should always be taken in conjunction with urethral smears on all new patients irrespective of symptoms. Repeat urethral cultures are required for follow-up of gonorrhoea but are not necessary in non-gonococcal urethritis. If direct plating is unavailable, use a sterile cotton wool swab and place into appropriate transport medium. Ensure that the specimen is transferred to the laboratory as soon as possible. The specimen should be left at room temperature.[9]

Identification of C. trachomatis
Many different identification kits are available. These are usually supplied with an appropriate swab which should be inserted at least 2–3 cm down the urethra. It is best to insert directly with no twisting or pushing action, which will cause a lot of discomfort. Straighten the penis with light tension before insertion and if the urethra is dry, moisten the swab with saline first. As the swab is withdrawn, a rotational movement may ensure a better specimen.

Urine tests

Two-glass urine test
The foreskin should be retracted fully and the patient asked to urinate into two clean specimen glasses, the first 10–20 ml into one glass, the rest into the second. If the urine is hazy, add sufficient 5% acetic acid to

dissolve the phosphate crystals which are responsible for the haze. When there is infection of the anterior urethra, the haze will persist in the first glass of urine due to the presence of pus cells, threads or flecks, but the second will be clear. If both glasses are abnormal, the infection also involves the posterior urethra, bladder or kidneys. This simple test may provide clear evidence of urethritis and is useful in follow-up of cases of non-gonococcal urethritis. Threads may be removed for microscopy either by Gram stain or as a wet preparation, to determine whether they contain mucous, sperm or pus cells. When the urine contains small threads or specks, a centrifuged deposit of urine can be examined in a similar manner. If pus cells are seen using this method, it is wise to request an early morning specimen, both to assess the degree of urethritis and to enable satisfactory urethral cultures to be taken.

This test is also invaluable in assessing patients who refuse to have a smear. 'False-negative' two-glass urine tests may be due to recent urination. If the urethral smear is positive and urine negative, the urethral sample may have been contaminated by subpreputial material.

The scientific evidence for use of a two-glass urine test has not been well documented. In some clinics the second specimen is omitted unless the symptoms are suggestive of prostatic or upper urinary tract problems. The first void urine can also be used for detection of *C. trachomatis* in some laboratories (see Chapter 3).

In all new patients urine should be tested by dipstick as described for women.

Early morning specimen
Urination removes accumulated urethral secretions. Ideally, patients should hold their urine for at least four hours before swabs are taken. If tests for urethritis are negative but the diagnosis is suggested by the history, the patient should be asked to reattend for an early morning specimen. Urethral smears and cultures and the two-glass urine test are performed after the patient has held his urine overnight. This can usually be achieved by restricting fluids for four hours and emptying the bladder before retiring. Emphasize that it is the swabs that are most important and that he should not collect a urine sample at home. This defeats the object of an early morning smear.

If it is impossible for the patient to attend the clinic in the morning, a similar result can be achieved by asking him to return after holding his urine for at least six hours.

EXAMINATION OF HOMOSEXUAL MEN

Homosexual men have specific risks for acquiring certain sexually transmitted diseases and have, as a group, an increased likelihood of infection by human immunodeficiency virus (HIV), hepatitis B, syphilis and anorectal infections. In order to meet the health needs of homosexual men, a knowledge of their sexual behaviour and appropriate examination is essential.

Physical examination
Skin and mucous membranes
All clothing should be removed to allow a good inspection of the skin, including soles of the feet and genitalia. Skin problems are common manifestations of HIV infection and syphilis (see Chapters 14,16,17 and 18). The mouth should be inspected for signs of ulceration, warts or other lesions.

Lymph nodes
Examine all sites for lymphadenopathy and check for hepatosplenomegaly.

Neurological examination
A brief neurological examination should be included in known HIV-positive patients or where syphilis is suspected. This should include fundoscopy, testing of reflexes and sensation. The need for more extensive neurological assessment depends on the history or any positive neurological finding on initial examination.

Genital examination
Genitalia and the perianal area should be examined as described above.

Investigations

Record the patient's weight. Take urethral and pharyngeal swabs as described above. Proctoscopy should always be undertaken at the initial attendance of a man who has been the receptive partner during anal intercourse. On proctoscopy there may be inflammation with blood, erythema, mucopus and mucosal oedema. In patients who have symptoms and/or signs of proctitis, culture should be taken for *N. gonorrhoeae*, *C. trachomatis* and herpes simplex virus. Lesions of the anal canal, such as warts, can be seen. Inflammatory changes associated with gonococcal, chlamydial or herpetic infection are generally limited to the distal 10 cm of the rectum. Take rectal samples as described above.

Sigmoidoscopy may be required to search for more proximal lesions and is indicated in cases of unexplained rectal bleeding, anorectal discharge, pain or incomplete defaecation. Rectal biopsy may be necessary.

In cases of diarrhoea, stool samples should be examined for bacterial pathogens, protozoa, ova and cysts. A minimum of three samples should be sent to the laboratory.

Blood tests

Universal precautions must be maintained. Remember to wear gloves when taking any blood samples and handle blood samples carefully. Never resheath the needle after use. Ensure that there is a satisfactory disposal bin close by. Clinics should have a clear policy for cases of spillage or needlestick injury.

Syphilis

Serological tests for syphilis (STS) should be carried out at the first visit in all new patients. For most patients a single screening test is required, e.g. TPHA or VDRL. Laboratories differ in the tests offered for screening; some may offer two serology tests routinely. In patients with genital ulcers or a previous history of syphilis, further serological tests are required, i.e. FTA-ABS, TPHA, in addition to VDRL/RPR (see Chapter 14).

HIV, hepatitis B and C

HIV tests should be available on request, but a number of points should be discussed before taking blood. These include partner notification, confidentiality, insurance implications, incubation period, latency and transmission. Before sending blood for hepatitis B status, establish whether the patient has ever had hepatitis before, whether their serological status is known or if a vaccine course has been given and if so, when.

Tests for viral infections should be available to all patients, as well as specifically targeted to those in higher risk behaviour groups (Table 2.2).

Table 2.2 Checklist for risk factors – HIV, hepatitis B and C

Group	HIV	Hep B	Hep C
Homo/bisexual	+	+	±
Intravenous drug user	+	+	+
Blood products	Pre-1985	–	+
Sex abroad, high-prevalence country	+	+	±
Sex with prostitute, esp. high-prevalence country	+	+	±

SPECIAL INVESTIGATIONS

Examination of ulcers

In all cases of genital ulceration, the first objective is to diagnose or exclude a sexually transmitted problem. The clinical examination can usually give enough information for a reasonably accurate diagnosis. The important features of genital ulcers are whether they are:

- multiple or single;
- large or small;
- painful;
- associated with lymphadenopathy;
- associated with oral ulceration;
- associated with oedema.

Note also the characteristics of the base and edge of the ulcer, whether there is surrounding erythema and evidence of scratching and superficial fissuring.

Chancre

Syphilis must be excluded in any patient presenting with a sore in the anal or genital region. For microscopy, clean the surface with a swab soaked in sterile saline. Induce bleeding by rubbing with a dry sterile gauze swab or scarifier, allowing the clot to retract before collecting the serum on to a glass slide. The serum can be squeezed out by applying pressure at the base of the lesion. Ideally, three specimens of serum should be obtained. Serum can be collected in a glass capillary or placed directly on to slides or the underside of the cover slip. Apply a cover slip and examine with dark-ground illumination or phase-contrast microscopy. Saline injected into the base of a lesion and aspirated and material scraped from a healing skin lesion are also worth examining by microscopy, as is the aspirate from a swollen lymph node following injection of 20 µl or less of saline. Spirochaetes other than treponemes can be seen in fluid from chronic ulcers, so serological tests for confirmation of diagnosis are mandatory. If clinical suspicion of syphilis is high, further repeat dark-ground examinations may be appropriate.

In secondary syphilis, treponemes can be sought in mucous patches and condylomata lata.

Herpes simplex virus (HSV)

HSV is the commonest cause of sexually transmitted genital ulceration in the developed world. The most specific and sensitive diagnostic method is isolation of the virus by cell culture. Apply firm pressure to the ulcer with a cotton or dacron-tipped swab. This can be exquisitely painful for the patient and he or she should be warned in advance. The swab should be transferred immediately to appropriate transport medium, squeezed out and discarded. If gel transport medium is used, the swab can be left in. The specimen should be sent to the laboratory as soon as possible and kept at +4°C if immediate analysis is unavailable.

In some units other methods of diagnosis are available, including electron microscopy to identify viral particles, and the use of monoclonal antibodies has made a direct immunofluorescence antigen test possible. There are also other antigen detection tests which may be especially useful if a patient presents after the lesions have crusted. If the blisters are intact, the fluid can be taken for analysis by deroofing the blister and using a capillary tube or allowing a cotton-tipped swab to absorb the fluid before transferring it to appropriate transport medium. In first-episode primary herpes genitalis, HSV can be isolated from the urethra and pharynx of the male and from the urethra, pharynx and cervix of the female.[10] It is common practice to take a cervical swab for herpes simplex virus in women although the vaginal/cervical examination will need to be deferred if the vulval area is too tender to permit speculum insertion.

Chancroid

Chancroid is a rare cause of genital ulceration in the United Kingdom, but a common cause worldwide. Diagnosis can be made by examination of stained films showing the organism *Haemophilus ducreyi* in typical configuration. The bacterium can be seen in specimens obtained from the ulcers or by aspiration of pus from an inguinal bubo (a confluent mass of lymph nodes). Material from the base of the ulcer should be inoculated directly on to suitable medium and transported to the laboratory as soon as possible for confirmation of the diagnosis.[11] Discuss with the microbiologists, who will need to prepare special culture medium. This may mean asking the patient to return the following day and delaying treatment.

Lymphogranuloma venereum

The most appropriate diagnostic test for *C. trachomatis* serotypes L1–3 is to culture pus aspirated from a bubo. Serological methods can also be used – the specific complement fixation test usually becomes positive 7–10 days after infection.

Granuloma inguinale

The diagnosis of this condition depends upon the demonstration of the infecting agent, *Calymmatobacterium granulomatis*. Biopsies or curettings of the ulcer base should be placed on a microscope slide and stained with Leishman or Giemsa stains. Donovan bodies can be seen (see Chapter 3).

Miscellaneous

Other viral causes of genital ulcers may need to be considered, e.g. cytomegalovirus, coxsackie virus. Specimens for viral culture should be sent in appropriate viral transport medium and specific requests discussed with the laboratory service. Some of these viruses can be more easily found in urine or stool samples.

Investigation of prostatic symptoms

The history or initial investigations may point to a prostatic cause. To establish a diagnosis of 'prostatitis', it is necessary to follow a methodical sequence of investigation – obtaining prostatic secretion alone is inadequate. The preferred procedure was described by Stamey.[12] Take a urethral swab then ask the patient to pass the first 10 ml void of urine into one glass (VB1) and 20 ml into a second glass (VB2). To obtain prostatic secretion, place a gloved finger in the rectum and massage the prostate gland. Firm pressure is required from proximal and lateral towards the distal end of the gland. The patient is often aware when the secretion starts to come down the urethra. Ensure that the foreskin, if present, is retracted. The prostatic secretion can be collected on the lid of a Petri dish. Then ask the patient to pass urine, collecting the first 10 ml (VB3).

Samples of the prostatic secretion can be placed on a microscope slide for Gram stain and inoculated directly on to selective medium for culture of gonorrhoea. Other organisms can be cultured by placing a cotton-tipped swab into the secretion and then placing in transport medium. Identification of *C. trachomatis* is rarely successful. The urine samples should be sent for microscopic examination for polymorphonuclear cells and cultured for organisms. The laboratory should do colony counts to estimate the concentration of bacteria in each urine sample. If the concentration of an organism in VB3 is more than 10-fold greater than in VB2, it is likely to be the cause of a bacterial prostatitis.

Examination of cerebrospinal fluid

Cerebrospinal fluid may be required for the diagnosis of conditions related to HIV infection and also to establish any involvement of the central nervous system in syphilis. CSF is obtained through a standard lumbar puncture technique and CSF pressure should be measured. Samples of CSF should be split for appropriate analysis by different laboratories. Estimations of the levels of protein, glucose (with simultaneous blood glucose) and cytological analysis are mandatory.

Investigation for bacteria, viruses and protozoa may be required. Two samples, each in a sterile screw-cap bottle, should be obtained. Immediate transport to the laboratory is important but when delay is unavoidable, the sample for bacterial culture should be held at room temperature or incubated. The viral sample should be kept at 4°C. A second specimen is taken to ensure that the cell count is not distorted by contamination with peripheral blood.

For the diagnosis of neurosyphilis, cell count and protein are more sensitive indicators, but serological tests of CSF for syphilis may be helpful. CSF may be required for the diagnosis of other diseases, e.g. cryptococcal antigen in cryptococcal infection (see Chapter 16).

INVESTIGATION OF SKIN DISEASES

Fungal infection

To exclude or confirm fungal infection, scrapings should be taken from the edge of the lesions. Collect the scrapings on to a folded slip of black paper. Microscopic examination is carried out by mounting skin scrapings in 10% potassium hydroxide solution.

Identification of scabies

To confirm the diagnosis of scabies, a mite should be located. Remove the mite from its burrow with a needle and examine it with magnification on a dry slide. Where it is impossible to remove the mite, a drop of liquid paraffin can be placed over the suspected site and the lesion scraped with a scalpel blade.

Pubic lice

The bites of lice may produce pinpoint red macules which can be seen in the hairy area of the genitals. Look for the eggs that protrude from the side of the hair; these may resemble dandruff. The adult lice can also be seen attached to hair or moving through the hair. When there is a heavy infestation axillary and chest hair and even eyebrows and eyelashes can be affected.[13]

Miscellaneous

Other skin conditions specific to the genitals may be recognized clinically but require histological confirmation. This is essential in malignant and premalignant conditions, e.g. vulval intraepithelial neoplasia (VIN) and erythroplasia of Queyrat, but is also helpful for definitive diagnosis of other skin diseases (e.g. psoriasis) (see Chapter 7).

Conjunctival specimens

Some sexually transmitted diseases involve the conjunctivae and other mucous membranes. When gonorrhoea is the likely diagnosis, a dry swab should be rubbed across the lower conjunctival surface and plated directly onto selective medium. For the diagnosis of chlamydial infection, wipe away excess pus then rub firmly with a swab. This should be squeezed into culture medium for *C. trachomatis*. Antigen detection tests can also be used.

MANAGEMENT OF CASES OF SEXUAL ASSAULT

The victims of sexual assault may present initially to a variety of physicians. For the genitourinary physician, it is vital to ascertain whether the victim has been assessed by a police surgeon to ensure that valuable evidence is not destroyed. Whatever the age of the victim, there are some important points to remember.

- Document both the history and the examination carefully. Even if the patient does not wish to report the assault at present, notes may be required at some future date.
- Establish whether the patient has reported the offence and had a forensic examination.
- If a forensic examination is required, ensure that a police officer is present. The Home Office provides a standard sexual assault examination kit.
- It is vital to comply with the principle of the 'chain of evidence'. The origin and history of any exhibit to be presented as evidence in a court of law must be clearly demonstrated to have followed an unbroken chain from its source to the court. All persons handling the sample, in addition to the place and conditions of storage, must be documented with a note of time, date, place and signatures where appropriate.
- Ensure that the best test is used for identification of sexually transmitted diseases if a court appearance is likely. Ensure that cultures are taken for *C. trachomatis*, either in addition to or instead of other methods. Ideally, discuss with the laboratory service before taking any swabs if a court case is likely and follow the chain of evidence.

Principles of the examination

The forensic examination and taking of specimens is directed towards the discovery of corroborative evidence that assault has taken place without consent. It may also provide evidence linking possible assailants to the complainant.

Table 2.3 Time limits for the detection of spermatozoa and seminal fluid

	Spermatozoa	Seminal fluid
Vagina	6 days	12–18 h
Anus	3 days	3 h
Mouth	12–14 h	
Clothing/bedding	Until washed	Until washed

Before commencing the examination, the doctor should obtain permission from the patient or guardian if appropriate. In the forensic examination all information given may be made available to the police.

Document the time, place and who was present for the examination. Ideally, joint examinations by police surgeons and genitourinary physicians (and/or paediatricians in the case of child sexual abuse) ensure that the forensic and infective aspects are dealt with properly. Ensure that the time and type of offence is documented so that the offence–examination interval is known. This is particularly relevant for assessing sexually transmitted infection when incubation periods may be important. Similarly, forensic samples need to be taken within a time limit, usually 3–4 days, although it is worth taking specimens up to seven days later (Table 2.3).

For forensic examination, ask the examinee to undress down to the underwear, standing on a sheet of paper. Evidence relating to the assault may drop from clothes and can be salvaged. The clothes are retained for forensic examination. Examine all areas of the body for trauma, which includes bruising and bite marks. Examination with a fluorescent light may facilitate the location of semen and saliva stains. The underwear can be removed and retained for examination.

Combings of pubic and head hair can be taken, along with control specimens plucked from the head and pubic region. Fingernail scrapings can be taken if circumstances suggest that blood or fibres could be present. Screening for sexually transmitted diseases or HIV infection is unhelpful within a few hours of the assault. Also, the identification of a sexually transmitted disease may count against the victim in court by undermining their character in the eyes of the jury and casting doubt on reliability of evidence.[14]

FEMALE VICTIMS OF ASSAULT

Examine carefully the upper arms and inner thighs for evidence of bruising, abrasions and petechial haemorrhages, where the offender may have used force to hold the victim down. Look for evidence of tears, especially at the posterior fourchette. Photographs are useful as a permanent record of injuries sustained. Take swabs from the external genitalia, including the anus if buggery is alleged. Use a cotton-tipped swab to remove samples of semen from the vagina. If assault occurred over 48 h before the examination, spermatozoa may survive longer in the endocervix and a swab should be taken from this site. The mouth is a common site of injury. If there is a history of oral penetration swabs should be taken from the gums and outer and inner aspects of the lips. These can be air dried or deep frozen. Take specimens of semen, seminal fluid or saliva from other areas of skin if present. These specimens are subject to DNA fingerprinting and determination of ABO secretor status.

If the screen for sexually transmitted disease is carried out at the same time as the forensic samples, forensic specimens should take priority. A screen for sexually transmitted disease can always be repeated at a later date. Ensure that the possibility of pregnancy has been raised. Postcoital contraception may be required. Blood samples will be required in forensic examination for blood grouping and DNA analysis (two EDTA bottles required). Blood may be required for drug and alcohol estimation. The question of acquisition of HIV or hepatitis B infection, which is thankfully a rare occurrence in the UK, should be discussed at some stage with the victim. A blood sample can be taken and saved without any evaluation at the initial examination pending a definitive decision.

The victim should be encouraged to return for follow-up examination. This opportunity can be used to discuss the chance of acquiring hepatitis B and HIV infection, if omitted previously. If a sexually transmitted disease is isolated, it is advisable, after treatment, to examine the patient to ensure that the infection has been cured. Referral to other agencies for psychological care may be required.

MALE VICTIMS OF RAPE

Reliable data on the incidence, nature and sequelae of sexual assault on males are lacking. Most of the reports come from the Untied States. Male victims are less likely to seek help following assault, but some seek medical help and fail to disclose the reason for requesting examination.[15,16]

It has been reported that multiple assailants are more common when the victim is male rather than female. Also, compared to female assault, male victims are more likely to sustain violent genital/non-genital trauma during the attack. The majority of male victims are homosexual but heterosexuals comprise a significant minority. The assailants can be heterosexual, but data available in the UK suggest that homosexual orientation is more likely. These epidemiological data are important when considering the management of male victims. The history should ask specifically about cleansing procedures performed after anal sex. Signs of extragenital injury are common in these cases and should be documented by drawing.[17]

Photographs also provide a permanent record of injuries. The risks of acquiring syphilis, hepatitis B and HIV are likely to be greater than for female victims of assault. Although the need for hepatitis B immunoglobulin immediately following assault is contentious, until further examination is available, this should be offered to all victims. Subsequently an active vaccination programme should be given and an accelerated course should be considered. Full screening for other sexually transmitted diseases is highly desirable. If these facilities are available prophylactic treatment is probably unnecessary. Psychological referral is likely to be necessary in these cases.

INVESTIGATIONS IN CASES OF CHILD SEXUAL ABUSE

Although most genitourinary physicians have very little experience of paediatric problems, sexual abuse is one area in which an increasing number are becoming involved. This is a very specialized area and examination of children should be done in conjunction with an experienced paediatrician. However, the expertise of the genitourinary physician in collecting adequate samples for sexually transmitted disease screening is invaluable. Referral to an experienced child sexual abuse team ensures that the legal aspects can be addressed and also cuts down the number of examinations required.[18] Children may present as victims of abuse in three main ways.

- Disclosure of direct allegations of abuse
- Behavioural problems
- Abnormal medical findings.

During the history, record verbatim what the child says. Obtain consent for examination from the parent and the child if the child is old enough to comprehend.

Abnormal medical findings

Physical evidence of abuse is present in less than 50% of girls and in only 43% cases of buggery. The presence of signs depends on type of abuse, length of time of abuse and whether abuse is acute or chronic. Very few signs, apart from the presence of semen in the genital

Table 2.4 Vulvovaginal signs of abuse

Diagnostic of penetration
- Laceration/scars extending into vaginal wall
- Attenuation of hymen with loss of hymenal tissue

Supportive of penetration
- Enlarged hymenal opening
- Notch associated with scarring
- Localized erythema and oedema
- Scarring of posterior fourchette

area, are diagnostic of sexual abuse (Table 2.4). It is important that the examination is systematic and the examiner has experience, understands the anatomy of prepubertal female genitalia and recognizes normal variations in the shape of the hymen.

Vulva and hymen

The vulva can be inspected with the child lying on the back with flexed and abducted hips and flexed knees. Another position favoured by some paediatricians is the knee/chest position (see Plate 4).[18] In the supine technique, the labia are separated with the tips of the fingers in a lateral and downward position until the introitus is exposed. In the knee/chest position the child rests her head on folded arms with the abdomen sagging downwards and the knees bent, with the buttocks in the air. The examiner can then press a thumb outward on the leading edge of the gluteus maximus. This position allows a better view of the posterior hymen.

The dimensions of the hymenal orifice are difficult to measure accurately unless taken with a colposcope or measured on a photograph. The dimensions are altered by position and the amount of hymenal tissue present. The orifice is generally larger in the knee/chest position. The surrounding skin should also be examined carefully.

Anal examination

The anus can be examined with the child in the lateral position. Perianal skin can be inspected by gentle separation of the buttocks. Bruising or bleeding without reasonable explanation must raise suspicion of abuse, but swelling and thickening of the skin may have other causes, e.g. perianal candidiasis, scratching due to threadworms, warts, lichen sclerosis.

The signs of buggery are listed in Table 2.5. Fissures from causes other than child sexual abuse are usually in the midline. Remember that a number of children suffer from constipation and have fissures as a result.

Table 2.5 Anal signs of abuse

Diagnostic of penetration
- Evidence of force/penetrating trauma. e.g. laceration/healed scar extending beyond the anal mucosa

Supportive of penetration
- Acute changes:
  erythema
  swelling
  bruising
  fissures
  venous congestion
- Chronic changes:
  thickening of anal verge skin
  increased elasticity
  reduction of sphincter tone
- Anal laxity
- Reflex and dilatation (RAD) >1 cm

Table 2.6 Sexually transmitted diseases in child sexual abuse

Infection	Incubation period	Risk of sexual abuse
Gonorrhoea	3–4 days	Probable (almost certain if child >2 yrs)
Chlamydia	7–14 days	Possible (probable if child >3 yrs)
Herpes	2–14 days	Possible
Trichomonas	1–4 weeks	Probable (if child >1 yr)
Warts	Weeks or months	Possible
Syphilis	Up to 90 days	Possible (exclude vertical transmission)
HIV	Within three months	Possible (exclude vertical transmission)
Hepatitis B	Within three months	Possible (exclude vertical transmission)

Multiple fissures away from the midline raise a high suspicion of abuse.

Veins in the perianal skin may become dilated, giving the appearance of bruising. These veins can also be seen in bowel disorders and therefore the presence of engorged veins should be interpreted with caution. Anal laxity can be caused by a number of medical conditions, e.g. neuromuscular disorders or severe constipation. It is usually unnecessary to insert a finger into the rectum and this should be avoided, as it is not useful as a diagnostic aid.

Reflex anal dilatation

When the buttocks are separated to display the anal area, the anus should be observed for 30 s. The external sphincter contracts briefly and is often seen to then relax. A positive test involves a relaxation of the internal sphincter as well as the external, so that the anus presents a cylindrical hole with a clear view into the rectum. As an isolated sign it is not pathognomonic of child sexual abuse, although when the dilatation is greater than 2 cm, no other diagnosis is likely.[19,20]

An assessment of sphincter tone can be made by gentle palpation of the anal verge.

IDENTIFICATION AND SIGNIFICANCE OF SEXUALLY TRANSMITTED DISEASES

Genital swabs should be taken to exclude STD.[21] Cotton-wool tipped swabs are more suitable for taking samples from the vaginal introitus in the absence of discharge and are more acceptable to children. Swabs should be taken from the vaginal introitus for Gram stain of exudate and culture for *N. gonorrhoeae*, *C. albicans* and other bacteria. One swab can be taken for wet slide/culture for *T. vaginalis*. The pH test is unhelpful in prepubertal children as the pH is alkaline anyway. A swab should be taken for culture of *C. trachomatis*. Blood samples can be taken if indicated.

When an STD is identified in a child, sexual abuse must be suspected but there are other explanations, which include vertical transmission and non-sexual transmission of STD agents. The presence of an STD in a child is rarely diagnostic of abuse although it can be used as corroborative evidence. An assessment of the risk of abuse in relation to sexually transmitted infection is given in Table 2.6. Ideally, any possible perpetrator should also be subjected to examination and investigation.

REFERENCES

1. Amsel, R., Totten, P.A., Spiegel, C.A. *et al.* (1983) Non-specific vaginitis: diagnostic criteria and microbial and epidemiological associations. *Am. J. Med.*, **74**, 14–21.
2. Szarewski, A., Cuzick, J., Nayagam, M. and Thin, R.N. (1990) A comparison of four cytological sampling techniques in a genitourinary medicine clinic. *Genitourinary Med.*, **66**, 439–443.

3. Szarewski, A., Cuzick, J. and Singer, A. (1990) Cervical smears following laser treatment; comparison of cervix brush versus cytobrush – Ayre spatula sampling. *Acta Cytol.*, **35**, 76–78.
4. Mahoney, J.B. and Chernesky, M.A. (1985) Effect of swab taking and storage temperature on the isolation of *Chlamydia trachomatis* from clinical specimens. *J. Clin. Microbiol.*, **22**, 865–867.
5. Dunlop, G.M.C., Goh, B.T., Darouger, S. and Woodland, R. (1985) Triple culture tests for diagnosis of chlamydial infection of the female genital tract. *Sexually Trans. Dis.*, **12**, 68–71.
6. Donegan, E.A. (1985) Laboratory methods in the diagnosis of gonococcal infection. In *Gonococcal Infection*, (eds. G. Brooks and E.A. Donegan), Arnold, London, p. 45.
7. Anon. (1992) Which pregnancy testing kits? *Pharmaceutical J.*, **249**, 232.
8. Stokes, E.J., Ridgway, G.L. and Wren, M.W.D. (1993) *Clinical Microbiology*, Arnold, London, p.13.
9. Stokes, E.J., Ridgway, G.L. and Wren, M.W.D. (1993) *Clinical Microbiology*, Arnold, London, 364.
10. Corey, L., Adams, H.C., Brown, A. and Holmes, K.K. (1983) Genital herpes simplex infections: clinical manifestations, cause and complications. *Ann. Intern. Med.*, **98**, 958–972.
11. Stokes, E.J. Ridgway, G.L. and Wren, M.W.D. (1993) *Clinical Microbiology*, Arnold, London, p.151.
12. Meares, E.M. and Stamey, T.A. (1968) Bacterial localisation patterns in bacterial prostatitis and urethritis. *Invest. Urol.*, **5**, 492.
13. Goldman, L. (1948) *Phthirius pubis* infestations of the scalp and cilia in young children. *Arch. Dermatol. Syphilol.*, **58**, 274.
14. Lacey, H. (1991) Rape, the law and medical practitioners. *Br. J. Sexual Med.*, **18**, 89–91.
15. Josephson, G.W. (1979) The male rape victim: evaluation and treatment. *J. Am. Coll. Emerg. Phys.*, **8**, 13–15.
16. Hillman, R.J., O'Mara, N., Taylor-Robinson, D. and Harris, J.R.W. (1990) Medical and social aspects of sexual assault of males: a survey of 100 victims. *Br. J. Gen. Pract.*, **40**, 502–504.
17. Schiff, A.F. (1980) Examination and treatment of the male rape victim. *South. Med. J.*, **73**, 1498–502.
18. Royal College of Physicians (1996) *Physical Signs of Sexual Abuse in Children*, RCP, London.
19. Hobbs, C.J. and Wynne, J.M. (1989) Sexual abuse of English boys and girls: the importance of anal examination. *Child Abuse Neg.*, **13**, 195–210.
20. Hobbs, C.J. and Wynne, J.M. (1986) Buggery in childhood: a common syndrome of child abuse. *Lancet*, **11**, 792–796.
21. Robinson, A.J. (1993) The management of suspected child abuse. *Br. J. Sexual Med.*, **20**, 16–20.

3 Diagnostic procedures in genitourinary medicine: practical laboratory aspects

D. Taylor–Robinson, B.J. Thomas and C. Ison

INTRODUCTION

The number of microorganisms to be considered in genitourinary medicine has increased remarkably over the past few years. The advent of the human immunodeficiency virus (HIV) and consequent acquired immunodeficiency syndrome (AIDS) has resulted in a proliferation of newly encountered microorganisms. In this chapter, however, we will not attempt to consider the diagnostic laboratory procedures for all microbes that might be encountered by genitourinary physicians. Thus, we make no mention of the hepatitis viruses, nor the opportunistic agents that are found as a result of HIV infection. Rather, we have concentrated on the main sexually transmitted diseases and their causative microorganisms, starting with the three traditional venereal diseases.

SYPHILIS

Treponema pallidum ssp. *pallidum*, a spirochaete, is the cause of the chronic infectious disease syphilis. The organism is transmitted during sexual intercourse or other intimate contact; it can also be transmitted from a pregnant woman to her fetus *in utero* or during birth. Three other treponemes – ssp. *pertenue*, ssp. *endemicum* and *T. carateum*, are also pathogenic for humans, causing yaws, bejel and pinta, respectively.[1] These three microorganisms are not sexually transmitted but they are related antigenically to *T. pallidum* and they produce serological test results that are indistinguishable from those found with syphilis.

Detection of *T. pallidum*

Examination of exudate from a genital ulcer is described in the previous chapter. A direct fluorescent antibody (DFA) test can be performed on a preparation which has been air dried and fixed with methanol or acetone.[2] The advantages of the DFA test over dark-field examination are that motile organisms are not required, pathogenic treponemes can be differentiated from non-pathogenic ones in oral lesions and tissue sections can be examined. An enzyme immunoassay (EIA) is available commercially (Visuwell® Syphilis Antigen EIA, ADI Diagnostics, Canada) as an alternative to microscopy. However, although it is sensitive, lack of specificity is a major problem at present.[3]

Polymerase chain reaction (PCR) assay

Although the PCR exhibits a high degree of sensitivity, the tests used so far will not differentiate between the subspecies of *T. pallidum*, namely *pallidum* and *pertenue*.[3] However, the technique has been used to detect *T. pallidum* DNA in the cerebrospinal fluid of patients with asymptomatic neurosyphilis and latent syphilis,[4] HIV-positive patients with central nervous system disease[5] and in specimens from patients with congenital infection.[6] Thus, the PCR assay may help to improve the diagnosis of neurosyphilis and congenital syphilis, conditions in which serology is difficult to interpret, as well as the diagnosis of early syphilis when a serological response may not have had time to develop.

Serological tests

These are divided into specific treponemal tests which use *T. pallidum* antigens and non-treponemal tests. In the UK both are used for screening, whilst in the USA treponemal tests are used only as confirmatory tests.

Non-treponemal tests

In early syphilis antilipid antibodies are produced in response to lipoidal material released from damaged tissues and to lipids from the treponemes. Non-treponemal antigens (reagins) are used in various tests to measure both the IgG and IgM responses. In the Venereal Disease Research Laboratory (VDRL) test and the unheated serum reagin (USR) test, flocculation is detected microscopically. In the rapid plasma reagin (RPR) card test, charcoal particles are added to the USR antigen. In the reagin screen (RS) test, a lipid-soluble dye is added to the antigen. Both of the latter tests are read by naked eye and since no microscope or water bath is required, they have some popularity. None of these tests is specific. A sample which is positive at a low titre, 1:8 or less, should prompt confirmation by use of a specific treponemal test. In primary syphilis, positive results are not obtained until 1–4 weeks after a chancre first appears and if there is no reaction, the test should be repeated at intervals up to three months. A negative test result, if accompanied by a negative specific treponemal test during this time, excludes the diagnosis of syphilis. The non-treponemal tests are reactive in secondary syphilis almost without exception, usually in titres of 1:16 or greater, and also in early latent syphilis. Thereafter, the titre decreases as the latent period increases.

Treponemal tests

The *T. pallidum* haemagglutination (TPHA) test uses sheep or turkey erythrocytes sensitized with sonicated *T. pallidum* extract as antigen. In the fluorescent treponemal antibody (FTA) test serum is added to treponemes fixed on a slide, followed by fluorescein-conjugated antihuman globulin. The test is made more specific by absorbing the group antibodies with *T. reiteri* (the FTA-abs test). A commercial EIA is available (Captia). At present it is not used widely, although the possibility of automation (see below) does make it attractive.[3]

The rationale of serological diagnosis

Syphilis serodiagnosis depends on the principle of dual level testing. Because of the low prevalence of infection, poor positive predictive values can result even when tests of high specificity (>99%) are used for screening. However, the use of such a test followed by a second confirmatory test of the same specificity increases the positive predictive value dramatically. Cardiolipin tests, such as the VDRL, are often used in areas of high prevalence and limited resources but, because of the prozone phenomenon, they should not be used alone to screen for untreated early infection. The TPHA test on its own is a good screen for syphilis at all stages apart, again, from the early primary stage. However, the activities of the VDRL and TPHA tests are complementary and their combined use provides an excellent screen for the detection or exclusion of syphilis at all stages, their combination being particularly important with regard to primary syphilis. Nevertheless, two tests require more labour, interpretation is subjective and the combination does not lend itself readily to automation. Considering these points, it is noteworthy that there is no evidence to suggest that, in the case of primary syphilis, screening for antitreponemal IgG by EIA is significantly less sensitive than the combination of VDRL and TPHA tests. Indeed, providing that a 1–2 week period of seronegativity is recognized during early primary infection and a high index of clinical suspicion is maintained with the facility to call for additional confirmatory tests, detection of early infection will not be compromised by using the antitreponemal IgG EIA as a single screening test.[3]

The FTA-abs test is used currently as the standard confirmatory method. It becomes reactive around the third week of infection and has a sensitivity ranging from 86% to 100% in primary infection, 100% in secondary cases and 96% to 100% in late stage infections. The Captia EIA has also been recommended as a confirmatory test but the possibility for automation endows it with a greater screening than confirmatory potential. The diagnostic requirements of the commonly used serological tests and their sensitivity in the various stages of syphilis are presented in Table 3.1.

Effect of treatment on serological tests

After successful treatment of early syphilis, the titre of non-treponemal tests should fall and eventually become negative. However, successful treatment of later stages of the disease may result in the persistence of antibody, with titres remaining constant or gradually

Table 3.1 Serological tests for syphilis

Stage of disease	Requirement for presumptive diagnosis[a]	Test positive in indicated % (range)
Early syphilis		
Primary		VDRL 80% (75–87%)
		TPHA 75% (70–90%)
		FTA-abs 85% (70–100%)
With no syphilis history	Positive VDRL test	
	Positive confirmatory TPHA or FTA-abs test	
With syphilis history	≥4-fold increase in VDRL titre compared with previous titre	
Secondary		VDRL 100%
		TPHA 100%
		FTA-abs 100%
With no syphilis history	Positive VDRL test (titre ≥ 1:16)[b]	
	If <16, confirmatory TPHA or FTA-abs test	
With syphilis history[c]	≥4-fold increase in VDRL titre compared with previous titre	
Early latent syphilis		VDRL 95% (90–100%)
		TPHA 98% (95–100%)
		FTA-abs 100%
With no syphilis history	Positive VDRL and TPHA or FTA-abs tests following negative VDRL in preceding year	
With syphilis history[c]	≥4-fold increase in VDRL titre compared with previous titre	
Late syphilis		FTA-abs 95%
		VDRL 70% (35–95%)
		TPHA 95%
With or without cardiovascular disease	Positive TPHA or FTA-abs test	
neurosyphilis	Positive serum TPHA or FTA-abs test. Positive CSF VDRL test[d]	
Neonatal congenital syphilis	Positive serum TPHA test or FTA-abs test together with positive CSF VDRL test	

[a]Presumptive unless confirmed
[b]False-negative reactions may occur occasionally due to the prozone phenomenon
[c]Other than in the current case
[d]The CDC (USA) considers that this meets the laboratory criteria for diagnosis

falling, but never rising. In contrast, treponemal tests are of little value in monitoring responses to therapy since they usually remain positive for life, even after successful treatment. Very rarely, such tests become negative if treatment is given early in the disease. It is clear that great care should be taken in interpreting the results of tests because, for example, the finding of a negative non-treponemal test and a positive specific test may be related to previous infection with *T. pallidum* which has been treated successfully.

Aspects of diagnosis requiring emphasis

Wherever feasible, dark-field microscopy should be exploited in making a diagnosis of syphilis. However, serological tests play a crucial role. The results should be interpreted according to the stage of the disease and with the knowledge that false-positive results are possible. Ideally, sera should be examined by using a non-treponemal test (VDRL) together with a treponemal test (TPHA or FTA-abs). When clinical and epidemiological evidence for the diagnosis of syphilis is questionable, a confirmatory treponemal test should always be used. A case has been made for using a single EIA test instead of a combination in primary syphilis. This could happen in the future, as could greater use of the PCR technique.

GONORRHOEA

Neisseria gonorrhoeae is the causative agent of gonorrhoea. The organism colonizes mucosal surfaces of the lower genital tract, the urethra in men and the cervix in women. In homosexual men, colonization of the rectum and pharynx is also common. Such uncomplicated gonococcal infection (UGI) may become complicated in a small proportion of patients when infection spreads to the upper genital tract and presents as epididymoorchitis or prostatitis in men and salpingitis or other pelvic inflammatory disease in women. Disseminated gonococcal infection, where the organism invades and causes a systemic infection, presents as septicaemia, rash and arthritis. This is a distinct condition that is not caused by untreated or repeated infection and is now rarely encountered.

N. gonorrhoeae colonizes only humans and has no other natural reservoir. It causes a purely sexually transmitted infection which can be controlled by prompt diagnosis and effective therapy to prevent further spread. UGI produces symptoms of discharge and/or dysuria in the majority (>90%) of men, which prompts them to seek treatment. UGI can be asymptomatic in more than 50% of women and is often detected by tracing contacts of infected men. As a consequence, women are colonized for longer than men and act as a reservoir of infection. The likelihood of untreated infection is also greater in women and leads to a higher incidence of complicated infection among women than men.

Sites to be tested

For the diagnosis of UGI in men, the urethra should always be sampled, as should the rectum in homosexual men and the cervix in women. However, it is common to isolate the organism from the female urethra which may have become colonized by leakage of infected vaginal secretion. Sampling of both the cervix and urethra in women is routine practice in most clinics and provides the highest isolation rate. In women it is also possible sometimes to isolate *N. gonorrhoeae* from the rectum, but again this often appears to be due to contamination from vaginal secretions. In addition, the pharynx should be sampled in homosexual men and in heterosexuals who report orogenital contact.

Detection of gonococci

Presumptive diagnosis

Direct visual detection of intracellular Gram-negative diplococci is the best method for the presumptive diagnosis of gonorrhoea. It is inexpensive, simple to perform and specific and a result can be obtained while the patient is in the clinic. Gram stain of a urethral smear has a sensitivity in symptomatic men of >95% in comparison with culture and when used by experienced personnel, it is highly specific. The sensitivity is lower (40–50%) when used on specimens from women and asymptomatic men where it is probable that fewer gonococci are present.[7] The sensitivity of the Gram stain procedure is also reduced when used to test specimens from the rectum, but this is caused by the many different types of bacteria present masking the Gram-negative cocci. Antigen detection through an immunoassay (Gonozyme, Abbott) in endocervical specimens is insufficiently sensitive[8] and the same comment may be made about a DNA probe (PACE, GenProbe) in the case of pharyngeal and rectal specimens.[9] Indeed, all attempts to increase sensitivity based on immunological or molecular techniques have so far failed to provide a test that is as readily acceptable as the Gram stain.[10] There is an undoubted need for a more sensitive, rapid test for the presumptive diagnosis of gonorrhoea in women and this may be addressed by new PCR and LCR kits.

Culture

Isolation of *N. gonorrhoeae* by culturing is still considered the 'gold standard' for the diagnosis of gonorrhoea. Its use is particularly important for the detection of Gram-negative, culture-positive patients and to provide an organism to be tested for antibiotic susceptibility. Culture is also important in medicolegal cases to confirm the diagnosis and to identify the infecting organisms.

The key to successful isolation of *N. gonorrhoeae* is the collection of an adequate specimen, efficient transport to the laboratory and provision of a suitable medium. Material collected with a disposable loop and inoculated directly onto culture medium that is incubated at 37°C in the clinic before transfer to the laboratory gives the highest isolation rate. This approach is not always possible and then collection by using a swab made of non-toxic material which is placed into a suitable transport medium, such as Stuart's or Amies, is very satisfactory. Swabs in transport medium should be stored at 4°C until sent to the laboratory and placed on culture media within 24 h for the best results.[10] *N. gonorrhoeae* is a fastidious organism, with an absolute requirement for iron, that colonizes warm, moist mucosal surfaces. Thus, for successful cultivation, an enriched agar base medium supplemented with haemin, glucose and amino acids, high humidity (90%) and incubation in 7% carbon dioxide are required. The enriched agar base most commonly used is GC agar base, which consists of peptones and starch and is supplied by many manufacturers (Difco Laboratories, BBL, Unipath). Columbia agar base has been used previously and is an adequate alternative. A supplement of horse blood, either chocolatized (heated at 80°C) or lysed, has been used predominantly. Haemoglobin was used originally by Thayer and Martin[11] but it is difficult to solubilize and has no advantage over horse blood. It is also possible to use a serum-free supplement such as IsoVitaleX or Kellogg's supplements which contain glucose, cysteine and haemin and hence remove the need for blood.

The addition of antibiotics to suppress the growth of other bacteria in the anogenital tract is often helpful[12] but they should be used judiciously as they may retard or inhibit to some extent the growth of N. gonorrhoeae itself. The most common combination in use is vancomycin, which inhibits Gram-positive organisms, colistin and trimethoprim, which inhibit other Gram-negative organisms, and nystatin or amphotericin which inhibit Candida spp. The need for a selective medium is dependent on the site to be tested. Indeed, urethral discharge from symptomatic men can be cultured on non-selective medium because few other bacteria are present, whereas the isolation of N. gonorrhoeae from the cervix, rectum or pharynx requires the use of a selective medium. This problem can be overcome for N. gonorrhoeae by the use of non-selective and selective media together.

Inoculated culture medium should be incubated at 36–37°C in 5–10% carbon dioxide and in high humidity for a minimum of 48 h before it is discarded as negative. The use of a CO_2 incubator is most convenient because each newly inoculated agar plate can be incubated immediately. However, if such an incubator is not available, CO_2 packs used in an anaerobic jar without a catalyst, or a candle jar, are suitable alternatives but require the medium to be batched for incubation.

Quality control
The efficiency of the isolation procedure can be monitored easily by comparing the results of Gram staining smears of urethral discharge from symptomatic men with those of culturing. Gram staining and culturing for N. gonorrhoeae should provide concurrent positive results in at least 95% of cases. Failure to achieve this usually indicates a problem with the culture medium but all aspects of the isolation procedure should be checked. It is also useful to monitor the agreement between the result of Gram staining endocervical smears and that of culture. The smears are usually positive in 40–50% of cases, but if the agreement falls below this, it may be helpful to recheck the original smears. These are simple parameters to measure and will help to maintain a high isolation rate.

Identification

The presumptive identification of N. gonorrhoeae isolated from genital specimens is made by determining the presence on selective media of typical colonies which are oxidase-positive and contain Gram-negative cocci. In some parts of the world, where laboratory facilities are limited, this is considered predictive of N. gonorrhoeae and no further tests are performed. When confirmation of the identity is required, this can be achieved using biochemical or immunological techniques which distinguish N. gonorrhoeae from other species of Neisseria, particularly N. meningitidis and N. lactamica.

Historically, carbohydrate utilization has been used as the method of choice. The ability of N. gonorrhoeae to utilize only glucose differentiates it from other Neisseria species including N. meningitidis, which utilizes both glucose and maltose, and N. lactamica, which utilizes glucose, maltose and lactose. The production of acid by viable organisms is detected by inoculating cystine tryptic agar (CTA), containing 1–2% carbohydrate, with a pure growth of organisms and examining after incubation at 37°C for 24 h.[13] It is imperative that the inoculum is pure and this is best achieved by an initial subculture on non-selective medium. A modification of this method in which preformed enzymes are detected in non-viable organisms involves the use of carbohydrates in a buffered solution.[14] This is inoculated with a heavy pure culture followed by incubation at 37°C in a waterbath, a procedure that will give a result in four hours.

Detection of aminopeptidases with chromogenic substrates has also been used to aid the identification of N. gonorrhoeae.[15] The production of γ-glutamyl aminopeptidase is characteristic of N. meningitidis and this is particularly useful for differentiating it from N. gonorrhoeae, which always provides a negative result. While the production of hydroxyprolyn aminopeptidase is characteristic of N. gonorrhoeae, it is also produced by N. lactamica. These tests should be used with care because such aminopeptidases are also produced by the non-pathogenic species of Neisseria.

Immunological approaches to the identification of N. gonorrhoeae use antibodies that give a positive reaction with N. gonorrhoeae and a negative reaction with all other species. Two tests are widely used: a coagglutination test (Phadebact GC OMNI Reagent, Launch

Diagnostics)[16] and a fluorescent antibody test (Micro Trak *N. gonorrhoeae* Culture Confirmation Test, Syva).[17] Application of these tests allows the identification of organisms directly from the primary isolation medium or those that are no longer viable. *Staphylococcus aureus* is used in the coagglutination test. This organism is rich in protein A, a surface protein which non-specifically binds to the Fc portion of IgG 2 and 4, so allowing the Fab portion to react with the antigen. Thus, when staphylococci are coated with specific antibodies to *N. gonorrhoeae* and mixed with a boiled suspension of gonococci, visible agglutination occurs. In the fluorescent antibody test, antibodies are used that are conjugated directly to fluorescein. A thin smear of gonococci on a slide is stained with the reagent, the antigen/antibody reaction resulting in green fluorescing cocci detected by fluorescence microscopy. All other species of *Neisseria* and other bacteria appear dark red. Both reagents contain a mixture of monoclonal antibodies that are directed at specific epitopes on the major outer membrane protein, Por (formerly known as P1). Both reagents have a high sensitivity and specificity, but false-negative results do occur occasionally and these should be confirmed by an alternative method.

Antibiotic susceptibility

The extensive use of penicillin for the treatment of gonorrhoea has resulted in the development of resistance. Chromosomal resistance is low level and the result of mutations at multiple loci which together increase the minimum inhibitory concentration (MIC) from ≤ 0.06 mg/l to ≥ 1 mg/l. Plasmid-mediated resistance is high level (MIC of 2–16 mg/l) and is the result of the acquisition of a low molecular weight plasmid of 3.2 or 4.4 MD that encodes for a TEM-1 type β-lactamase. The increase in penicillin resistance has led to the use of alternative antibiotics including spectinomycin, ciprofloxacin and ceftriaxone. Resistance to spectinomycin has occurred sporadically but has not created a therapeutic problem. Resistance to ciprofloxacin is increasing and is currently found most frequently in isolates from the Western Pacific region, whilst resistance to ceftriaxone is undocumented.

The primary aim of antibiotic susceptibility testing is to predict the possible outcome of therapy. If first-line treatment results in few failures, there may be no need to test every isolate. For instance, ciprofloxacin and ceftriaxone are known to provide a therapeutic success approaching 100% and, therefore, testing the susceptibility of gonococcal isolates will give no additional information. If this approach is employed, however, it may be useful to store all isolates of *N. gonorrhoeae* for a short period of time for testing in the event of a therapeutic failure, for surveillance purposes and for testing for β-lactamase (penicillinase) production. Plasmid-mediated resistance in *N. gonorrhoeae* can be detected easily. The most convenient way of detecting penicillinase activity is to use the chromogenic cephalosporin (Nitrocefin, Unipath) test.[18] This is based on the hydrolysis of the amide bond which results in a colour change from yellow to red and can be performed on a few colonies taken directly from the primary isolation plate. Tetracycline resistance is evident if there is no zone of inhibition around a 10 μ disc or there is growth on GC agar containing 10 mg/l. Either of these findings is a good predictor of the presence of the *tet*-M determinant.[19]

It is more difficult and time consuming to detect chromosomal resistance. Most laboratories use disc diffusion tests but these should be used with caution as the result depends on the disc content chosen, media used and the inoculum and strain of *N. gonorrhoeae* being tested. Determination of the MIC is seldom required and is best performed by a reference laboratory. Alternative methods that give an indication of the MIC are the E-test[20,21] and the breakpoint agar dilution technique.[22] The E-test consists of a plastic carrier strip with a predefined continuous exponential antibiotic gradient on one side which is applied to the surface of a previously seeded agar plate. The MIC is recorded at the point where the zone interacts with the strip. The methodology for a full MIC determination is that used in the breakpoint technique, but selected concentrations of antibiotic are used to allow categorization of strains into those that are susceptible, have reduced susceptibility or are resistant.

Aspects of diagnosis requiring emphasis

Despite many efforts to improve diagnostic techniques, microscopic visualization of intracellular Gram-negative diplococci remains an important way of diagnosing gonorrhoea, being most sensitive in symptomatic men and least sensitive in women and asymptomatic men. Microscopy should therefore be allied to culture and even if molecular techniques eventually improve the detection rate, isolation will remain an essential prelude to antibiotic sensitivity testing.

CHANCROID

Haemophilus ducreyi is the causative agent of chancroid or soft chancre. Chancroid, together with syphilis, is the most common cause of genital ulceration in the tropics. In contrast, chancroid is uncommon in Europe and is usually the result of an imported infection. In the United States, there has been an increasing number of small epidemics since 1981, but it is still considerably less common than in tropical countries.[23] In Africa, it has assumed greater importance because of the association between genital ulcer disease and increased transmission of HIV. *H. ducreyi* is a fastidious Gram-negative coccobacillus. The organisms enter at a break in the epithelium and this is the site of the ulcer. Chancroid ulcers appear 3–10 days after contact with an infected individual and are the most painful of the genital ulcerations. The disease is found more commonly among men than women. The role of asymptomatic carriage in the spread of infection is unclear because of the difficulty in making a definitive diagnosis. In many parts of the world where chancroid is prevalent, the diagnosis is made on the basis of clinical presentation and exclusion of other causes of genital ulceration. However, this approach lacks sensitivity and specificity because mixed infections with the agents of genital herpes and syphilis are common. Chancroid presents as painful, purulent ulcers or a painful inguinal lymphadenopathy. Three types of chancroid ulcer have been described: the giant ulcer where several small ulcers have merged; the follicular type which is superficial and often originates in a hair follicle, initially forming a pustule which then breaks down into an ulcer; and the dwarf ulcer which has an irregular base and clear haemorrhagic borders and resembles an herpetic ulcer. In men, the ulcers form most commonly on the coronal sulcus, frenulum and foreskin and in women, at the entrance to the vagina.[24]

Detection of *H. ducreyi*

Microscopy
Exudate from an ulcer is smeared on a slide, fixed and Gram stained. The smear may show a distinctive picture of Gram-negative coccobacilli arranged in patterns resembling 'schools of fish'.[25] However, the specificity of this approach is variable[26] because of the large number of other organisms which may be present. Attempts to produce *H. ducreyi*-specific antibodies for use in immunofluorescence tests show promise, but none of these antibodies is available commercially.

Culture
Isolation of *H. ducreyi* from ulcers or buboes is still regarded as the definitive method for the diagnosis of chancroid. However, culture of this fastidious organism is difficult and even in laboratories with appropriate facilities, the sensitivity may reach only 80%. Early workers used fresh or heat-inactivated clot media for isolation, but it is now possible to obtain satisfactory isolation rates using solid media. Various media have been described which provide an enriched base supplemented primarily with blood but also with foetal calf serum and/or IsoVitaleX (Table 3.2). The greatest sensitivity for culture has been achieved by using two or three different media in combination.[24] This probably occurs because strains of *H. ducreyi* differ in their preference for GC agar, Müller Hinton or Columbia agar as a basal medium. In recent years, a charcoal-based medium (Table 3.2) has been described which may overcome the need for using multiple media (Y. Dangor, personal communication).

The choice of antibiotics to aid selection of *H. ducreyi* from other bacteria at first proved difficult because of its susceptibility to many antibiotics. However, the incorporation of vancomycin into medium (3 mg/l) has greatly enhanced the isolation of *H. ducreyi*.[27] It is imperative that fresh medium is used, within 1–2 weeks of preparation, and that it is allowed to warm to room temperature before inoculation. Inoculated culture plates should be incubated at 33°C and not at 37°C, which can be lethal. A high humidity and 5% CO_2 are also necessary for growth. *H. ducreyi* may be isolated after 48 h but it is slow growing and negative cultures should be incubated for a further three days at least before being discarded.

Specimens for culture should be obtained from the ulcer using either a dry or moist swab. The ulcer does not need to be cleaned but excess exudate should be removed to reach the base. Specimens from buboes should be obtained by aspiration with a wide-bore

Table 3.2 Isolation media for *Haemophilus ducreyi*

---

*Gonococcal foetal calf serum agar**
GC agar base
1% IsoVitaleX
1–2% haemoglobin
5% foetal calf serum

*Müller Hinton chocolatized horse blood agar**
Müller Hinton agar base
5% chocolatized horse blood
1% IsoVitaleX

Rabbit blood agar
Heart infusion agar
5% defibrinated horse blood
1% IsoVitaleX

Charcoal agar
Columbia agar
1% haemoglobin
2% activated charcoal
1% IsoVitaleX
5% foetal calf serum

---

*GC agar or Müller Hinton agar can be used
NB: Vancomycin 3 mg/l should be added to each medium

needle. Direct inoculation of culture medium gives optimum results. Transport media have not been used extensively because they have led to inferior detection. However, a medium described by the National Centre for Sexually Transmitted Diseases in Johannesburg, South Africa,[28] has overcome many of the problems and may be an alternative where direct inoculation is difficult.

The features of *H. ducreyi* colonies are characteristic. They are grey to yellow, friable and can be pushed intact across the surface of the agar. The colonies are also oxidase positive when tested with a 1% solution of N-N-N-N-tetra-methyl-*p*-phenylenediamine dihydrochloride. These characteristics, together with the presence of pleomorphic Gram-negative coccobacilli, are indicative of *H. ducreyi*. The production of alkaline phosphatase and nitrate reductase by *H. ducreyi* also differentiates it from other species of *Haemophilus*, but tests for these enzymes are seldom used routinely.

Many isolates of *H. ducreyi* produce β-lactamase and may be insusceptible to penicillin. Detection of penicillinase by the chromogenic cephalosporin test or suitable alternative is recommended. Testing the susceptibility of isolates to other therapeutic antibiotics is difficult and is rarely performed routinely, but patterns of resistance determined in reference laboratories give a good prediction of therapeutic success.

Aspects of diagnosis requiring emphasis

Microscopic examination of Gram-stained smears lacks specificity. The latter could be improved by the use of monoclonal antibodies to *H. ducreyi* in fluorescence tests, but this approach is not yet routine. Molecular techniques show promise but detection depends currently on culture which is difficult because of the fastidious nature of the organisms.

GRANULOMA INGUINALE (DONOVANOSIS)

Granuloma inguinale is a chronic ulceration of the genital region, caused by the bacterium *Calymmatobacterium granulomatis*. It is a disease predominantly of tropical regions although imported cases are seen in developed countries. Interest in this and similar conditions has been heightened recently following the observation of increased transmission of HIV in patients with genital ulcer disease.[29]

The disease starts with a subcutaneous nodule at the site of infection, which enlarges to form a bright red, slowly extending ulcer with a characteristically smooth, raised, rolled edge. The most common sites for such primary lesions are the distal penis in men and the introitus in women, although many other sites, such as the groin, the anus and the cervix and, more remotely, the neck and mouth, may also be involved. As the disease progresses, the lesions may become sclerotic and may be secondarily infected. Hopefully, recent attempts to culture *Calymmatobacterium granulomatis* will prove successful because there have been no reports of success since the 1960s.[30] Consequently, the diagnosis is still made by demonstrating the pathognomonic groups of intracellular bacteria, or Donovan bodies, in large histiocytes in smears from tissue deep in the ulcer base and in biopsies.[31]

Collection of specimens

Smears

An active area of ulceration is chosen and cleaned with saline. The optimum site for sampling is the leading edge of a progressive lesion, although the base of an ulcer is also suitable. Cells may be obtained in three ways. A direct impression may be made on a glass slide from the diseased area, although excess debris and other contaminating bacteria sometimes make this a less than adequate specimen. Alternatively, the lesion is cleaned gently to remove such debris and a second cotton-tipped swab is then rolled firmly across the edge. A larger amount of material may be obtained by scraping away a small piece of friable tissue from the edge of the lesion with a scalpel or by removing it with forceps. Smears from swabs or dabs from the cut edges of tissue should be made immediately and small pieces of tissue may then be crushed between two microscope slides. Methanol fixing a smear, preferably before sending it to the laboratory for staining, will preserve the integrity of the Donovan bodies for most staining methods. However, a rapid-staining technique (Rapi-Diff) requires treatment of the smear with a fixative supplied by the manufacturer.

Biopsy specimens

Punch biopsies are recommended for early lesions where the bacterial load may be small, for areas of sclerosis from which adequate smears cannot be made and for heavily superinfected ulcers. In addition, the histological features of granuloma inguinale may help to differentiate this condition from other genital ulcerations. Optimum use may be made of biopsy material prior to fixation by rubbing the cut edge on a microscope slide to make a smear. The biopsy specimen is then placed in 10% buffered neutral formalin fixative for transportation to the laboratory.

Detection of organisms

Smears

Donovan bodies may be demonstrated in smears by a variety of techniques and their appearance may vary between patients and according to the stain used. The application of Giemsa, Leishman or Wright stains produces deep blue-purple staining pleomorphic bacteria, usually with, but occasionally without, a non-staining capsule. Dark bipolar staining gives the organisms a 'closed safety pin' appearance. They occur in clusters in cytoplasmic vacuoles of large mononuclear cells, the nuclei of which are often oval and eccentric. A rapid-staining technique, RapiDiff, in which a mixture of eosin and a thiazine dye is used, takes less than one minute to complete and has been shown to be as sensitive as a standard Giemsa method. RapiDiff-stained bodies are slightly larger than those stained by Giemsa and have an irregular outline, an unstained capsule and pink-purple cytoplasm.[32]

Biopsy specimens

For the demonstration of Donovan bodies in fixed, paraffin-embedded tissue, sections are cut at 6 μm. Giemsa or silver stains are preferred to haematoxylin and eosin, silver staining especially producing a characteristic black colour. Staining of semithin sections with a slow Giemsa technique or with thionine azure II basic fuchsin has also produced good results.

Histological examination of stained biopsy sections may aid in the differential diagnosis of granuloma inguinale from other genital ulcer diseases. A specimen from an ulcer in which few lymphocytes are seen and in which there is a mixed inflammatory response of plasma cells, neutrophils and histiocytes and Donovan bodies stained black by a silver stain is very characteristic. Donovan bodies can be demonstrated in 60–80% of well-taken and properly processed samples from infected patients. Although the bodies are recognized more easily in smears than in sections, examination of sections may reveal organisms in patients with negative smears.

Recent demonstrations of groups of fluorescing organisms in cervical lymph node biopsies stained with specific antisera and the development of an indirect immunofluorescence serological test using paraffin-embedded sections indicate that, in the future, fluorescein-labelled monoclonal antibodies may provide a useful additional specific staining method for the identification of the organisms.[33]

Serological tests

The complement fixation test has never become established for diagnosing granuloma inguinale, despite some early indications of its sensitivity and specificity.[34] More recently, an indirect immunofluorescence technique has been used in which Donovan bodies in protease-treated, dewaxed and rehydrated tissue sections comprised the antigen.[35] It was found to have a sensitivity of 100% and a specificity of 98% when sera from patients with proven granuloma inguinale and other genital ulcerations were tested at a dilution of 1 in 160. In the current absence of culture methods, this technique may provide a useful adjunct for diagnosing individual cases of granuloma inguinale.

Aspects of diagnosis requiring emphasis

Detection of Donovan bodies by one of several staining techniques is the method of diagnosis. Serology may gain more prominence but Donovan body detection will remain the diagnostic choice until the mystery of successful culturing has been fathomed.

CHLAMYDIA TRACHOMATIS INFECTION

Chlamydia trachomatis is an obligate intracellular bacterium which most commonly causes infections of the urethra in men, the cervix and/or urethra in women and the conjunctivae in both (serovars D-K). Ascending infection may result in epididymitis or pelvic inflammatory disease and the organism is also linked to the arthritis which follows non-gonococcal urethritis in about 1% of men and cervical infection in women (sexually acquired reactive arthritis; SARA). In addition, serovars L1–3 of *C. trachomatis* are the cause of lymphogranuloma venereum (LGV) which is seen most often in developing countries.

Culture of *C. trachomatis* organisms (chlamydiae) in cell monolayers has been the 'gold standard' for the diagnosis of chlamydial infections, and remains so especially where there are legal implications. This is because culture is deemed to have high specificity, despite being only about 75% sensitive.[36] In cell culture, chlamydiae form cytoplasmic inclusions comprising individual elementary bodies (EBs) which can be identified by a variety of staining techniques. Alternative methods of detection include direct fluorescent antibody (DFA) staining of smears, enzyme immunoassays (EIAs), nucleic acid probes[37] and the Amplicor (Roche) polymerase chain reaction (PCR)[38] and ligase chain reaction (LCR; Abbott)[39] technologies.

The best DFA reagents contain anti-MOMP monoclonal antibodies which produce bright, evenly distributed fluorescence over the EB surface. This staining method can be very sensitive and specific when fluorescence is interpreted by an experienced observer, but its subjectivity and labour intensiveness make it unsuitable in laboratories that deal with a large number of samples.

Most commercially available EIAs detect chlamydial group-specific lipopolysaccharide antigen, which is extracted from the organism by heating. The EIAs have a wide range of sensitivity[40,41] and their use may result in infection being detected in only about 60% of women who are chlamydia-positive by a more sensitive method.[42] Specificity is less of a problem, but DFA confirmation of reactive samples or the use of the blocking antibody provided in the kits should eliminate any false-positive results. The 'desk-top' assays, which are designed to give an instant result, are insensitive[43] and the most widely evaluated commercially available nucleic acid probe assay is no more sensitive than some EIAs.[44]

The PCR and LCR assays provide a hitherto unequalled combination of sensitivity and detection in a wide range of samples, including urine, vulval and vaginal. Both have been shown to be more sensitive than most other diagnostic methods for genital tract samples[45,46] and both should be invaluable for screening large populations using non-invasive[47] or minimally invasive sampling methods.

Collection of specimens

Genital tract and conjunctival specimens

The distance a swab needs to be inserted into the male urethra for optimal recovery of chlamydiae has never been determined, but 3–4 cm is recommended to emphasize that deep urethral and not meatal swabbing is required. Swabbing of the squamocolumnar epithelial junction of the cervix in women, after removal of excess mucus, should be firm, as should that of the conjunctivae after excess exudate has been removed. Vaginal swabs may be taken without use of a speculum. Cotton-tipped swabs are superior to calcium alginate or Dacron and, for isolation, aluminium shafts are less likely to interfere with the culture, especially if the swab is left in the transport medium.[49]

Samples destined for commercial immunoassays should be taken with the swab provided or recommended by the manufacturer; however, these swabs may be toxic if the sample is intended for culture. The cytobrush is a highly efficient alternative collecting tool for cervical samples[50] but the excess mucus and cell debris which it removes may also be toxic for cell cultures; the consensus is that it offers little, if any, advantage over the much cheaper cotton-tipped swab.

If the sample is destined for a DFA test, the swab should be rolled firmly on the clear area of a coated slide and the adequacy of the smear checked microscopically. After drying at room temperature, the smear is fixed with methanol or acetone. Ideally, smears should be stored at 4°C before despatch to the laboratory, although fixed smears may be sent by post. Material for culture should be placed immediately in a cryopreservative solution, such as 0.2 M sucrose-phosphate (2SP) buffer or glutamate contained in a vial designed specially for low-temperature storage. The swab should be agitated vigorously in the liquid, expressed against the side of the tube and discarded. Viability is preserved best by snap-freezing the sample without delay in liquid nitrogen. Samples for testing by an EIA should be placed immediately in the transport medium provided in the kit and the swab removed or retained, according to the manufacturer's instructions.

The procedure for collection of samples for DNA amplification-based methods of detection varies according to whether the test is a commercially produced one, such as Amplicor (Roche) and the LCR-based assay kit (Abbott), or an 'in-house' assay. In the former case, only swab specimens collected and transported using the manufacturers' specimen collection and transport kits are acceptable, as the constitution of the transport and diluent buffers is crucial to subsequent stages of the assays. Samples for 'in-house' assays are usually collected in a small volume of phosphate-buffered saline and the conditions for transportation and storage must be agreed with the laboratory staff who perform the test.

Urine samples

Attempting to culture chlamydiae from urine has never been recommended because of the lack of sensitivity of the method, together with the toxicity of the sample for cell monolayers. EIAs, however, have been much more promising, in that testing a centrifuged deposit from a first-catch urine from symptomatic men may be as sensitive as testing a urethral swab by culture.[51] Urines from women are less useful,[52] only the DFA and LCR tests so far being sensitive enough to detect the small number of EBs shown to be present.[53] An early morning urine sample has no advantage over one collected from the patient in the clinic and similarly, there is no correlation between the period of time since previous micturition and DFA- or EIA-positive results.[51] For DFA testing or an EIA, the first-passed 20 ml of urine (FPU) should be collected in a sterile container and stored at 4°C until despatch to the laboratory. For the Amplicor assay, 10–50 ml of FPU are collected in a plastic pot at least two hours after previous micturition. For the LCR assay, 20 ml of FPU are collected at any time of the day and the sample may be frozen at −20°C before testing.

Detection of chlamydiae

Genital tract and conjunctival swabs

DFA staining

Fluorescein-labelled monoclonal antibodies are applied to a fixed smear at either room temperature or 37°C for 15–30 min. After washing in distilled water and drying, smears are examined by fluorescence microscopy; chlamydial EBs are seen as bright apple-green discs of characteristic appearance. Depending upon the skill of the microscopist, a single elementary body can constitute a positive result, although most manufacturers recommend that a minimum of 10 are seen. If cellular material is very sparse and no EBs are seen, the smear should be recorded as inadequate and another sample requested.

Culture
Specimens stored in liquid nitrogen should be thawed rapidly and processed by centrifuging them onto 24-h-old monolayers of cycloheximide-treated McCoy cells. These cells may be grown in either shell vials or microtitre plates, although use of the latter carries some risk of cross-contamination between samples. After incubation at 37°C for 48 h, the monolayers are fixed with methanol, stained with Giemsa reagent and examined by dark-ground microscopy. A positive culture contains bright green, autofluorescing cytoplasmic inclusions which may appear wrapped around the cell nucleus. As alternatives, staining with a monoclonal fluorescent antibody or Lugol's iodine may be used. The former is very specific but the latter, which stains the glycogen in the inclusions brown, may give rise to artefacts, especially with cervical samples.

EIAs
Samples for testing by an EIA should be processed according to the manufacturer's instructions. The results of some studies have suggested that lowering the cut-off level for a positive result might increase the sensitivity of some assays, although specificity may suffer as a result.[54]

PCR and LCR
For DNA amplification assays, the reagent-preparation area and the sample-processing area should be remote from the amplification and detection areas, to prevent contamination by amplified DNA. The Amplicor assay, however, incorporates a step in which any contaminating amplified product is destroyed at the first thermal cycling temperature. Specimens for testing by Amplicor or LCR may be stored at 2–25°C and processed within four days of collection. Assays should be performed according to the manufacturer's instructions, although there is some evidence that decreasing the initial dilution of samples for the Amplicor assay may increase sensitivity.[55]

Urine samples
Before any procedure, urines that have been stored at 4°C should be warmed at 37°C to dissolve any precipitate. For commercially produced assays, the volume of urine recommended by the manufacturer should be used and the instructions in the package insert followed.

DFA staining
The whole volume should be vortexed to distribute the contents evenly and then 1 ml removed and centrifuged at 13 000 rpm for 15 min in a microfuge. The resulting deposit is resuspended in a small volume of deionized water and a volume that will make a smear of reasonable thickness is dried on the clear well of a coated slide. The smear is fixed in acetone and stained with the DFA reagent. Characteristic apple-green EBs may be seen by fluorescence microscopy. In some urine deposits, EBs may not be distributed evenly throughout the smear, but may adhere preferentially in groups to a few cells.

EIAs
Only the volume recommended for the assay should be centrifuged since the deposit from a larger volume may contain excess substances which are inhibitory. Indeed, it may be an advantage to store urines at 4°C for up to four days before testing to allow the breakdown of inhibitors of the assay. False-positive results may be recorded for urines in some assays in which polyclonal antibodies are used because of Gram-negative bacteria cross-reacting with common LPS epitopes.[56]

PCR and LCR
Seven millilitres of urine are centrifuged and processed for the Amplicor assay and 1 ml for the LCR.

Serology
The genus-specific complement fixation test makes no useful contribution to the diagnosis of chlamydial genital infections, with the possible exception of LGV.[57] The more sensitive microimmunofluorescence (MIF) test is useful for serological surveys, but its role in the diagnosis of individual infections is limited to patients with LGV and SARA[58] and to neonates with pneumonia in whom specific IgM antibody may be detected.[59] EIAs are available commercially for detecting IgM, IgG and IgA but experience with them is limited.

The antigen used in the MIF test comprises EBs grown in the yolk sac of embryonated eggs and acetone fixed as small dots on a slide. A single serovar may be used or multiple serovars may be clustered, especially when serovar-specific IgM antibody responses are being sought. After incubation of the antigens with serial dilutions of serum and a fluorescein-labelled antihuman globulin, slides are counterstained with Evan's blue and brightly fluorescing EBs are sought by microscopy. The significance of antibody titres is contentious but IgM and IgG levels of 1 in 8 or more have been considered to indicate a current and a past chlamydial infection, respectively.[60]

Aspects of diagnosis requiring emphasis

Isolation of *C. trachomatis* in cultured cells can no longer be regarded as the 'gold standard' for diagnosis. Furthermore, EIAs and DNA probes are insufficiently sensitive to detect small numbers of EBs and even the best of the EIAs will have to give way to either PCR or LCR, and particularly the latter, as they prove themselves ready to take over in routine diagnosis. Unfortunately, there is no comfort to be gained by turning to serology.

MYCOPLASMAL INFECTIONS

Many conditions have been associated with infection by mycoplasmas but few can be considered to have a mycoplasmal cause. The diseases and the mycoplasmas with which they are likely to be associated aetiologically are (i) non-gonococcal urethritis in men – *Ureaplasma urealyticum* organisms (hereafter referred to as ureaplasmas) and *Mycoplasma genitalium*; (ii) acute epididymitis – ureaplasmas; (iii) pelvic inflammatory disease in women – *M. hominis*; (iv) urethritis and arthritis in hypogammaglobulinaemic patients: – ureaplasmas. Other mycoplasmas, for example *M. fermentans, M. pirum, M. primatum, M. penetrans, M. spermatophilum* and even *M. pneumoniae*, have the capacity to cause sexually transmitted disease but there is insufficient unequivocal evidence to know that this is so.

Collection of specimens

The detection of mycoplasmas and ureaplasmas in the male urogenital tract may be accomplished by collecting and testing a urethral swab specimen and/or a FPU specimen. The latter is likely to prove less sensitive unless the specimen is centrifuged and the cellular deposit is tested. High vaginal and endocervical swab specimens are likely to be more satisfactory than urethral swab or urine specimens for detecting the organisms in women. Swabs from whatever site should be expressed immediately in mycoplasmal or other transport medium, not broken off into the medium, and should never be allowed to dry. If specimens are to be used for DNA detection, they may be taken and transported in phosphate-buffered saline (pH 7.2). Specimens should be kept at 4°C and ideally should reach the laboratory within 24 h. If transportation cannot be undertaken within a few days (five at a maximum), the medium containing the specimen should be frozen to –70°C or placed in liquid nitrogen and transported in the frozen state. Concern over speed of transportation is less if the specimens are to be examined by non-cultural methods. Specimens received unfrozen should be examined promptly, if possible, rather than being frozen, because some loss of viability is inevitable through the freezing process. It is feasible but undesirable to keep specimens at 4°C for a week or longer but if testing is not possible within this time, they should be frozen at –70°C or in liquid nitrogen, but not at –20°C.

Detection of organisms

Culture

Of the various media used for the culture of mycoplasmas and ureaplasmas,[61,62] SP4 medium, developed originally to cultivate spiroplasmas, has in particular improved the isolation not only of the more fastidious mycoplasmas[63] but also those more easily isolated, such as *M. hominis*. Inoculation of specimens into liquid medium, which is diluted serially, followed by subculture to liquid or agar media, provides the most sensitive method for isolation.[64] Ureaplasma colonies, in particular, often fail to develop after putting a specimen directly onto agar medium, whereas a colour change (yellow to red) occurs in liquid medium containing 0.1% urea because it is metabolized to ammonia (alkaline) by the urease enzyme of ureaplasmas. In addition, the specimen is diluted in medium containing arginine (0.1%); *M. hominis* metabolizes arginine to ammonia, so that a similar colour change occurs. Ureaplasmas usually produce a change within 24–48 h or less and rarely thereafter, and *M. hominis* less rapidly, but usually well within a week.

Serial dilution of specimens is valuable for various reasons,[64] most notably for diluting antibodies, antibiotics and other substances which otherwise inhibit isolation. To confirm that a mycoplasma has been isolated, small aliquots of medium, taken from liquid cultures which are just changing colour, are introduced into fresh liquid medium and/or onto agar medium. On the latter, colonies of the genital mycoplasmas develop best in an atmosphere of 95% N_2 plus 5% CO_2. Those of *M. hominis* have the classic 'fried egg' appearance and are up to 200–300 μm in diameter. Ureaplasmas produce the smallest colonies, 15–30 μm in diameter, which usually do not have a 'fried egg' morphology. However, on agar medium containing either manganous sulphate or calcium chloride,[64] ureaplasmas produce colonies that are slightly larger and dark brown so that they are easier to detect.

Commercial kits designed for the isolation and identification of *M. hominis* and ureaplasmas are available. Specialist laboratories are likely to have media of superior quality than in these but successful use of the kits has been reported[65] and they may be of particular value where the need to detect these microorganisms arises infrequently.

Cultured organisms require specific identification. Incorporation of specific antisera in filter-paper discs on agar to inhibit colony development (agar growth inhibition)[66] is widely used but distinguishing between *M. genitalium* and *M. pneumoniae* may require several techniques, more than one antiserum and perhaps monoclonal antibodies allied to Western blotting.[67] Epiimmunofluorescence or immunoperoxidase techniques used to stain colonies are advantageous in enabling different ureaplasmal serovars or, indeed, mycoplasmal species to be distinguished in mixtures.[68]

Antigen or DNA detection

Non-cultural procedures have a particular place in the detection of mycoplasmas that are impossible, or almost impossible, to culture, for example, *M. genitalium*. Antigen detection tests are insensitive and despite some success,[69] so too are straightforward DNA probes, with the inevitable move towards the use of the PCR technique. In this regard, DNA primers specific for human ureaplasmas[70] and *M. genitalium*,[71] among others, have been developed and the method has been of considerable value diagnostically. However, isolation of organisms through culture remains a worthy goal since quantitative results are more difficult to achieve with the PCR technique and it does not permit an assessment of antibiotic sensitivity or other biological features.

Serology

Detection of organisms, together with an antibody response to them, is the ultimate in diagnosis. Detection of an antibody response alone carries less weight and even less if reliance is placed on the existence of antibody in a single serum sample. Many serological tests have been used to measure antibodies to mycoplasmas[64] but the metabolism inhibition test has greater specificity than most and has been used to detect antibody responses in various clinical settings.[64] It does, however, require special expertise, a feature not needed with EIAs which are becoming more widely employed. Thus, by use of a cell lysate ureaplasmal antigen and commercially available alkaline phosphatase conjugates, ureaplasmal responses were detected in two-thirds of patients with non-gonococcal urethritis. Furthermore, a modified EIA was used to measure changes in the levels of *M. hominis* antibody classes occurring in women with acute salpingitis.[64] Most recently, an EIA based on lipid-associated membrane proteins (LAMP) of *M. penetrans* has been used to measure antibody to this mycoplasma[72] and the same method has been applied to detecting antibody to *M. genitalium*.

The direct immunofluorescence test has gained in popularity and is rapid, reproducible and quite sensitive and specific for measuring antibody to *M. genitalium*,[64,73] there being less cross-reactivity with *M. pneumoniae* than seen with some other methods. The test has been used to detect antibody responses to *M. genitalium* in men with non-gonococcal urethritis and in women with salpingitis.[64,74] Responses may develop slowly and collection of a second 'convalescent-phase' serum sample too early could result in failure to detect a rise in the titre of antibody.

Aspects of diagnosis requiring emphasis

Serology has some merit but alone it takes second place to detection of the organisms or a combination of the two approaches. Ureaplasmas are cultured quickly and easily, *M. hominis* more slowly and fastidious mycoplasmas, such as *M. genitalium*, can be detected usually only by use of the PCR.

BACTERIAL VAGINOSIS

Bacterial vaginosis (BV) is a condition in which there is an imbalance in the microbial flora of the vagina. A working group in 1984[75] was of the opinion that it would be appropriate to use the term 'bacterial vaginosis' because of the association with various and many bacteria and because of the lack of an inflammatory response, 'vaginosis' rather than 'vaginitis' denoting this. The discussion here comprises a brief resumé of the methods that are least helpful and those that are most useful in routine practice for diagnosing BV.

Diagnostic methods that are least helpful

Culture techniques

Gardner and Dukes[76] described a close association between BV and the isolation of *Gardnerella vaginalis* and as a consequence, over the next 25 years or more, isolating and identifying *G. vaginalis* became a means of diagnosing vaginosis. However, with improved methods of culture, it became clear that *G. vaginalis* could often be found in the vagina of women without BV,[77] as could other microorganisms associated with the condition. Thus, although culture may be required as part of some research investigations, clinicians should not otherwise request it. It has no purpose and can be misleading in routine clinical practice.

Non-culture techniques

Of the non-culture procedures, several require examination of vaginal specimens by relatively sophisticated laboratory techniques which are not rapid.[78] These comprise detection of (i) non-volatile fatty acids by gas-liquid chromatography; (ii) diamines by thin-layer chromatography; and (iii) proline aminopeptidase. These methods may have their place in research on vaginosis, but not in routine clinical practice. However, Gram staining of vaginal smears is another non-culture diagnostic technique that is of value (see below).

Diagnostic methods that are most helpful

The signs of BV that can be detected by the clinician were described in 1983 as the composite criteria.[79] Fulfilment of at least three of the following four criteria is required to make the diagnosis: a thin, homogeneous vaginal discharge, a vaginal pH value of more than 4.5, a positive KOH test, the presence of clue cells in a wet mount preparation. Clearly, some of these signs need microscopical or other support to be fully appreciated. Individually, a discharge has a sensitivity of only about 50–70% and is not a specific indicator of BV. A raised pH value has a sensitivity of 92–97% but is of low specificity. A vaginal malodour may be accentuated in the 'whiff' or amine test in which a drop of 10% KOH is added to a sample of vaginal fluid. The test has a sensitivity of 40–80%, but it is usually specific for BV. Of the signs mentioned, the detection of clue cells is the single most sensitive (80% or more) and specific (90% or more) criterion for BV, but it is, of course, observer dependent.

Gram-stained smear

Secretion or discharge taken from the high vagina, but not from the ecto- or endocervix, is smeared on a glass slide, air dried, Gram stained and examined microscopically. In the case of women with BV, clue cells may be seen but it is not necessary to see such cells to make the diagnosis. The key feature is the absence of typical large Gram-positive bacilli (lactobacilli) and their replacement with Gram-variable or Gram-negative rods. Scoring systems for describing and recording the microbial flora exist[80,81] but the simplest and perhaps the most satisfactory system is to grade smears as follows: grade 1 refers to the normal flora which consists predominantly of lactobacillus morphotypes (Fig. 3.1a); grade 3, which is consistent with BV, is recorded when there are very few or no lactobacillus morphotypes, but the presence of many other bacteria and sometimes clue cells (Fig. 3.1b); grade 2 refers to an intermediate flora in which the number of lactobacillus morphotypes is obviously reduced but they are not absent and there is some increase in the number of other bacteria.

Aspects of diagnosis requiring emphasis

Attempting to fulfil the composite clinical criteria for diagnosing BV requires a microscope. The use of microscopy to examine a Gram-stained smear made from swabbing the high vagina is an alternative to be recommended. Attempting to detect sexually transmitted pathogens in women with BV by culturing or other tests may be appropriate, depending on the circumstances, but the habit of requesting that vaginal specimens should be cultured for *G. vaginalis*

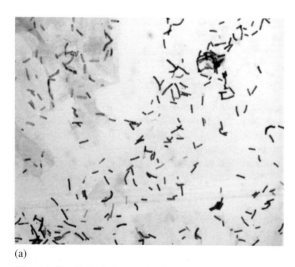
(a)

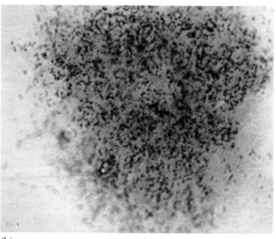
(b)

Fig. 3.1 Grading of smears. (a) Grade 1. (b) Grade 3.

or other bacteria associated with BV has no place, unless such microbiology is part of a research investigation.

TRICHOMONAS VAGINALIS INFECTION

Trichomonas vaginalis is a protozoan that commonly infects humans. In women it causes a vaginal discharge which is characteristically frothy, green and malodorous. The majority of patients are symptomatic but asymptomatic infection does occur. Infection is more common in women of lower socioeconomic status and can occur together with other sexually transmitted diseases. In men, symptomatic infection is rare, although *T. vaginalis* will survive in the urethra and can be transmitted by sexual contact. *T. vaginalis* causes a characteristic vaginal discharge but it is not always present so that this cannot be used alone to make the diagnosis. Laboratory diagnosis is usually made either by microscopy and/or by culturing vaginal discharge for the trichomonads. Molecular methods have been tried as alternatives[82] but conventional methods are more often used.[83]

Detection of trichomonads

Microscopy
Direct microscopy is the simplest method for detection of the trichomonads. A suspension of vaginal fluid is made in saline on a slide, covered with a cover slip (a 'wet mount') and examined under low-power microscopy. The trichomonads are easily identified in a fresh specimen, particularly by their characteristic jerky motility. If the specimen is not examined immediately, they will lose their motility and will be difficult to distinguish from polymorphonuclear leucocytes. The sensitivity of a saline wet mount has been reported to be between 38% and 92% and is highly specific.

Dried smears stained with acridine orange or fluorescein-labelled antibodies have been used as alternatives to the wet mount.[84] Trichomonads stained with acridine orange and examined by fluorescence microscopy appear brick-red with a yellowish-green, elongated or round nucleus.[85] Yeasts also stain red but are smaller and have a distinctive morphology. The sensitivity of staining with acridine orange, in comparison with culture, is between 60% and 70%. The use of a pool of monoclonal antibodies labelled with fluorescein to produce a cross-reactive reagent has been shown to have a sensitivity of 83–86%.[84,86] However, *T. vaginalis* is known to exhibit antigenic heterogeneity and this could lead to false-negative results. Both tests are considered to be 100% specific.

Enzyme immunoassay (EIA)
Such assays have been used to detect antigens in vaginal fluids[86] but they have not found a place in routine detection.

Culture
This is the most sensitive method for detecting *T. vaginalis* but requires complex media and incubation for up to 6–7 days.[87,88] Vaginal secretions are collected with a swab which is placed immediately in culture medium, such as Diamond, which has been warmed to room temperature. Medium contained in a plastic envelope, available commercially, may be used for convenience.[89] The inoculated medium is incubated at 35–37°C for seven days and an aliquot examined microscopically for motile trichomonads at intervals, for example after 2–3 days and 6–7 days. Although the combination of culture with microscopy in this way provides the most sensitive method for the laboratory diagnosis of *T. vaginalis*, culture is expensive and time consuming and is used infrequently. If fresh specimens are obtained, screening a wet saline mount will provide a high degree of sensitivity and is often used successfully as the sole method of diagnosis.

Aspect of diagnosis requiring emphasis

In most clinical situations, the simplicity of wet mount microscopy is such that culture, even though it is the recognized 'gold standard' with superior sensitivity, is seldom used.

CANDIDIASIS

The lower genital tract of many women of childbearing age is colonized in small numbers by *Candida* species which are considered part of the normal flora. However, changes in the vaginal ecology can give the yeasts an advantage, resulting in increased numbers and the symptoms of vulvovaginitis. Factors that may initiate the overgrowth of *Candida* species include increased oestrogen levels, as seen, for example, in pregnancy, the use of broad-spectrum antibiotics and corticosteroids, diabetes mellitus and immunosuppression. *Candida albicans* is the predominant species isolated and can account for more than 75% of yeast vulvovaginitis. The discharge produced is cream coloured and curdy and is usually accompanied by intense itching. Although this is very characteristic, the presence of yeasts is normally confirmed by microscopy and/or culture.[90]

Detection of organisms

Microscopy

Yeasts can be detected by using a wet mount or a Gram-stained preparation. A wet mount is made by collecting vaginal fluid with a swab and preparing a suspension in saline on a microscopy slide. An alternative is to Gram stain a smear of vaginal material. The presence of pseudohyphae or budding yeast cells in either of these preparations is indicative of *C. albicans*. The wet mount is simple to examine and this may be done quickly by experienced staff, whereas the Gram-stained preparation can be useful for less experienced personnel.

Culture

Isolation of *C. albicans* is the most sensitive method for the detection of yeasts in the vagina. However, vaginal fluid from women with no signs or symptoms of vaginitis may yield positive cultures. Hence, isolation of *C. albicans* should be interpreted in conjunction with the results of microscopy and the symptoms of the patient.[91] In most instances, the wet mount preparation is adequate to make a diagnosis in symptomatic women. However, if culture is required, vaginal secretions should be collected with a swab and inoculated into Sabouraud's agar supplemented with chloramphenicol (50 mg/l). After incubation at 30°C for 48 h, any colonies should be identified by growth in serum followed by microscopy for the detection of germ tubes.

Aspects of diagnosis requiring emphasis

As for *T. vaginalis*, wet mount or Gram stain microscopy prevails in routine diagnosis. Detection by this approach may have a greater clinical significance than by culture, the superior sensitivity of the latter allowing detection in healthy women.

HERPES SIMPLEX VIRUS INFECTION

Sixty to 80% of genital herpes simplex virus (HSV) infections are caused by HSV-2 and the remainder by HSV-1. The classic presentation of primary genital infection by either serotype is multiple vesicular lesions on the external genitalia, progressing to pustules and ulcers that gradually crust over and heal. Virus is also shed from the cervix in about 90% of women with primary infection. Although typical symptomatic primary genital HSV infection can be diagnosed reliably on clinical grounds, there is some clinical overlap between HSV-induced and other genital ulcerations and laboratory confirmation of the diagnosis can be of benefit in management of the patient.

Many primary HSV infections and reactivations are entirely asymptomatic and little virus is shed. In these circumstances, rapid and sensitive laboratory diagnosis may be crucial; for example, to allow asymptomatic women who are about to go into labour a delivery by caesarean section. Rapid diagnosis and treatment with antiviral chemotherapy might also prevent dissemination of the virus in asymptomatic patients who are immunosuppressed or immunodeficient as a result of HIV infection. A variety of tests is available for the diagnosis of HSV infection. The method chosen will depend upon the urgency of making the diagnosis and the site to be tested.

Collection of specimens

Vesicle fluid and to a lesser extent pustules are most likely to be virus positive, while only 25% of crusted lesions yield virus.[92] Fresh vesicles should be aspirated with a small gauge needle on a tuberculin syringe. Any remaining fluid is absorbed with a premoistened cotton-tipped swab. The base of the lesion is swabbed firmly with a cotton-tipped swab to obtain epithelial cells. Calcium alginate swabs should not be used as they may interfere with viral isolation.

For direct examination, vesicle material should be spread on a glass slide and fixed appropriately for the subsequent staining method and, for typing procedures, two smears should be fixed on each slide for direct comparison. For culture, the fluid is removed into 1–2 ml of cold viral transport medium (VTM), the syringe being rinsed to recover the maximum amount of virus. Several transport media have been shown to preserve HSV infectivity at both 4°C and at ambient temperature.[93] Material for EIA should be put into appropriate specimen transport buffer.

Cervical samples should be collected without the use of lubricants, which might interfere with isolation. In addition to the endocervical surface, it is advantageous to swab the external os, the vagina and the labia and put the material into VTM prior to culture. For direct examination, smears are fixed immediately on glass slides with the appropriate fixative and for EIAs, material is put into the appropriate transport medium.

Detection of virus

Direct staining

Direct-staining methods are efficient for detecting virus in fresh vesicles but their overall lack of sensitivity, especially when the viral load is small, necessitates confirmatory culture. Intact HSV-infected cells may be identified rapidly by Papanicolaou (Pap) or Wright–Giemsa (Tzank) staining, which relies on the presence of enlarged cells and sometimes syncytia and intranuclear inclusions. Cytological techniques are, however, only 30–80% as sensitive as culture and the cytopathic effect (CPE) may not be distinguishable from that of other viruses.[94] Direct fluorescent antibody or immunoperoxidase staining is much more specific and sensitivity values of 88–100% have been recorded.[95,96,97]

Culture

Isolation and identification of the virus in cell culture remains the 'gold standard' for diagnosis. It is used to confirm a clinical suspicion of symptomatic HSV infection and especially to reveal asymptomatic shedding. Human diploid fibroblast lines, such as MRC-5 and WI-38, as well as HEp-2 and Vero cells, are often used for HSV culture. Most procedures incorporate a 1-h absorption stage before incubation at 37°C and in the case of genital lesions, a viral CPE can usually be observed in cultures in 3–4 days. Primary cell lines, such as mink lung cells, and rhabdosarcoma cells are more sensitive and develop a CPE more rapidly, but they are not used routinely for diagnosis.[98]

Centrifugation of the sample onto cell monolayers on cover slips in shell vials reportedly increases the speed and sensitivity of viral detection,[99] as does pretreatment of cells with dexamethasone[100] and the use of cells in suspension.[101]

Virus can be identified early in cultured cells or simply confirmed by an EIA, by *in situ* DNA hybridization or by immunological staining with biotin-avidin complexes, staphylococcal protein A and immunoperoxidase. Identification can also be combined with typing of the isolate by a direct fluorescent antibody technique. Such staining techniques can reveal foci of infected cells before a typical CPE becomes apparent. Sensitivities up to 100% have been reported when these methods are applied to regular tissue culture systems, but these values may be lower when 24-well-plate cultures or shell vial cultures are tested after incubation for only 16–24 h.[102]

EIAs

Solubilized viral antigens can be detected by EIAs, such as Herpchek and IDEIA, without amplification in cell culture. Sensitivity values of up to 99% are reported when these are used for symptomatic infections, but only around 60% for asymptomatic infections.[103,104,105] A rapid 'desk-top' EIA (Surecell, Kodak) has had limited testing: sensitivity values ranging from 64% to 100% have been reported, depending on whether the lesion is crusted or a fresh vesicle.[106,107] The specificity of all these assays is very high and standard culture techniques can be used to test EIA-negative samples to ensure maximum detection.

PCR

The detection of HSV DNA by PCR techniques which amplify a specific DNA sequence, for example a target on the DNA polymerase gene[108] or the thymidine kinase gene[109] is proving to be rapid and sensitive. Use of this technique will be invaluable for diagnosing potentially life-threatening HSV infections and asymptomatic relapses, for which less sensitive techniques are unreliable.[110]

Serology

The major requirement of HSV serology is to detect current or past infection, particularly with HSV-2, and thus predict a potential relapse, for example in pregnant women when they reach the time of delivery, or in immunocompromised patients.

Current infection is diagnosed only by demonstrating seroconversion between acute- and convalescent-phase sera during primary episodes. Recurrences cannot be diagnosed serologically, since the antigenic load is insufficient to stimulate an anamnestic response. Thus, stable, high antibody titres may be the result of a previous infection with either serotype, with or without a current relapse. Such titres may also indicate recurrence of a previously asymptomatic episode or a current infection in which seroconversion has already occurred.

Testing for IgG antibody is achieved most efficiently by the use of commercially available EIAs or latex agglutination assays, but these cannot distinguish between the two serotypes. Detection of IgM antibody by current methods does not improve the specificity of the diagnosis.

An assay, in which glycoprotein G(gG-2) derived from HSV-2-infected cells is used in a conventional ELISA format, enables HSV-1 to be differentiated from HSV-2, IgG antibody to be demonstrated in a past HSV-2 infection and IgM antibody in a current primary infection.[111] A similar assay, using recombinant DNA derived gG-2,[112] had a sensitivity of 93% and a specificity of 99% for the detection of IgG antibody, compared to confirmed HSV-2 culture results.

Aspects of diagnosis requiring emphasis

If vesicle fluid is available and a rapid diagnosis is required, direct immunofluorescence or immunoperoxidase microscopy is likely to provide it. Of the other methods available, the PCR technique is the most likely to find a permanent niche, but isolation and identification of HSV in cell culture has not yet been displaced as the 'gold standard' for diagnosis. Some commercially available serological kits are of limited value because of the crossreactivity of HSV-1 and HSV-2. Full commercial development of tests which distinguish between the two viruses on the basis of their glycoproteins is likely to be of benefit to the field.

HUMAN PAPILLOMA VIRUS INFECTION

About 80 different human papillomavirus (HPV) genotypes have now been cloned from clinical material and at least 28 of these affect the genital tract.[113] HPV types 6 and 11 are found predominantly in benign genital warts (condylomata acuminata) and to a lesser extent in premalignant lesions of various genital sites, including the cervix, vulva and penis. The lesions are usually of low grade, for example cervical intraepithelial neoplasia grades I and II. These HPV types have not been found in malignant lesions of the cervix, whereas HPV types 16, 18 and some others have been found associated with all grades of intraepithelial neoplasia and with malignant disease of the cervix and penis. Although infectious HPV virions have been produced by transfecting primary keratinocytes with HPV type 18 DNA,[114] HPVs have not been grown reliably in cell cultures. However various procedures outside and in the laboratory are available to detect infections caused by them.

Diagnostic procedures

Clinical examination with the unaided eye is the oldest diagnostic technique. However, many HPV infections, especially those of the cervix, produce lesions that are flat and grossly invisible and so require microscopic examination for detection. The intraepithelial neoplastic lesions that occur on the transformation zone of the cervix are best seen with the aid of a colposcope (a microscope) after painting the cervix with 5% acetic acid.[115] Colposcopy has been extended to include microscopic examination of the whole female anogenital tract, and of the male anogenital tract too, with or without the application of acetic acid.[116,117] The sensitivity and specificity of the procedure depends on the operator and on the area studied. Well-defined criteria have now been established for the interpretation of acetowhite changes of the cervix.[118] However, if women have not been preselected on the basis of cytology, colposcopic examination, combined with acetic acid treatment, lacks specificity and, in comparison with the molecular techniques mentioned below, it also lacks sensitivity for detecting HPV infections.[115] The same comment may be made about the sensitivity and specificity of the technique when applied to the male genitalia, especially for scrotal lesions.[116]

At colposcopy, for lesions that are not otherwise visible, specimens may be taken for examination by cytological, histological and molecular techniques, the latter being detailed below. Of course, the same procedures may be applied to lesions that are visible and do not require colposcopy. Cytological and histological examinations reveal cellular changes produced by HPV infection. In contrast, electron microscopy may be used to detect virus directly. However, it is costly, time consuming, insensitive and not suitable for detecting specific types and direct detection is now based entirely on molecular technology.

Molecular procedures

The specimens to be tested (cell scrapes, biopsies) are placed preferably in a small volume of phosphate-buffered saline. In addition, it is possible to sample the superficial layers of the skin to detect HPV DNA by using the 'Superglue' (cyanoacrylate glue) technique.[119] One drop of glue is placed on a clean glass slide which is applied immediately to the skin, kept in place for 1 min and then removed.

The technology used to detect HPV DNA is hybridization of the DNA with either DNA or RNA probes directed against specific types of HPV DNA. Amplification of the DNA by PCR technology before hybridization greatly increases the sensitivity of detection and this procedure is accepted and widely used.[120] All DNA, both human and viral specific, is extracted from the clinical sample and then general primers are used to amplify DNA of several HPV types (for example, 6, 11, 16, 18, 31 and 33). This is done by using a reaction mixture containing the DNA, the primers and *Taq* polymerase. The mixture is subjected to 40 or more cycles of amplification using a PCR processor.[119,121] If PCR amplification is not used, the DNA is cut with one or more restriction enzymes chosen to best demonstrate characteristic banding patterns. The fragments of DNA so produced, or the products of the PCR, are then separated electrophoretically in an agarose gel, denatured and transferred and fixed to a nylon membrane for Southern hybridization with one or more type-specific labelled oligonucleotide probes[119,121] (Fig. 3.2).

Attempts are made to avoid cross-contamination of DNA products by preparing reagents, processing patient material and undertaking the PCR assay and Southern blotting in separate rooms. Positive displacement pipettes or single-use Pasteur pipettes are used to distribute reagents for the PCR and to add the DNA samples to the reaction mixtures. Positive and negative controls are included in each assay.

The PCR enzyme immunoassay is a simple and rapid method for detecting multiple HPV types and may prove useful for large epidemiological studies.[122]

Diagnostic procedures in genitourinary medicine 37

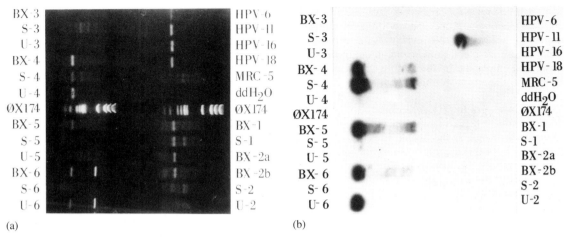

Fig. 3.2 HPV DNA detection. DNA fragments are separated electrophoretically in an agarose gel (a). Once denatured, transferred and fixed to a nylon membrane, specific fragments are visualized through Southern hybridization with one or more labelled oligonucleotide probes (b).

Hybrid capture, in which hybrids formed between HPV DNA and an RNA probe are detected by antibody binding and subsequent chemiluminescence, is of similar sensitivity to the PCR.[123]

Serology

Circulating antibodies in patients with genital HPV infections are mainly to antigens L1 and E4, E6/7 and E2, in that order. Such antibodies are also detected in individuals without overt signs of disease, although presumably with subclinical infections. Western blot (immunoblot) and enzyme-linked immunosorbent assays have been developed for measuring antibodies to HPV using, as antigens, bacterially expressed fusion proteins, synthetic peptides or HPV propagated in a mouse xenograft system.[124] However, reproducibility has been a problem. Tests that are based on capsid proteins produced by cloning HPV in vaccinia virus or baculovirus, the expressed capsids containing conformationally correct L1 and L2 epitopes, seem to be more promising. The results of studies using HPV type 16 L1/L2 virus-like particle EIAs have suggested that current or recent exposure to HPV stimulates a clear antibody response which gradually declines.[125] Seroconversion occurs about 8 months after detection of HPV DNA at the cervix and antibody persists for at least 40 months.[126]

Aspects of diagnosis requiring emphasis

Cytology and histology are the basis of routine diagnosis. However, these methods point only indirectly to an HPV infection. Since it is not possible to isolate HPVs in cultured cells, their detection and typing depend entirely on molecular techniques, HPV DNA amplification by the PCR before hybridization with specific probes being the method of choice. Serology will probably have an increasing role in the future.

HUMAN IMMUNODEFICIENCY VIRUS (HIV) INFECTION

HIV is classified in the family Retroviridae, genus *Lentivirinae*, along with viruses that cause immunosuppression, arthritis and autoimmune disease in a variety of animal hosts. HIV exists as two antigenically distinguishable subtypes, HIV-1 and HIV-2, which differ in the efficiency of their transmission and replication. The HIV genome comprises two copies of a single-stranded RNA molecule.

The natural history of HIV infection is one of progression from an acute, symptomatic but antibody-negative primary phase to an extended asymptomatic period with minimal clinical and immunological abnormalities. At the end of this stage, the appearance of opportunistic infections marks profound immunodeficiency and progression to AIDS. Efficient diagnosis of HIV infection must take account of the lag phase between infection and antibody production, the antigenic variation between HIV strains and the very low levels of virus that exist in some samples.

Features of HIV infection

Antigens in HIV infection

Three genes containing highly conserved regions code for the core proteins p24 and p55 (*gag* gene), the

RNA-dependent DNA polymerase, protease and endonuclease (*pol* gene) and the two major envelope proteins, gp 120 (gp 140 in HIV-2) and gp41 (*env* gene). These structural proteins are strongly immunogenic and important in the serological response to HIV.

Antibody response in HIV infection

There is a phase between infection and the appearance of antibody, known as the 'window' period. Antibody appears 2–8 weeks after the acute, symptomatic stage of infection, which is marked by a transient viraemia. Antibodies to the major core (*gag*) proteins, p25 and p55, appear first, accompanied by a corresponding decrease in viraemia[127] and *env* and *pol* antibodies may develop at the same time or slightly later. As viral replication and disease progress, p24 antibody declines as immune complexes develop. Anti-*env* antibody, however, remains raised.[128]

Specific IgM, IgG and IgA antibodies develop during HIV infection. The IgM antibody response has been shown to peak at about one week after seroconversion and thereafter to decline, so it may be useful for monitoring specific incidents of seroconversion. The IgG antibody response is the best characterized, rising steadily and reaching a plateau after 3–5 weeks. Levels of IgA antibody are generally low and variable.[129]

Prognostic indicators in HIV infection

Surrogate markers can be used to monitor the progress of HIV infection. The most useful is the absolute number of lymphocytes carrying the CD4 marker, the major T-lymphocyte receptor for the viral surface glycoprotein gp120. Depletion of these cells, which are important in the control of pathogens and neoplasms, marks the onset of immunodeficiency and AIDS.[130] Viral load, measured by the level of p24 antigen, infectious virus or proviral DNA and RNA, is also a good indicator of the progress of HIV infection[131] and low levels of the immune activation marker β_2-microglobulin correlate with rapid progress to AIDS.[132]

Samples and methods for HIV testing

Detection of HIV antibodies in serum is the most efficient way to determine whether an individual has been exposed to HIV infection and many tests are available for this purpose. Antibodies can also be detected in saliva, oral mucosal transudates and urine[135,136] but as test performances may vary,[137] these are not yet the samples of choice.

The definitive test for active infection is the demonstration of virus or viral antigens in cells or body fluids. Virus can be cultured from infected peripheral blood mononuclear cells (PBMCs) or body fluids and identified in cell monolayers by a p24 ELISA, immunofluorescence, Western blotting, a reverse transcriptase assay and *in situ* hybridization. The most sensitive method of detecting viral antigens is the PCR assay which can be adapted to detect integrated or free episomal HIV, proviral DNA or HIV RNA.

Collection of samples

Blood samples for antibody tests are collected by venepuncture. Plasma, serum and other body fluids are stored at −20°C. Virus isolation from heparinized or EDTA-treated blood samples and from other body fluids ideally should be undertaken immediately. If this is impractical, virus in whole blood will survive for 36 h at room temperature or, if a longer period is envisaged, storage should be at 4°C after the first 24 h. PBMCs should be separated from blood by Ficoll-Hypaque gradient centrifugation. They can be maintained in medium containing serum and IL-2 for up to 10 days before culturing or stored in liquid nitrogen in culture medium containing 10% dimethyl sulphoxide (DMSO). Samples for antigen detection can be stored at −70°C. The results of a recent study have shown that frozen-thawed total white cells and buffy coat samples may be as useful as PBMCs for detection of HIV by PCR.[138]

Antibody screening tests

ELISAs

The ELISA is most commonly used to screen for HIV antibody because it is inexpensive, reliable and rapid. It exists in many formats and modifications are constantly being devised to boost sensitivity and specificity and to enable simultaneous screening for HIV-1 and HIV-2.[139] Commercially available assays are well characterized[140] and their standardization provides consistency between laboratories. They are designed to detect antibodies to the multiple antigens produced during the course of infection, such as p24, p17, gp160, gp120 and gp41, and tests range in sensitivity from 93% to 100%.[141]

Indirect ELISAs use either HIV structural proteins purified by sucrose gradient centrifugation or recombinant and synthetic peptides as antigen. These antigens are absorbed onto beads or microtitre plates or, for greater specificity, 'captured' by a monoclonal antibody. HIV antibodies are bound and detected after being complexed with enzyme-labelled antihuman IgG. The more recently developed antigen sandwich technique assays, in which a second labelled antigen is added after the first antibody, detect both IgM and IgG antibodies so that they are useful at the time of seroconversion.

Competitive ELISAs are based on cruder antigen preparations and specific HIV antibody in the sample competes with labelled antibody in the test. They have a lower sensitivity but greater specificity than indirect methods.[142]

Agglutination assays
Whole blood or serum is mixed with latex beads coated with recombinant or synthetic antigen, rotated on a card and evaluated by eye for agglutination. No electrical equipment is needed so they can be used when facilities for performing ELISAs are not available. Sensitivity and specificity are variable, ranging from 71% to 99% and 93% to 99%, respectively.[141]

Dot-blot assays
Recombinant or synthetic antigen is blotted passively onto a nitrocellulose support or 'captured' on microparticles and antibody is detected by a colour reaction. The method is very rapid but expensive. Good sensitivity has been reported by some investigators[143,144,145] but caution in early infection is urged by others.[146]

Particle adherence test
IgG antibody in a test sample is captured by rabbit antihuman IgG antibody on a microtitre plate; when a suspension of HIV-coated gelatine particles is added, they will adhere to any specific antibody present. Excellent results have been reported with urine specimens.[147]

Confirmatory assays

False-positive and negative results may occur in any HIV antibody test. These may be due, for example, to variability in the sensitivity and specificity of the test system, to the sample being taken too early or too late in the course of the infection, to the inability of the assay to detect antibodies to the infecting subtype or to human error. It is important, therefore, to perform a confirmatory assay of high specificity.

Western blot (WB) assay
This is used most widely for confirmation. Electrophoresed viral proteins from virus cell lysates, or synthetically produced, are transferred to nitrocellulose membranes and reacted with the serum sample. The profile of antibodies bound to the viral proteins is identified by a conjugated second antibody–substrate reaction. Although this method is still considered to be the 'gold standard', the existence of different criteria for a positive result and problems with weak bands can make interpretation difficult.[148]

Indirect immunofluorescence assays (IFA)
These simple, rapid and cheap assays can be used to resolve equivocal results of Western blotting. However, they are unsuitable for examining large numbers of samples and require experienced staff to interpret the fluorescent staining of infected cells by positive sera.

Line immunoassays (LIA)
These involve the application of bands of recombinant proteins or synthetic polypeptides to plastic or nitrocellulose strips without electrophoresis, so that antibody responses to optimal amounts of uncontaminated antigen are detected. They are considerably cheaper than WBs and are reliable for confirmation.[149]

Virus culture and direct antigen tests

Virus culture
Between 3 and 6×10^6 PBMCs to be tested are cocultured with phytohaemagglutinin-stimulated PBMCs from seronegative individuals, in a medium containing IL-2, in 5% CO_2 at 37°C. The medium is replaced every 3–4 days, filtered or centrifuged and the supernatant is tested for virus by one of the following assays.

ELISA
Infected cells are treated with detergent to disrupt the virions, p24 antigen is captured with bound antibody and detected with a labelled monoclonal antibody in an enzyme–substrate reaction. A positive reaction is verified by a reduction in the optical density reading after treating the original sample with anti-p24 antibody.

Reverse transcriptase (RT) assays
These are performed on virus pelleted from 1 ml of culture fluid, which is disrupted and placed in a reaction mixture with a poly(a) RNA template primer, unlabelled dNTPs and a radioactively labelled dTTP. The presence of RT and hence virus is indicated by the detection of labelled reaction products by autoradiography.[150] The assay is not specific for HIV, so it is used to confirm HIV infection in conjunction with other evidence.

WB assay
Different HIV gene products can be distinguished by analysing infected cell lysates or culture supernatants with high titre control antisera to all HIV proteins. Reactive bands may be detected by an enzyme–substrate system or, when very low levels of antigen are involved, by autoradiography. The WB assay is particularly useful for identifying HIV-1 and

HIV-2 infections if the results of other tests are ambiguous.

Indirect IFA
Infected cells are identified by reaction with known positive and negative HIV antisera and a fluorescein-labelled second antibody. The test is cheap but experience is needed for interpretation.

In situ hybridization
Hybridization is achieved by applying a denatured ^{35}S-labelled HIV DNA probe to suitably denatured nucleic acids in cells or a biopsy sample. Following application of photographic emulsion and development, the presence of autoradiographic grains over cells indicates intracellular HIV. This technique detected virally infected cells 2–5 days before an RT assay on the culture supernatant was positive.[151]

PCR
A rapid culture/PCR method has been described for the quantitation of infectious HIV from plasma.[152] The results correlated closely with those of longer culture-based assay methods.

Direct antigen tests
The major advantage of the p24 antigen ELISA and the *in situ* hybridization and highly sensitive PCR technologies is that virus can be detected without propagation in cell culture.

ELISA
Antibody-captured p24 antigen is reacted with an enzyme-conjugated antibody, which produces a colour change when a substrate is added. Disassociation of antigen–antibody complexes using low pH prior to testing increases sensitivity.[153]

Table 3.3 Summary of diagnostic methods currently available

Disease/infection	Microscopy	Culture	Other tests	Serology	Minimum requirement
Syphilis	+ (Dark ground)	–	–	+	Serology
Gonorrhoea	+ (Gram stain)	+	–	–	Gram stain
Chancroid	±* (Gram stain)	+	–	–	Gram stain
Granuloma inguinale	+ (Giemsa stain)	–	–	±*	Giemsa stain
C. trachomatis infection	+ (Fluorescence)	+	+ (EIA; PCR; LCR)	±*	Fluorescence stain
Mycoplasmal infection	–	+	+ (PCR)	+	Culture/PCR
Bacterial vaginosis	+ (Gram stain)	–	+ (Composite criteria)	–	Gram stain
T. vaginalis infection	+ (Wet mount; fluorescence)	+	–	–	Wet mount
Candidiasis	+ (Wet mount; Gram stain)	+	–	–	Wet mount
Herpes simplex virus infection	+ (Fluorescence)	+	+ (EIA)	–	Fluorescence stain/EIA
Human papillomavirus infection	+ ('Pap' smear: koilocytes)	–	+ (PCR)	–	'Pap' smear
Human immunodeficiency virus infection	–	+	+ (EIA; PCR)	+	Serology

*Poor option

In situ hybridization
Using the technique outlined previously, HIV antigen can be identified in many tissues, including the lymphatic system.[154]

PCR
Proviral DNA sequences can be amplified by the PCR from only 1 μg of cellular DNA which is equivalent to 20 μl of blood, but the many different strains of HIV are efficiently detected only if the primers target highly conserved regions of the genome, such as those in the *gag, env* and *pol* genes and the long terminal end repeat. A commercially available PCR assay which targets the *gag* gene has been shown to have 100% sensitivity and between 96% and 100% specificity.[155,156] In comparison with detection of proviral DNA in PBMCs by PCR, viral culture from PBMCs and plasma has been shown to be only 89% and 75% sensitive, respectively.[157]

Aspects of diagnosis requiring emphasis

Detection of antibody to HIV by an ELISA is only presumptive evidence of infection. Because non-specific reactions may produce false-positive results, a positive test must be confirmed by a second ELISA and an independent confirmatory test, such as WB. Furthermore, the absence of antibodies should not be considered as conclusive proof of lack of infection. Antibody testing should be repeated at intervals in individuals who are at high risk to exclude testing during the antibody-negative 'window' period, which may be extensive.

If antibody test results are inconclusive, infection by HIV-2 may be revealed with type-specific antisera or WB analysis of a viral lysate. The p24 antigen ELISA may detect the transient viraemia of early infection with both viral subtypes and, more likely, the increasing viral load of late disease. Detection of viral antigen by culture or by PCR are the most reliable methods.

Measures are now available to measure the plasma viral load, which is a good predictor of progression to AIDS, and to assess the development of antiviral drug resistance.

CONCLUSIONS AND RECOMMENDATIONS

There are many options in diagnosing sexually transmitted infections/diseases caused by specific microorganisms. We recognize that what is done may be outside the control of individual genitourinary medicine physicians but they should at least know what the options are and, furthermore, not be reluctant to question those in the laboratory if test results do not match up with clinical observation or suspicion. In these days of cost cutting and so-called cost-effectiveness, what is best for the patient, which of course should be the primary concern, may not equate with what can be offered on a limited budget. We recognize, therefore, that some of the more sophisticated procedures may not be considered, simply because they are too expensive or, sometimes, because they are outside the capabilities of the laboratory.

In Table 3.3 we present a summary of the diagnostic methods currently available for diagnosing a particular disease or infection, together with the minimal requirements for making such a diagnosis. It is noteworthy that the latter may often be met by the availability of simple microscopy (rarely, fluorescence microscopy). If this is not possible, then syndromic diagnosis and treatment is the only way forward. While this may have to be the case in some developing countries, it is unlikely to be so in developed countries. Nevertheless, sensible use is not always made of the facilities available. This is particularly so in investigations of infections of the female genital tract when often there is a lack of appreciation of what is needed. Our recommendations, shown in Table 3.4, should find acceptance by both clinicians and laboratory personnel since they are designed to prevent excessive specimen taking and inappropriate laboratory work.

Table 3.4 Recommendations for investigating sexually transmitted genital tract infections in women

Lower genital tract infection
  Investigations, for example, of vaginal discharge, vulvovaginitis, vulval itching*

  Endocervical infection
    Specimen required: endocervical swab for smear and culture.
    Examination for *N. gonorrhoeae* and *C. trachomatis*

  Vaginal infection
    Specimen required: high vaginal swab for smear
    Examination for *T. vaginalis*, *Candida* spp., bacterial vaginosis

Upper genital tract infection
  Investigation, for example, of lower abdominal pain
  Specimen required: endocervical swab for smear and culture.
  Examination for: *N. gonorrhoeae* and *C. trachomatis*

*A swab from the vulva, especially the skin, is usually unhelpful

REFERENCES

1. Young, H. and Penn, C.W. (1990) Syphilis, yaws and pinta, in *Topley and Wilson's Principles of Bacteriology, Virology and Immunology*, (eds G.R. Smith and C.S.F. Easmon), Arnold, Kent, pp. 588–604.
2. Lukehart, S.A., Tam, M.R., Hom. J. *et al.* (1985) Characterization of monoclonal antibodies to *Treponema pallidum*. *J. Immunol.*, **134**, 585–592.
3. Young, H. (1992) Syphilis: new diagnostic directions. *Int. J.STD & AIDS*, **3**, 391–413.
4. Hay, P.E., Clarke, J.R., Strugnell, R.A. *et al.* (1990) Use of the polymerase chain reaction to detect DNA sequence to pathogenic treponemes in cerebrospinal fluid. *FEMS Microbiol. Letts*, **58**, 233–238.
5. Hay, P.E., Clarke, J.R., Taylor-Robinson, D. and Goldmeier, D. (1990) Detection of treponemal DNA in the CSF of patients with syphilis and HIV infection using the polymerase chain reaction. *Genitourinary Med.*, **66**, 428–432.
6. Sanchez, P.J., Wendel, G.D., Grimpnel, E. *et al.* (1993) Evaluation of molecular methodologies and rabbit infectivity testing for the diagnosis of congenital syphilis and neonatal nervous system invasion by *Treponema pallidum*. *J.Infect. Dis.*, **167**, 148–157.
7. Goodhart, M.E., Ogden, J., Zaidi, A.A. and Kraus, S.J. (1982) Factors affecting the performance of smear and culture tests for the detection of *Neisseria gonorrhoeae*. *Sex.Trans.Dis.*, **9**, 63–69.
8. Schachter, J., McCormack, W.M., Smith, R.F. *et al.* (1984) Enzyme immunoassay for diagnosis of gonorrhoea. *J. Clin. Microbiol.*, **19**, 57–59.
9. Lewis, J.S., Fakile, O., Foss, E. *et al.* (1993) Direct DNA probe assay for *Neisseria gonorrhoeae* in pharyngeal and rectal specimens. *J. Clin. Microbiol.*, **31**, 2783–2785.
10. Ison, C.A. (1990) Methods of diagnosing gonorrhoea. *Genitourinary Med.*, **66**, 453–459.
11. Thayer, J.D. and Martin, J.E. (1964) A selective medium for the cultivation of *Neisseria gonorrhoeae* and *Neisseria meningitidis*. *Public Health Report*, **79**, 49–57.
12. Phillips, I., Humphrey, D., Middleton, A. and Nicol, C.S. (1972) Diagnosis of gonorrhoea by culture on selective medium containing vancomycin, colistin, nystatin and trimethoprim (VCNT). A comparison with Gram-staining and immunofluorescence. *Br. J. Vener. Dis.*, **48**, 287–292.
13. Flynn, J. and Watkins, S.A. (1972) A serum-free medium for testing fermentation reaction in *Neisseria gonorrhoeae*. *J.Clin.Pathol.*, **25**, 525–527.
14. Brown, W.J. (1974) Modification of the rapid fermentation test for *Neisseria gonorrhoeae*. *Appl. Microbiol.*, **27**, 1027–1030.
15. D'Amato, R.F., Eriquez, L.A., Tomforde, K.M. and Singerman, E. (1978) Rapid identification of *Neisseria gonorrhoeae* and *Neisseria meningitidis* by enzymatic profiles. *J. Clin. Microbiol.*, **7**, 77–81.
16. Young, H. and Moyes, A. (1989) Utility of monoclonal antibody co-agglutination to identify *Neisseria gonorrhoeae*. *Genitourinary Med.*, **65**, 8–13.
17. Ison, C.A., Tanna, A. and Easmon, C.S.F. (1988) Evaluation of a fluorescent monoclonal antibody reagent for the identification of cultured *Neisseria gonorrhoeae*. *J. Med. Microbiol.*, **26**, 121–123.
18. O'Callaghan, C.H., Morris, A., Kirby Sy, S.M. and Shingler, A.H. (1972) Novel method for the detection of beta-lactamases by using a chromogenic cephalosporin substitute. *Antimicrob. Agents Chemother.*, **1**, 283–288.
19. Ison, C.A., Tekki, N. and Gill, M.J. (1993) Detection of the *tetM* determinant in *Neisseria gonorrhoeae*. *Sexually Trans. Dis.*, **20**, 329–333.
20. Van Dyck, E., Smet, H. and Piot, P. (1994) Comparison of E-test with agar dilution for antimicrobial susceptibility testing of *Neisseria gonorrhoeae*. *J. Clin. Microbiol.*, **32**, 1586–1588.
21. Yeung, K-H., Ng, L-K. and Dillon, J.R. (1993) Evaluation of E-test for testing antimicrobial susceptibilities of *Neisseria gonorrhoeae* isolates with different growth media. *J. Clin. Microbiol.*, **31**, 3053–3055.
22. Ison, C.A., Branley, N.S., Kirtland, K. and Easmon, C.S.F. (1991) Surveillance of antibiotic resistance in clinical isolates of *Neisseria gonorrhoeae*. *Br. Med. J.*, **303**, 1307.
23. Schmid, G.P., Sanders, L.L., Blount, J.H. and Alexander, E.R. (1987) Chancroid in the United States. Re-establishment of an old disease. *J. Am. Med. Assoc.*, **258**, 3265–3268.
24. Morse, S.A. (1989) Chancroid and *Haemophilus ducreyi*. *Clin. Microbiol. Revs.*, **2**, 137–157.
25. Messing, M., Sottnek, F.O., Biddle, J.W., *et al.* (1983) Isolation of *Haemophilus* species from the genital tract. *Sexually Trans. Dis.*, **10**, 56–61.
26. Choudhary, B.P., Kumari, S., Bhati, R. and Agarwal, D.S. (1982) Bacteriological study of chancroid. *Indian. J. Med. Res.*, **76**, 370–385.

27. Hammond, G.W., Lian, C.J., Wilt, J.C. and Ronald, A.R. (1978) Comparison of specimen collection and laboratory techniques for the isolation of *Haemophilus ducreyi*. *J. Clin. Microbiol.*, **7**, 39–43.
28. Dangor, T., Radebe, F. and Ballard, R.C. (1993) Transport media for *Haemophilus ducreyi*. *Sexually Trans. Dis.*, **20**, 5–9.
29. Piot, P. and Laga, M. (1989) Genital ulcers, other sexually transmitted diseases, and the sexual transmission of HIV. *Br. Med. J.*, **298**, 623–624.
30. Senghal, V.N. and Prasad, A.L. (1986) Donovanosis – current concepts. *Int. J. Dermatol.*, **5**, 8–16.
31. Van Dyck, E. and Piot, P. (1992) Laboratory techniques in the investigation of chancroid, lymphogranuloma venereum and donovanosis. *Genitourinary Med.*, **68**, 130–133.
32. O'Farrell, N., Hoosen, A.A., Coetzee, K. and van den Ende, J. (1990) A rapid stain for the diagnosis of granuloma inguinale. *Genitourinary Med.*, **66**, 200–201.
33. Freinkel, A.L. (1988) Granuloma inguinale of cervical lymph nodes simulating tuberculous lymphadenitis: two case reports and review of published reports. *Genitourinary Med.*, **64**, 339–343.
34. Goldberg, J., Weaver, R.H., Packer, H. and Simpson, W.G. (1953) The complement fixation test in the diagnosis of granuloma inguinale. *Am. J. Syphilis, Gonorrhea & Vener. Dis.*, **37**, 71–76.
35. Freinkel, A.L., Dangor, Y., Koornhof, H.J. and Ballard, R.C. (1992) A serological test for granuloma inguinale. *Genitourinary Med.*, **68**, 269–272.
36. Schachter, J. (1984) Biology of *Chlamydia trachomatis*, in *Sexually Transmitted Diseases, 2nd edn*, (eds K.K. Holmes et al.), McGraw-Hill, New York, pp. 243–257.
37. Taylor-Robinson, D. and Thomas, B.J. (1991) Laboratory techniques for the diagnosis of chlamydial infections. *Genitourinary Med.*, **67**, 256–266.
38. Dutilh, H., Bébéar, C., Rodriguez, P. et al. (1989) Specific amplification of a DNA sequence common to all *Chlamydia trachomatis* serovars using the polymerase chain reaction. *Res. Microbiol.*, **140**, 7–16.
39. Dille, B.J., Butzen, C.C. and Birkenmeyer, L.G. (1993) Amplification of *Chlamydia trachomatis* DNA by ligase chain reaction. *J. Clin. Microbiol.*, **31**, 729–731.
40. Taylor-Robinson, D. (1994) Immunochemical techniques for the detection of chlamydial species, in *Rapid Methods and Automation in Microbiology and Immunology*, (eds R.C. Spencer, E.P. Wright and S.W.B. Newson), Intercept. Andover, pp. 361–363.
41. Thomas, B.J., MacLeod, E.J. and Taylor-Robinson, D. (1993) Evaluation of sensitivity of ten diagnostic assays for *Chlamydia trachomatis* by use of a simple laboratory procedure. *J. Clin. Pathol.*, **46**, 408–410.
42. Thomas, B.J., MacLeod, E.J., Hay, P.E. et al. (1994) Limited value of two widely used enzyme immunoassays for detection of *Chlamydia trachomatis* in women. *Eur. J. Clin. Microbiol. Infect. Dis.*, **13**, 651–655.
43. Ferris, D.G., Martin, W.H., Fischer, P.M. and Petry, L.J. (1990) A comparison of rapid enzyme immunoassay tests for the detection of *Chlamydia trachomatis* cervical infections. *J. Fam. Pract.*, **31**, 597–601.
44. Clarke, L.M., Sierra, M.F., Daidone, B.J. et al. (1993) Comparison of the Syva Micro Trak enzyme immunoassay and Gen-Probe PACE 2 with cell culture for diagnosis of cervical *Chlamydia trachomatis* infection in a high prevalence female population. *J. Clin. Microbiol.*, **31**, 968–971.
45. Bianchi, A., Scieux, C., Brunat, N. et al. (1994) An evaluation of the polymerase chain reaction Amplicor for *Chlamydia trachomatis* in male urine and female urogenital specimens. *Sexually Trans. Dis.*, **21**, 196–200.
46. Story, A., Najim, B. and Lee, H.H. (1997) Vulval swabs as alternative specimens for ligase chain reaction detection of genital chlamydial infection in women. *J. Clin. Microbiol.*, **35**, 836–838.
47. Chernesky, M.A., Jong, D.J., Lee, H. et al. (1994) Diagnosis of *Chlamydia trachomatis* infections in men and women by testing first void-urine by ligase chain reaction. *J. Clin. Microbiol.*, **32**, 2682–2685.
48. Thomas, B.J., Pierpoint, T., Taylor-Robinson, D. et al. (1998) Sensitivity of the ligase chain reaction assay for detecting *Chlamydia trachomatis* in vaginal swabs from women who are infected at other sites. *Sexually Trans. Inf.*, **74**, 140–141.
49. Mahoney, J.B. and Chernesky, M.A. (1985) Effect of swab type and storage temperature on the isolation of *Chlamydia trachomatis* from clinical specimens. *J. Clin. Microbiol.*, **22**, 865–867.
50. Lees, M.I., Newnam, D.M., Plackett, M. et al. (1990) A comparison of cytobrush and cotton swab sampling for the detection of *Chlamydia trachomatis* by cell culture. *Genitourinary Med.*, **66**, 267–269.

51. Hay, P.E., Thomas, B.J., Gilchrist, C. et al. (1991) The value of urine samples from men with non-gonococcal urethritis for the detection of Chlamydia trachomatis. Genitourinary Med., 67, 124–128.
52. Kellog, J.A., Vanderhoff, B.T., Seiple, J.W. and Hick, M.E. (1994) Comparison of first-voided urine specimens with endocervical swab specimens for enzyme-linked immunosorbent assay detection of Chlamydia trachomatis in women. Arch. Fam. Med., 3, 672–675.
53. Thomas, B.J., Pierpoint, T., Taylor-Robinson D. et al. (1998) Quantification of Chlamydia trachomatis in cervical and urine specimens from women attending a genitourinary medicine clinic: implications for screening strategies. Int. J. STD & AIDS, 9, 448–451.
54. Zceberg, B., Thelin, I. and Schalen, C. (1992) Chlamydia trachomatis antigen detection by Chlamydiazyme combined with Chlamydia Blocking Agent verification. Int. J. STD & AIDS, 3, 355–359.
55. Thomas, B.J., MacLeod, E.J. and Taylor-Robinson, D. (1995) Evaluation of a commercial polymerase chain reaction assay for Chlamydia trachomatis and suggestions for improving sensitivity. Eur. J. Clin. Microbiol. Infect. Dis., 14, 719–723.
56. Goudswaard, J., Sabbe, L. and van Belzen, C. (1989) Interference by Gram-negative bacteria in the enzyme immunoassay for detecting Chlamydia trachomatis. J. Infect. Dis., 18, 94–96.
57. Puolakkainen, M., Koussa, M. and Saikku, P. (1987) Clinical conditions associated with positive complement fixation serology for Chlamydiae. Epidemiol. Infect., 98, 101–108.
58. Keat, A., Thomas, B.J. and Taylor-Robinson, D. (1983) Chiamydial infection in the aetiology of arthritis. Br. Med. Bull., 39, 168–174.
59. Schachter, J., Grossman, M. and Azimi, P.H. (1982) Serology of Chlamydia trachomatis in infants. J. Infect. Dis., 146, 530–535.
60. Bowie, W.R., Wang, S-P., Alexander, E.R. et al. (1977) Aetiology of non-gonococcal urethritis. J. Clin. Invest., 59, 735–742.
61. Freundt, E.A. (1983) Culture media for classic mycoplasmas, in Methods in Mycoplasmology, Vol 1, (Eds S. Razin and J.G. Tully), Academic Press, New York, pp. 127–135.
62. Shepard, M.C. (1983) Culture media for ureaplasmas, in Methods in Mycoplasmology, Vol 1, (Eds S. Razin and J.G. Tully), Academic Press, New York, pp. 137–146.
63. Tully, J.G., Rose, D.L., Whitcomb, R.F. and Wenzel, R.P. (1979) Enhanced isolation of Mycoplasma pneumoniae from throat washings with a newly modified culture medium. J. Infect. Dis., 139, 478–482.
64. Taylor-Robinson, D. (1989) Genital mycoplasma infections. Clin. Lab. Med., 9, 501–523.
65. Renaudin, H. and Bébéar, C. (1990) Evaluation des systèmes Mycoplasma PLUS et SIR Mycoplasma pour la détection quantitative et l'étude de la sensibilté aux antibiotiques des mycoplasma genitaux. Pathol. Biol., 38, 431–435.
66. Clyde, W.A. (1964) Mycoplasma species identification based on growth inhibition by specific antisera. J. Immunol., 92, 958–965.
67. Morrison-Plummer, J., Jones, D.H., Daly. K. et al. (1987) Molecular characterization of Mycoplasma genitalium species-specific and cross-reactive determinants: identification of an immunodominant protein of M. genitalium. Isr. J. Med. Sci., 23, 453–457.
68. Taylor-Robinson, D. (1983) Serological identification of ureaplasmas from humans, in Methods in Mycoplasmology, Vol 1, (eds S. Razin and J.G. Tully), Academic Press, New York, pp. 57–63.
69. Hooton, T.M., Roberts, M.C., Roberts, P.L. et al. (1988) Prevalence of Mycoplasma genitalium determined by DNA probe in men with urethritis. Lancet, i, 266–268.
70. Blanchard, A., Hentschel, J., Duffy, L. et al. (1993) Detection of Ureaplasma urealyticum by polymerase chain reaction in the urogenital tract of adults, in amniotic fluid, and in the respiratory tract of newborns. Clin. Infect. Dis., 17, (suppl 1), 148–153.
71. Palmer, H.M., Gilroy, C.B., Furr, P.M. and Taylor-Robinson, D. (1991) Development and evaluation of the polymerase chain reaction to detect Mycoplasma genitalium. FEMS Microbiol. Letts., 61, 199–203.
72. Wang, R.Y-H., Shih, J.W-K., Grandinetti, T. et al. (1992) High frequency of antibodies to Mycoplasma penetrans in HIV-infected patients. Lancet, 340, 1312–1316.
73. Furr, P.M. and Taylor-Robinson, D. (1984) Microimmunofluorescence technique for detection of antibody to Mycoplasma genitalium. J. Clin. Pathol., 37, 1072–1074.
74. Møller, B.R., Taylor-Robinson, D. and Furr, P.M. (1984) Serologic evidence implicating Mycoplasma genitalium in pelvic inflammatory disease. Lancet, i, 1102–1103.
75. Weström, L., Evaldson, G., Holmes, K.K. et al. (1984) Taxonomy of vaginosis: bacterial vaginosis – a definition, in Bacterial Vaginosis, (eds P.-A. Mårdh and D. Taylor-Robinson), Almqvist and Wiksell International, Stockholm, pp. 259–260.

76. Gardner, H.L. and Dukes, C.D. (1955) *Haemophilus vaginalis* vaginitis. A newly defined specific infection previously classified as 'nonspecific' vaginitis. *Am. J. Obstet. Gynecol.*, **69**, 962–976.
77. Totten, P.A., Amsel, R., Hale, J. et al. (1982) Selective differential human blood bilayer media for isolation of *Gardnerella (Haemophilus) vaginalis*. *J. Clin. Microbial.*, **15**, 141–147.
78. Easmon, C.S.F., Hay, P.E. and Ison, C.A. (1992) Bacterial vaginosis: a diagnostic approach. *Genitourinary Med.*, **68**, 134–138.
79. Amsel, R., Totten, P.A., Spiegel, C.A. et al. (1983) Nonspecific vaginitis: diagnostic criteria and microbial and epidemiologic associations. *Am. J. Med.*, **74**, 14–22.
80. Spiegel, C.A., Amsel, R. and Holmes, K.K. (1983) Diagnosis of bacterial vaginosis by direct Gram stain of vaginal fluid. *J. Clin. Microbiol.*, **18**, 170–177.
81. Nugent, R.P., Krohn, M.A. and Hillier, S.L. (1991) Reliability of diagnosing bacterial vaginosis is improved by a standardized method of Gram stain interpretation. *J. Clin. Microbiol.*, **29**, 297–301.
82. Briselden, A.M. and Hillier, S.L. (1994) Evaluation of Affirm VP Microbial Identification Test for *Gardnerella vaginalis* and *Trichomonas vaginalis*. *J. Clin. Microbiol.*, **32**, 148–152.
83. Meysick, K. and Garber, G.E. (1995) *Trichomonas vaginalis*. *Curr. Opin. Infect. Dis.*, **8**: 22–25.
84. Bickley, L.S., Krisher, K.K., Punsalang, A. et al. (1989) Comparison of direct fluorescent antibody, acridine orange, wet-mount and culture for detection of *Trichomonas vaginalis* in women attending a sexually transmitted diseases clinic. *Sexually Trans. Dis.*, **16**, 127–131.
85. Greenwood, J.R. and Kirk-Hillaire, K. (1981) Evaluation of acridine orange stain for detection of *Trichomonas vaginalis*. *J. Clin. Microbiol.*, **14**, 699.
86. Krieger, J.N., Tam, M.R., Stevens, C.E. et al. (1988) Diagnosis of trichomoniasis. Comparison of conventional wet-mount examination with cytologic studies, cultures and monoclonal antibody staining of direct specimens. *J. Am. Med. Assoc.*, **259**, 1223–1227.
87. Schmid, G.P., Matheney, L.C., Zaidi, A.A. et al. (1989) Evaluation of six media for the growth of *Trichomonas vaginalis* from vaginal secretions. *J. Clin. Microbiol.*, **27**, 1230–1233.
88. Thomason, J.L., Gelbart, S.M., Sobun, J.F. et al. (1988) Comparison of four methods to detect *Trichomonas vaginalis*. *J. Clin. Microbiol.*, **26**, 1869–1870.
89. Beal, C., Goldsmith, R., Kotby, M. et al. (1992) The plastic envelope method, a simplified technique for culture diagnosis of trichomoniasis. *J. Clin. Microbiol.*, **30**, 2265–2268.
90. McCormack, W.M., Starko, K. and Zinner, S.H. (1988) Symptoms associated with vaginal colonization with yeast. *Am. J. Obstet. Gynecol.*, **158**, 31–33.
91. Odds, F.C., Webster, C.E., Riley, V.C. and Fisk, P.G. (1987) Epidemiology of vaginal *Candida* infection: significance of number of vaginal yeasts and their biotypes. *Eur. J. Obstet. Gynaecol. Reprod. Biol.*, **25**, 53–66.
92. Moseley, R.C., Corey, D., Benjamin, A.D. et al. (1981) Comparison of viral isolation, direct immunofluorescence and direct immunoperoxidase techniques for detection of genital herpes simplex virus infection. *J. Clin. Microbiol.*, **13**, 913–918.
93. Jensen, C. and Johnson, F.B. (1994) Comparison of various transport media for viability maintenance of herpes simplex virus, respiratory syncytial virus and adenovirus. *Diag. Microbiol. Infect. Dis.*, **19**, 137–142.
94. Fife, K.H. and Corey, L. (1984) Herpes simplex virus, in *Sexually Transmitted Diseases*, 2nd edn, (eds K.K. Holmes et al.), McGraw-Hill, New York, pp. 941–952.
95. Pouletty, P., Chomel, J.J., Thouvenot, D. et al. (1987) Detection of herpes simplex virus in direct specimens by immunofluorescence assay using a monoclonal antibody. *J. Clin. Microbiol.*, **25**, 958–959.
96. Sabil, D., Othman, S.K. and Isahak, I. (1990) Comparison of direct immunoperoxidase and direct immunofluorescence for the detection of herpes simplex virus antigen in cell culture. *Malays. J. Pathol.*, **12**, 35–38.
97. Gleaves, C.A., Rice, D.H., Bindra, R. et al. (1989) Evaluation of a HSV-specific monoclonal antibody reagent for laboratory diagnosis of herpes simplex virus infection. *Diag. Microb. Infect. Dis.*, **12**, 315–318.
98. Johnston, S.L.G., Wellens, K. and Siegel, C.S. (1990) Rapid isolation of herpes simplex virus by using mink lung and rhabdosarcoma cell cultures. *J. Clin. Microbiol.*, **28**, 2806–2807.
99. Woods, G.L. and Mills, R.D. (1988) Conventional tube cell culture compared with centrifugal inoculation of MRC-5 cells and staining with monoclonal antibodies for detection of herpes simplex virus in clinical specimens. *J. Clin. Microbiol.*, **26**, 570–572.
100. West, P.C., Aldrich, B., Hartwig, R. and Haller, C.J. (1989) Increased detection of herpes simplex

virus in MRC-5 cells treated with dimethyl sulphoxide and dexamethasone. *J. Clin. Microbiol.*, **27**, 770–772.
101. Luker, G., Chow, C., Richards, D.F. and Johnson, F.B. (1991) Suitability of infection of cells in suspension for detection of herpes simplex virus. *J. Clin. Microbiol.*, **29**, 1554–1557.
102. Espy, M.J., Wold, A.D., Jespersen, D.J. *et al.* (1991) Comparison of shell vials and conventional tubes seeded with rhabdomyosarcoma and MRC-5 cells for the rapid detection of herpes simplex virus. *J. Clin. Microbiol.*, **29**, 2701–2703.
103. Baker, D.A., Pavan-Langston, D., Gonik, B. *et al.* (1990) Multicenter clinical evaluation of the Du Pont Herpchek HSV ELISA, a new rapid diagnostic test for the direct detection of herpes simplex virus. *Adv. Exp. Med. Biol.*, **263**, 71–76.
104. Verano, L. and Michalski, F.J. (1990) Herpes simplex virus antigen direct detection in standard virus transport medium by Du Pont Herpchek enzyme-linked immunosorbent assay. *J. Clin. Microbiol.*, **28**, 2555–2558.
105. Cone, R.W., Swenson, P.D., Hobson, A.C. *et al.* (1993) Herpes simplex virus detection from genital lesions: a comparative study using antigen detection (Herpchek) and culture. *J. Clin. Microbiol.*, **31**, 1774–1776.
106. Dorian, K.J., Beatty, E. and Atterbury, K.E. (1990) Detection of herpes simplex virus by the Kodak SureCell Herpes Test. *J. Clin. Microbiol.*, **28**, 2117–2119.
107. Zimmerman, S.J., Moses, E., Sofat, N. *et al.* (1991) Evaluation of a visual, rapid, membrane enzyme immunoassay for the detection of herpes simplex virus antigen. *J. Clin. Microbiol.*, **29**, 842–845.
108. Shimizu, C., Shimizu, H., Mitsuda, T. *et al.* (1994) One-step determination of herpes simplex virus types 1 and 2 by polymerase chain reaction. *Mol. Cell Probes*, **8**, 193–198.
109. Kimura, H., Shibata, M., Kuzishima, K. *et al.* (1990) Detection and direct typing of herpes simplex virus by polymerase chain reaction. *Med. Microbiol. Immunol. Berlin*, **179**, 177–184.
110. Cone, R.W., Hobson, A.C., Brown, Z. *et al.* (1994) Frequent detection of genital herpes simplex virus DNA by polymerase chain reaction among pregnant women. *J. Am. Med. Assoc.*, **272**, 792–796.
111. Ho, D.W., Field, P.R., Sjogren-Jansson, E. *et al.* (1992) Indirect ELISA for the detection of HSV-2-specific IgG and IgM antibodies with glycoprotein G (gG-2). *J. Virol. Methods*, **36**, 249–264.
112. Spiezia, K.V., Dille, B.J., Mushahwar, I.K. *et al.* (1990) Prevalence of specific antibodies to herpes simplex virus type 2 as revealed by an enzyme-linked immunosorbent assay and Western blot analysis. *Adv. Exp. Med. Biol.*, **278**, 231–242.
113. Sonnex, C. (1998) Human papillomavirus infection with particular reference to genital diseases. *J. Clin. Pathol.*, **51**, 643–648.
114. Meyers, C., Mayer, T.J. and Ozbun, M.A. (1997) Synthesis of infectious human papillomavirus type 18 in differentiating epithelium transfected with viral DNA. *J. Virol.*, **71**, 7381–7386.
115. Paavonen, J., Stevens, C.E. and Critchlow, C.W. (1988) Colposcopic correlates of cervical human papillomavirus infection. *Obstet. Gynecol. Surv.*, **43**, 323.
116. Schultz, R.E. and Skelton, H.G. (1988) Value of acetic acid screening for flat genital condylomata in men. *J. Urol.*, **139**, 777–779.
117. Wikström, A., Hedblad, M-A., Johansson, B. *et al.* (1992) The acetic acid test in evaluation of subclinical genital papillomavirus infection: a comparative study on penoscopy, histopathology, virology and scanning electron microscopy findings. *Genitourinary Med.*, **68**, 90–99.
118. Cartier, R. (1984) *Practical Colposcopy*, 2nd edn. Laboratoire Cartier, Paris.
119. Hillman, R.J., Ryait, B.K., Botcherby, M. and Taylor-Robinson, D. (1993) Changes in HPV infection in patients with anogenital warts and their partners. *Genitourinary Med.*, **69**, 450–456.
120. De Villiers, E-M. (1992) Laboratory techniques in the investigation of human papillomavirus infection. *Genitourinary Med.*, **68**, 50–54.
121. Hillman, R.J., Botcherby, M., Ryait, B.K. *et al.* (1993) Detection of human papillomavirus DNA in the urogenital tracts of men with anogenital warts. *Sexually Trans. Dis.*, **20**, 21–27.
122. Clavel, C., Rihet, S., Masure, M. *et al.* (1998) DNA-EIA to detect high and low risk HPV genotypes in cervical lesions with E6/E7 primer mediated multiplex PCR. *J. Clin. Pathol.*, **51**, 38–43.
123. Farthing, A., Masterson, P., Mason, W.P. *et al.* (1994) Human papillomavirus detection by hybrid capture and its possible clinical use. *J. Clin. Pathol.*, **47**, 649–652.
124. Galloway, D.A. (1992) Serological assays for the detection of HPV antibodies, in *The Epidemiology of Human Papillomavirus and Cervical Cancer*, (eds N. Munoz, F.X. Bosch, K.V. Shah and A. Meheus), ARC Scientific Publication No. 119, IARC Lyon.
125. Wideroff, L., Schuffman, M.H., Hoover, R. *et al.*

(1996) Epidemiologic determinants of seroreactivity of human papillomavirus type 16 virus-like particles in cervical HPV 16 DNA-positive and -negative women. *J. Infect. Dis.*, **174**, 937–943.
126. Carter, J.J., Koutsky, L.A., Wipf, G.C. et al. (1996) The natural history of human papillomavirus type 16 capsid antibodies among a cohort of university women. *J. Infect. Dis.*, **174**, 927–936.
127. Graziosi, C., Pantaleo, G., Butini, L. et al. (1993) Kinetics of human immunodeficiency virus type 1 (HIV-1) DNA and RNA synthesis during primary HIV-1 infection. *Proc. Natl Acad. Sci. USA*, **90**, 6405–6409.
128. Cheingsong-Popov, R., Panagiotidi, C., Bowcock, S. et al. (1991) Relation between humoral responses to HIV *gag* and *env* proteins at seroconversion and clinical outcome of HIV infection. *Br. Med. J.*, **302**, 23–26.
129. Gallarda, J.L., Henrard, D.R., Liu, D. et al. (1992) Early detection of antibody to human immunodeficiency virus type 1 by using an antigen conjugate immunoassay correlates with the presence of IgM antibody. *J. Clin. Microbiol.*, **30**, 2379–2384.
130. De Wolf, F., Lange, J.M., Houweling, J.T. et al. (1989) Appearance of predictors of disease progression in relation to the development of AIDS. *AIDS*, **3**, 563–569.
131. Saag, M.S., Crain, M.J., Decker, W.D. et al. (1991) High level viraemia in adults and children infected with human immunodeficiency virus in relation to disease stage and CD4+ lymphocyte levels. *J. Infect. Dis.*, **164**, 72–80.
132. Lacey, C.J.N., Forbes, M.A., Waugh, M.A. et al. (1987) Serum B_2 microglobulin and human immunodeficiency virus infection. *AIDS*, **1**, 123–127.
133. Mellors, J.W., Rinaldo, C.R., Gupta, P. et al. (1996) Prognosis in HIV-1 infection predicted by the quantity of virus in plasma. *Science*, **272**, 1167–1171.
134. Notermans, D.W., Goudsmit, J., Danner, S.A. et al. (1998) Rate of HIV-1 decline following antiretroviral therapy is related to viral load at baseline and drug regimen. *AIDS*, **12**, 1483–1490.
135. Desai, S., Bates, H. and Michalski, F.J. (1991) Detection of antibody to HIV in urine. *Lancet*, **337**, 183–184.
136. Van den Akker, R., van den Hoek, J.A.R., van den Akker, W.M.R. et al. (1992) Detection of HIV antibodies in saliva as a tool for epidemiological studies. *AIDS*, **6**, 953–957.
137. Major, C.J., Read, S.E., Coates, R.A. et al. (1991) Comparison of saliva and blood for immunodeficiency virus prevalence testing. *J. Infect. Dis.*, **163**, 699–702.
138. Adams, M., Lee, T.H., Busch, M.P. et al. (1993) Rapid freezing of whole blood of buffy coat samples for polymerase chain reaction and cell culture analysis: application to detection of human immunodeficiency virus in blood donor and recipient repositories. The Transfusion Safety Study Group. *Transfusion*, **33**, 504–508.
139. Ayres, L., Avillez, F., Garcia-Benito, A. et al. (1990) Multicenter evaluation of a new recombinant enzyme immunoassay for the combined detection of antibody to HIV-1 and HIV-2. *AIDS*, **4**, 131–138.
140. World Health Organization (1991) *Operational Characteristics of Commercially Available Assays to Determine Antibodies to HIV-1 and/or HIV-2*. Report 4. GPA/MBR/91.6. WHO, Geneva.
141. Van Kerckhoven, I., Vercauteren, G., Piot, P. and van der Groen, G. (1991) Comparative evaluation of 36 commercial assays for detecting antibodies to HIV. *Bull. WHO*, **69**, 753–760.
142. Ferns, R.B., Tedder, R.S. and Lloyd Donoghue, J. (1988) Comparison of a monoclonal anti-HIV/*gag* solid phase with a polyclonal anti-HIV solid phase for detecting anti-HIV-1 in a competition ELISA. *J. Virol. Methods.*, **20**, 143–153.
143. Kelen, G.D., Bennecoff, T.A., Kline, R. et al. (1991) Evaluation of two rapid screening assays for the detection of human immunodeficiency virus-1 infection in emergency department patients. *Am. J. Emerg. Med.*, **9**, 416–420.
144. Constantine, N.T., Zhang, X., Li, I. et al. (1994) Application of a rapid assay for detection of antibodies to human immunodeficiency virus in urine. *Am. J. Clin. Pathol.*, **101**, 157–161.
145. Asihene, P.J., Kline, R.L., Moss, M.W. et al. (1994) Evaluation of rapid test for detection of antibody to human immunodeficiency virus type 1 and type 2. *J. Clin. Microbiol.*, **32**, 1341–1342.
146. Lyons, S.F. (1993) Evaluation of rapid enzyme immunobinding assays for the detection of antibodies to HIV-1. *S. Afr. Med. J.*, **83**, 115–117.
147. Parry, J.V. and Mortimer, P.P. (1989) An immunoglobulin G antibody-capture particle-adherence test (GACPAT) that allows economical large-scale screening. *AIDS*, **3**, 173–176.
148. Centers for Disease Control (1989) Interpretation and use of the Western blot assay for serodiagnosis of human immunodeficiency virus type 1 infection. *Morbid. Mortal. Wkly Rpt*, **38**, 1–7.

149. Fransen, K., Pollet, D.E., Peeters, M. *et al.* (1991) Evaluation of a line immunoassay for simultaneous confirmation of antibodies to HIV-1 and HIV-2. *Eur. J. Clin. Microbiol. Infect. Dis.*, **10**, 936–946.
150. Hoffman, A.D., Banapour, B. and Levy, J.A. (1985) Characterization of the AIDS-associated retrovirus reverse transcriptase and optimal conditions for its detection in virions. *Virology*, **147**, 326–335.
151. Busch, M.P., Beckstead, J.H., Hollander, H. and Vyas, G.N. (1988) A histomolecular approach to the detection and study of human immunodeficiency virus infection, in *HIV Detection by Genetic Engineering Methods*. (eds Luciw, P.A. and Steiner, K.S.), Marcel Dekker Inc, New York, pp. 209–242.
152. Ariyoshi, K., Bloor, S., Bieniasz, P.D. *et al.* (1994) Development of a rapid quantitative assay for HIV-1 plasma infectious viraemia-culture-PCR (CPID). *J. Med. Virol.*, **43**, 28–32.
153. Miles, S.A., Balden, E., Magpantay, L. *et al.* (1993) Rapid serologic testing with immune-complex-dissociated HIV p24 antigen for early detection of HIV infection in neonates. Southern California Paediatric AIDS Consortium. *N. Engl. J. Med.*, **328**, 297–302.
154. Harper, M.E., Marselle, L.M., Gallo, R.C. and Wong-Stahl, F. (1986) Detection of lymphocytes expressing human T-lymphotropic virus type III in lymph nodes and peripheral blood from infected individuals by *in situ* hybridization. *Proc. Natl. Acad. Sci. USA*, **83**, 772–776.
155. Zaaijer, H.L., Cuypers, H.T., Reesink, H.W. *et al.* (1994) Detection of HIV-1 DNA in leucocytes using a commercially available assay. *Vox. Sang*, **66**, 78–80.
156. Whetsell, A.J., Drew, J.B., Milman, G. *et al.* (1992) Comparison of three nonradioisotopic polymerase chain reaction-based methods for detection of human immunodeficiency virus type 1. *J. Clin. Microbiol.*, **30**, 845–853.
157. Kwok, S., Mack, D.H., Sninsky, J.J. *et al.* (1988) Diagnosis of human immunodeficiency virus in seropositive individuals: enzymatic amplification of HIV viral sequences in peripheral blood mononuclear cells, in *HIV Detection by Genetic Engineering Methods*, (eds Luciw, P.A. and Steiner, K.S.), Marcel Dekker Inc, New York, pp. 243–255.

4 Principles and problems of organizing clinical trials in genitourinary/HIV medicine

V.S. Kitchen and G.P. Taylor

INTRODUCTION

Since the mid 1980s, genitourinary medicine (GUM) has been arguably the most rapidly changing specialty in the UK. These changes have occurred primarily as a result of the advent of HIV as a new and lethal, sexually acquired infection. In this context, the clinical management of HIV-related disease has proved extremely challenging and has resulted in a wider recognition of the value of well-planned and executed clinical trials as a means of defining optimal therapy.

Clinical trials of novel therapeutic agents are divided into distinct phases, each serving a specific purpose. Phase I studies examine the pharmacokinetics and early safety of the agent in healthy volunteers and require only a small number of subjects. Phase II studies are conducted in a patient population and are designed to determine the safety and tolerance of the agent in a clinical setting and optimal dosing regimens. These studies also require relatively small patient numbers.

Where it is considered inappropriate or unethical to administer the agent to healthy volunteers, as with anticancer chemotherapy, phase I and II studies are combined. The urgency and competition to develop new antiretroviral agents for the treatment of people with HIV infection and AIDS has also led to the combination of phase I and II studies. Additionally, in recent years there has been a move towards including some efficacy measurements in these studies. However, these trials tend to involve small numbers of patients and numerous patient groups and remain an inappropriate means of evaluating drug efficacy.

Phase III studies involve much larger numbers of patients and are designed to investigate the efficacy of the agent in addition to further establishing drug safety and tolerability. In phase III studies, with few exceptions, it is necessary to introduce some form of blinding in order to eliminate bias when assessing the results of the novel agent against those of the comparator drug, which may be an alternative (usually licensed) medication or placebo. For logistical reasons, phase I/II studies are often preferentially undertaken as single-centre trials and therefore tend to be conducted in centres with substantial numbers of patients. However, owing to the need for much larger numbers of study subjects, phase III studies will usually be conducted as multicentre trials. Phase I/II studies are usually initiated by industry, but phase III studies may be instigated by industry or by a research coordinating body such as the Medical Research Council (MRC) in the UK or the National Institutes of Health (NIH) in the USA. Phase IV studies take place after the agent has received a marketing licence and will not be considered further here.

In some instances it may be feasible for physicians to instigate studies independently to determine the comparative efficacy of given drug regimens or management strategies. However, in planning such studies, it is important to determine the running costs of the proposed investigation at the outset and to establish the source of any required funding. The financing of clinical studies is usually easier where the study is instigated by an external body (see Budget, below).

In this chapter we will discuss primarily the principles and pitfalls encountered in undertaking phase I, II and III clinical trials in HIV/GU medicine; however, many of the issues discussed are applicable to clinical research in all its forms.

THE OUTSET

Investigational compound

Before agreeing to collaborate in the clinical trial of a new compound, you will wish to see the available toxicology data. The investigational drug will have been tested *in vitro*, against a number of human cell lines and bacterial species, at concentrations considerably higher than are proposed for clinical usage, in order to detect potential cytotoxicity and mutagenesis respectively.

Having passed these tests, mammalian toxicity studies will have been undertaken using rodents, dogs and occasionally primates. These studies provide data on acute and subacute organ toxicity and are usually extended to address the issue of teratogenicity. In larger mammals, e.g. primates, the pharmacokinetic data obtained may help to determine the initial dosing regimens for clinical trials. Depending on the stage of drug development, limited human data may be available also.

Having reviewed the toxicology data, you should examine the data on drug efficacy. In the case of antiretroviral agents, this may be restricted to an *in vitro* assessment.

It is worth recalling at this stage that the compound will constitute more than the active pharmaceutical agent and adverse reactions may occur to subsidiary agents (e.g. the antacid buffer of dideoxyinosine).

Study design

The study objectives will depend to a large extent on the purpose of the investigation. A clinical trial designed by industry will usually focus on the collection of data required for licensing purposes, whereas an investigation instigated by a scientific body is likely to be different, e.g. to reexamine the use of an established treatment.

The study design should address four essential components: the objectives, the study population, the therapeutic intervention and the study endpoints.

Having defined these, the major consideration is whether or not the study objectives require the inclusion of a control arm and whether randomization and double blinding of the planned interventions are necessary. Not all studies can be double blind or need to be placebo controlled. You may find it useful to seek the advice of an experienced statistician at this stage, in order for these issues to be addressed adequately. Additionally, the statistician should be able to determine, using power calculations, the requisite subject numbers and study duration which should be defined at the outset. These calculations should take into account the perceived efficacy of the comparative therapies, the characteristics of the study population, the natural history of the disease in question and the chosen endpoints. Consideration should also be given to the clinical relevance of any potential observed differences in effect between the planned interventions.

Many clinical trials in HIV disease have been confounded by marked differences between *in vitro* and *in vivo* drug activity, inadequate study duration and uncertainty surrounding the relevance of chosen endpoints. A further confounding factor in terms of study design is the high percentage dropout of subjects seen particularly in studies of longer duration. This needs to be addressed, both in terms of the power calculations and in terms of subsequent data analysis (see Statistical analysis, below).

Subject eligibility

The selection of eligibility criteria should balance the need to define a sufficiently homogeneous population (to reduce the likelihood of patient bias in the study groups), the need to keep the study applicable to a broad population base and the practicalities of recruitment. If particular groups of patients are likely to exhibit different responses to a given intervention, it is helpful to predetermine these subgroups and incorporate them into the study design.

The concept of disease staging is very familiar in oncology and staging systems have been drawn up over the years and changed as knowledge advances. In HIV medicine two approaches to staging exist in parallel: the clinical system – asymptomatic, ARC, advanced ARC, AIDS, advanced AIDS; and a system based on laboratory markers, particularly the CD4+ lymphocyte count and plasma HIV RNA copy number. These approaches are usually integrated as a means of stratifying trial subjects. Unfortunately, due to the nature of HIV disease, these definitions may not be ideal for determining trial subgroups. For example, many patients not fulfilling the criteria for an AIDS diagnosis can be clinically far less healthy than some who have established AIDS. Eligibility criteria are usually precise if based upon laboratory results but clear guidance should be given in less certain areas, for instance, on what constitutes alcohol or other substance abuse, if this is likely to be an area of importance.

Study endpoints

Study endpoints may be based on measured changes in clinical or laboratory parameters and relate to both drug efficacy (or lack of it) and toxicity. In terms of toxicity, it is usual to define four grades of severity; mild, moderate, severe and life-threatening. For mild and moderate (grade 1/2) toxicity, observation or dose reduction may suffice; however, the more severe (grade 3/4) toxicities usually warrant at least temporary withdrawal of treatment. Guidelines should be provided for quantifying and responding to toxicity and for reintroducing treatment in patients in whom the toxicity was not severe or was suspected, but not confirmed, to be drug related.

The role of surrogate markers in trials of antiretroviral therapy was brought into question, particularly by the results of the Concorde Study[1]. As a result, there was a resurgence of enthusiasm for clinical endpoints including quality of life measurements. However, clinical endpoints in trials of antiretroviral therapy require long follow-up and the measurement of well-being is beset with the problems of subjectivity. These problems existed, with regard to HIV treatment, to a large degree because of the limited efficacy of the antiretroviral compounds available. However, as treatment becomes clinically more effective and produces greater and more reliable changes in laboratory markers, these markers became acceptable surrogates of clinical outcome.

A number of studies have shown that patients attending HIV clinics and participating in trials are often using a broad array of complementary medicines and practices.[2,3] Whilst this may not influence laboratory markers and clinical endpoints, it may bias assessment of well-being and should perhaps be included in the documentation of therapy.

Statistical analysis

The approach to be taken in the statistical analysis of a clinical trial should be specified at the outset. For a study with small numbers of patients taking curative treatment for a short period, an interim analysis may not be necessary. However, where a study is projected to run for months or years, a preplanned interim analysis by independent persons provides important data on the safety profile of the trial regimens and is a reassuring mechanism for both participants and investigators. The timing of interim analyses will in part be determined by the length of the study and the perceived likelihood of reaching a firm conclusion at a date prior to trial conclusion but should also ensure that unexpected differences, especially toxicities, are not missed for long periods. The level of statistical significance required needs to be specified at study outset and this value will determine the number of interim analyses performed throughout the trial.[4]

Analysis of the trial data should be performed in accordance with the initial patient randomization. In addition, the method of analysis will need to take account of the study objectives. For phase I/II studies which focus on safety and tolerability issues, an 'on treatment analysis' may be more appropriate than the 'intention to treat' analysis most commonly employed to determine comparative efficacy in phase III trials. If a large proportion of patients drop out of a given arm of the protocol, this should be investigated further as a potentially important trial outcome, as this may be due to unreported side effects.

All subjects, as far as possible, should be included in all the statistical analyses. Sufficient data on potential confounding factors should be collected throughout the trial in order for these to be controlled for adequately in the final analysis. The effect of confounding factors should be assessed using multivariate analysis and not by the *post hoc* definition of subgroup, which can generate misleading results.

PRACTICAL ISSUES

Patient base

If you are asked to undertake a clinical trial it will be on the understanding that your centre has the appropriate patient base to fulfil the allotted recruitment quotient. If you are concerned that you may have too few suitable patients to recruit within the prescribed timescale, it is often wise to make this plain to the study instigators at the outset. At this stage it may be feasible to alter the entry criteria (which may be unduly restrictive), to reduce the patient allocation to your centre or to increase the duration of recruitment. Taking an early proactive role allows for discussion between the sponsor and yourself regarding a mutually acceptable approach to the problem and helps to build an atmosphere of trust between you.

It is also important to take into consideration the pressure exerted on your patient base by undertaking more than one study in the same patient group at the same time. This is a particularly common scenario in the context of clinical trials of new antiretroviral agents, where there may be intense competition for a relatively small patient group. It is important to remember that the enthusiasm of potential study subjects for the intervention proposed will dramatically influence the number of recruits to your study. The views of patients can be profoundly influenced by media attitudes and by those of their peer group and developments in HIV medicine have attracted more attention and controversy than most. The attitudes of patients to a given intervention are subject to frequent change, particularly in the area of antiretroviral therapy, and this often presents the greatest confounding variable in recruitment. As a general principle, the studies most attractive to patients are those that offer a therapeutic option not otherwise available, where the study visits are not unduly frequent nor the investigations unduly arduous.

Space

It is important at the outset to identify the clinical space at your centre in which the proposed trial can be conducted. The provision of an identified area enhances subject recruitment and facilitates the smooth running of the study.

Additional space is required intermittently for visits from the trial monitor, who will need access to the subjects' clinical notes in order to verify data entries in the case report forms (CRFs). It is therefore worthwhile discussing the likely times and frequency of these visits to ensure the availability of suitable space for this activity. Similarly, storage facilities are required in order to keep the CRFs safely until the study is officially closed at your site. Current European guidelines stipulate that CRFs should be stored securely for 15 years following study closure. It may be possible to negotiate with individual companies for the provision of offsite storage, at a neutral location, for this period.

Space may also be required for the secure storage of study drugs. Ideally, study drugs should be kept in the hospital pharmacy and dispensed on prescription by the trial physician. This facilitates drug accountability, an important issue in terms of good clinical practice and trial audit (see later). It is therefore best to involve the lead pharmacist at an early stage in the proceedings. This is particularly important where the study requires the administration of intravenous drugs which may require preparation in the appropriate setting by pharmacy staff.

Budget

The funding required to undertake a study should be determined and finalized prior to subject recruitment. The level of financial support needed will be determined by the study protocol and should be broken down into staff costs (in terms of posts or sessions) and laboratory and other investigations. It is often useful to engage the help of the finance officer for your institution before finalizing your costings, as the clinician's tendency is to underestimate the level of funding required, an error that becomes increasingly obvious as the study progresses. This department may also undertake negotiations with the funding body on your behalf and may agree to manage the financial aspects of the study for its duration. However, their involvement will lead to an additional charge for overheads which is calculated as a percentage of the total study costs and which can therefore be considerable.

If you need to employ additional staff in order to run the study, it is important, where possible, to keep the costing for these salaries separate from the 'per patient' study costs. Poor patient recruitment or reduced subject evaluability may occur despite the considerable efforts of well-motivated research staff. Under these circumstances, a study costed solely on a 'per patient' basis will result in the liability of your department for financing the resultant shortfall in research staff salaries.

Ethics committee submission

All clinical trials require ethics committee approval and for multicentre studies, this means local approval at each individual centre although regional ethics committees are being introduced. The submission to the ethics committee will usually include a synopsis of the study design, a statement concerning patient confidentiality, copies of the patient information sheet and consent form, the clinical trials exemption certificate where the investigational agent is unlicensed (or a drug data sheet where the drug is licensed) and a copy of the indemnity agreement with the pharmaceutical company involved (see Good clinical practice, below). Ethics committees may well have their own customized questionnaires, in which case obtaining this on disk is a worthwhile first step.

The safety of the compound(s) should be adequately addressed in the protocol but in seeking ethical approval, you will also need to address specifically the question of discomfort, if any, incurred by the patient/volunteer as a result of the study investigations or administration of the study drug.

Personnel

Principal investigator
The principal investigator for a clinical trial at a given site carries overall responsibility for the conduct of the study. The role begins with an appraisal of the proposed trial protocol to ensure he or she is

comfortable with the ethical issues surrounding the study design and is convinced that the question posed is of sufficient practical or academic merit to warrant investigation. If the principal investigator has reservations regarding the study design, it may be possible for this to be altered in collaboration with the sponsor. Protocol amendment most commonly occurs in relation to patient selection criteria. If, however, there is a more fundamental concern with the study protocol that is not amenable to revision, it may be necessary for the principal investigator to decline further participation. It would be unethical for the investigator to proceed with a trial against his or her better judgement, regardless of the ethics committee's position on the matter.

The principal investigator is also responsible for monitoring study recruitment and for ensuring adherence to good clinical practice. He or she needs to maintain good communication with clinical colleagues to ensure a high profile for the study and should be prepared, if study enrolment is significantly below expected levels, to modify recruitment methods (see The trial physician and nurse, below). The principal investigator is also required to update the ethics committee on the progress of the study and to inform them of amendments to the trial protocol and of any new reports of adverse events associated with the study medication.

On study completion, the investigator should request access to the raw data for independent analysis in order to ensure that the interpretation of the trial results has not been influenced unduly by any commercial interest of the sponsoring company. He or she should also ensure that the results are published promptly in an appropriate medical journal or, failing this, are at least released into the public domain and specifically to study participants and collaborating clinicians (see Publication, below).

The trial physician and nurse
The best clinical trials staff are enthusiastic about the process of medical research, have good communication and organizational skills and are both flexible and methodical. They should have not only the ability to enrol patients in sufficient numbers but also the temperament to maintain their enthusiasm over what invariably turns out to be a more prolonged study period than envisaged initially. Whilst reasons for trial entry are variable, most trial participants cite as their principal motivation the rapport they develop with these investigators and the opportunity it affords them for increased medical attention and a greater understanding of their condition. The willingness of study staff to adopt flexible hours of work facilitates the participation of patients who remain in full-time employment, for whom 'out of hours' study visits may be not only convenient but also helpful in maintaining confidentiality.

The research staff need to understand the significance of the role they play in the trial's process. A great deal rests on the accuracy of the data they record and their attention to detail. At best, their errors may result in a minor protocol violation; at worst, they may endanger life. In all cases the interests of the patient override the study and the clinician should always discontinue treatment for changes which may not be included in the trial criteria for toxicity.

Short, single-centre studies may provide sufficient part-time research interest for a registrar and/or nurse with predominantly NHS commitment. However, where the staff are primarily involved in clinical trials work, they may wish to pursue additional research interests of their own, which may be related to or independent from the primary study. Where possible, time, facilities and supervision should be offered to support these objectives within reason. Similarly, if a member of the research staff wishes to obtain further clinical skills, it may be appropriate to integrate this clinical work into his or her timetable. It is important to balance time allocated to the study and to independent activities and not to underestimate the amount of 'hidden' time that a properly conducted study requires.

Recruitment

The objective is to recruit highly motivated, well-informed patients who fulfil all the entry criteria and who will complete the study. Recruitment problems will depend to some extent on the nature of the study to be conducted, but understanding what motivates patients to enter clinical studies is helpful.

There were a number of interesting findings in a study of the factors influencing HIV-positive patients to participate in antiretroviral drug trials.[5] The assumption is that many of these patients take part in set studies with an altruistic desire to find a cure for HIV infection and, secondarily, to have access to antiretroviral therapy as soon as possible and to be active in combating their infection. A questionnaire was administered to subjects who had been participating in antiretroviral studies for several months. From their responses, it was clear that they perceived that continuity of care, close contact with a small medical team and ready access to their trial nurse or doctor were the most significant personal benefits gained from study participation.

Comparisons with clinical oncology trials, where there is quite extensive literature on motivation and recruitment, reveal some differences between these and antiretroviral trials. Oncology trial participants seem to be motivated by a belief that entering a study allows them to access more effective therapy than would normally be available.[6] Subject motivation was different again amongst 36 women enrolled in a phase I/II study to assess vaginal virucides, with altruism (100%) and a particular interest in the disease (97%), although they were not themselves infected, the primary reasons given for participating.[7]

Recruitment to studies, particularly in HIV disease, is clearly influenced by the volunteer's perception of the medication under investigation. In one of our studies the common thread linking participants was the desire to take anything other than the only licensed medication which they perceived as toxic and without benefit. Thus they entered a phase I/II dose-finding study of a related compound which the investigators considered might be as toxic, or more so, than the established drug.

The perceived importance of early access to a potentially active drug was underscored by the very limited recruitment to the placebo arm of the Alpha Study.[8] There is no doubting the importance of the placebo or deferred treatment arm of the Concorde Study, but one of the selling points for the next MRC study, Delta, was the absence of a placebo arm and many current studies are comparing different combination therapies. Also important has been the provision of open label drugs to patients who have reached a study endpoint or have been intolerant of the monotherapy (in the case of the Delta Study, zidovudine).

Recruitment to studies is undoubtedly influenced by external pressures. Many potential recruits will discuss with friends and relatives whether to enter a study, especially a long-term study. The media can influence recruitment negatively by casting doubts on the merits of a therapy or positively effect enrolment by their reporting of a 'scientific breakthrough'. However, the patient's physician has probably the greatest influence in terms of a subject's decision to enter a therapeutic trial. His or her key role has been documented in several studies of recruitment of patients to cancer therapy trials, reviewed by Carolyn Cook Gotay,[9] where the non-entry of eligible patients was attributed to physician-related variables in 50–94% of cases. Conversely, the patients of a physician enthusiastic about the trial's process are most likely to enter a study.

Failure to promote a study may be due to lack of familiarity with the study design, insufficient time in clinics to discuss the study, study fatigue (being overwhelmed with potential studies for patients) or a professional lack of faith in the therapeutic regimens offered. Some of these problems can be circumvented by publicity, especially around the clinic, the ready availability of patient information sheets, immediate access to a staff member who can spend adequate time with the patient discussing the pros and cons of entering the study, regular personal reminders to colleagues of the essential study details and the targeting of eligible patients either directly or indirectly.

The principles of recruitment to clinical studies in genitourinary medicine and HIV medicine are the same, although there will be differences in the nature of the studies and the patient base, the short duration and safety of treatment in, for example, studies of gonorrhoea, non-specific urethritis or candida increasing the acceptability of the study to patients. However, the intensive reviews required in clinical trials compared to that indicated for standard treatment and the knowledge that such treatment is effective may act as a relative disincentive.

GOOD CLINICAL PRACTICE

Ethical considerations

Guidelines on good clinical practice for trials have been issued by the European Community[10] and these should be referred to. The principles upon which these guidelines are based are stated in the revised Declaration of Helsinki:

> *The design and performance of each experimental procedure involving human subjects should be clearly formulated in an experimental protocol which should be transmitted for consideration, comment and guidance to a specially appointed committee independent of the investigator and the sponsor provided that this independent committee is in conformity with the laws and regulations of the country in which the research experiment is performed.*

Essentially, the first objective of these guidelines is to protect the trial subject. Submission of all protocols to an independent ethics committee is the initial step and the guidelines advise on the nature and purpose of this committee. The ethics committee is expected to examine not only the protocol and whether the study is appropriate and safe but also the suitability of the investigator and the facilities. Particular attention is given to the manner in which information is given to trial subjects and to the method of obtaining informed consent (see Ethical considerations, below). A record of all patients approached to enter the study and reasons for non-entry should also be held. The guidelines also stipulate that information which becomes available during the study and is relevant to the trial participants must be made known to them by the investigator. This includes the results of other studies as well as data arising from their own study. To further safeguard the trial participant, it is important that provision is made for compensation in the unfortunate event of injury or death attributable to the trial intervention. It is standard practice for the sponsoring companies to provide an indemnity agreement to this end.

Informed consent

The ethics committee will need to be assured that written and informed consent will be obtained from all participants prior to study entry, by named members of the research staff. The patient information sheet will include details of possible side effects, discomfort or inconvenience. Particular emphasis should be given to the freedom of any patient to withdraw from the study, at any time, without having to give any reason and without any prejudice to their future care. It is useful to include a timetable of expected attendances and procedures in the patient information sheet.

The subject should have provided informed written consent prior to any involvement in the study, including phlebotomy for screening investigations.

In some cases where the study design is simple and of short duration, it may be appropriate to obtain informed consent immediately after discussing the relevant study information with the prospective subject. However, where the study involves a prolonged commitment from the subject (e.g. antiviral drugs or chemotherapy) it will be appropriate for the potential volunteer to consider his or her views regarding the trial over a longer period (i.e. 24 hours to 1 week). This tends to result in a reduced dropout rate as those who subsequently decide to take part in the study are likely to be well motivated and committed.

Confidentiality

The confidentiality of patient data is paramount in the context of all clinical trials. However, periodically throughout the study, it will be necessary for persons other than the medical and nursing personnel at the study site to view the patients' notes. These will include the trial monitors (company employees who verify the recorded data) and occasionally the trial auditors who may be appointed by the company, the trial coordinators or an external agency such as the FDA. In all cases the clinical records must be viewed on site in the presence of a member of the institute undertaking the study. Only anonymized data can leave the study site. It is normal practice to use a trial number with either a confidential hospital number, date of birth, initials or a soundex code. This latter code is a device whereby the patient's name is converted into numbers but then cannot be identified as the code can only be deciphered in one direction. It is essential to use at least two points of identification to ensure the safety of the system.

Trial maintenance

After patient recruitment there are three essential areas to consider: patient assessment, administration of compound(s) and data collection. The time required to assess a patient participating in a clinical study is often at least 50% longer than for a routine outpatient appointment and, depending on the nature of the study, may take two or three times longer. Assessment of the patient may be weekly, bimonthly or, in studies where early change is anticipated, for instance the treatment of oral candida, every other day (including weekends). Whilst it is ideal that a single observer assesses each subject, this may not be feasible for short-duration studies and two or more staff should be familiar with the study and known to the participants. In any case, provision should be made for the 'out of hours' assessment of all trial participants, including the provision of necessary information to the trial personnel and an adequate staff handover. We have found it useful to provide written instructions for trial participants on how to obtain assistance out of hours and have instituted a 24-h cover for clinical trials via a research nurse rota and a long-range pager. Many patients have their GPs involved in their care and where this is so, information concerning the study should be given to the GP.

Pharmacy

It is the role of the pharmacy in clinical trials to monitor drug supply and to be responsible for drug accountability. In long-term studies, ensuring the supply of the trial compound is potentially problematic. Whilst the sponsoring company should have no difficulty with supply, ensuring the arrival of up to 100 individually labelled three-month treatment packages without the pharmacy becoming overrun with tablets for ex-participants can be surprisingly difficult.

Once again, the system usually works best where this responsibility is clearly allocated. Pharmacy will also be accountable for all trial drugs received and dispensed, a favourite area for audit! Finally, pharmacy are likely to have some role in monitoring compliance, informally if not formally, and friendly communication between prescriber and dispenser can be very fruitful.

Records

In a clinical trial there are two parallel sets of records: the clinic notes, which are maintained as usual, although with some difference in emphasis, and which provide the hard data against which trial data are checked; and the CRF. Each item on the CRF has to be verified against another source, usually the clinic notes. This is the remit of the clinical trial monitor. To avoid the need for future clarification, clarity and consistency in the clinical notes are essential. For example, creps, crackles and râles recorded in the notes might generate three separate adverse event entries. The description of skin lesions can also prove very confusing. Dates, doses and duration of therapy take on greater significance in trials. Noting that your patient had an asthma attack in the 3/12 since last attendance which responded to his or her salbutamol inhaler may be a sufficient clinical record but for the CRF, dates of onset and recovery as well as more precise details of therapy taken are required. Patients often become very good at remembering such information or start diaries when they realize the detail that is needed. Proprietary medications should be recorded by their generic names and actual dosages. If the CRF is altered, the entry to be changed must be crossed out but still legible, the new entry made and all alterations signed and dated.

The collection of data on adverse events is of paramount importance in a clinical trial of a new agent. All symptoms and signs of disease, whether suspected to be compound related or attributable to a disease process, have to be recorded with rigour to ensure that increased occurrences of events that could be otherwise attributed elsewhere do not escape detection. There is often a large gulf in the understanding of what constitutes an 'adverse event' between a nurse or clinician and a trial monitor. In many studies, any hospitalization is considered, by definition, to be an adverse event and demands notification within 24 h. It is important that this occurs even where the cause of admission may appear to be entirely unrelated to the study intervention.

Audit

Audit is an essential part of good clinical trial practice. Sponsoring pharmaceutical companies will not only double or triple check every entry in the CRF and require an explanation of each anomaly but they will also institute their own audit, often using external auditors, to ensure there are no surprises when they are in turn audited by the regulatory authorities. This audit will again check that the data on the CRFs can be confirmed by entries in the clinical notes, that written informed consent was obtained prior to trial entry, that adverse events have been reported according to the trial protocol and that the pharmaceutical products under investigation are properly handled by an appropriate person and accurate records kept. The areas where it is easiest to find fault are often simple things such as results recorded on the CRF but not filed in the notes or a copy of the patient information sheet missing from the notes.

STUDY TERMINATION

Data monitoring committee

Most commonly, a clinical trial is closed after the requisite number of subjects has been recruited and followed up for the appropriate period. However, a study may be terminated prematurely if interim analysis of the data shows a strikingly significant benefit of one treatment modality over another or if a study treatment has a toxicity profile that suggests continuation of the trial would be unethical. Interim analysis of data should, where appropriate, be undertaken by the data monitoring committee, an independent body usually including one or more of the principal investigators, which meets at predetermined times throughout the duration of the study. Interim analysis of trial data should occur infrequently and in a closely regulated manner, to prevent distortion of the study results (see Statistical analysis, above).

Publication

Following trial closure, there is an inevitable delay whilst data are verified and analysed statistically. This process can take months, during which time there may be intense pressure from a number of different parties, including the study participants, the investigators themselves, the pharmaceutical company and the business world.[11] Once available, it is important that the results are released in a well-organized and appropriate manner. Ideally, the investigator should have access to the results first, in order for the preliminary release to the study patients to be formulated appropriately. Study participants should be ensured ready access to this document prior to the public release of information. Thereafter, the results should be simultaneously provided for all other interested parties, an issue which is paramount where results are likely to generate interest from the media or contain price-sensitive information which could impact on the stock market. Only few clinical trials yield such sensitive information, but this situation arises quite commonly in the area of antiretroviral therapy.

A preliminary publication in the medical press is usually the most appropriate means by which such study results are first aired publicly, as this helps prevent distortion of the information by the lay press. However, this approach has led to criticism from some physicians and scientists, who feel that the results of the clinical trials should only be published following full statistical analysis of the trial in its entirety.

What happens, however, if the results of your clinical trial are deemed unsuitable for publication or if your submission fails to be accepted into the medical press? At present, there is considerable concern that pharmaceutical companies may be able to influence the exposure of less-than-promising data to the public domain. The guidelines on good clinical practice formulated by the Association of the British Pharmaceutical Companies state that study results may be either confidential to the company or intended for publication, but that the intention should be to publish results 'where this is warranted'. However, a statement from the Royal College of Physicians further clarifies the position of the clinical investigators in this matter by stating that it is their responsibility 'to ensure that there is prior agreement with any financial sponsor that the results of the research may be submitted to journals of the investigator's choice and that the sponsor may not seek to influence the publication of the results of the research.[12]

It has been suggested that a computer database of the findings of unpublished clinical trials should be established under the jurisdiction of a public body, with ethics committees undertaking a wider role in quality assurance.[13] Clinical investigators would then submit study reports to this database, thus allowing ready access to the results by any interested party. The absence of such a centralized database leads to wasted resources in terms of participants' hours, investigators' efforts and sponsors' support for those studies which currently fail, for whatever reason, to be published. Such a system would also ensure that similar research was not needlessly repeated, wasting further resources, and would provide an invaluable source for the metaanalysis.

In 1997 the medical media attempted to address this problem by announcing an 'amnesty' for unpublished clinical trials allowing authors to register such data which could then be included in systematic reviews.[14] These trials would be included in the Cochrane Controlled Trials Register which already includes reports published in conference proceedings as well as peer-reviewed papers. An HIV/AIDS review group was registered in March 1998 and an application from a Sexually Transmitted Diseases Group is pending (September 1998). Access to this information is available on the World Wide Web (http://www.cochrane.co.uk).

The scope and need for good clinical trials in HIV/GUM seems likely to continue to expand in the future, both in conjunction with industry and independent of commercial interests. Participation in such studies can be extremely rewarding for patients and health-care workers alike, although the time and commitment required to see a study through from conception to published report should not be underestimated.

COURSES

Membership of the Association of Clinical Research in the Pharmaceutical industry (tel: 01628 29617; fax: 01628 21230) is open to any person involved in the running of clinical trials for the pharmaceutical industry. They aim to enhance the standards of clinical research carried out on behalf of the pharmaceutical industry and organize workshops and courses.

REFERENCES

1 Concorde Coordinating Committee (1994) Concorde: MRC/ANRS randomised double-blind controlled trial of immediate and deferred zidovudine in symptom-free HIV infection. *Lancet*, **343**, 871–881.
2 Anderson, W., O'Connor, B.B., MacGregor, R.R. and Schwartz, J.S. (1993) Patient use and assessment of conventional and alternative therapies for HIV infection and AIDS. *AIDS*, **7**, 561–566.
3. Barton, S.E., Davies, S., Schroeder, K. *et al.* (1994) Complementary therapies used by people with HIV infection. *AIDS*, **8**, 661.
4. Armitage, P. and Berry, G. (1987) *Statistical Methods in Medical Research*, Blackwell Scientific, Oxford.
5. Gay, V.I., Hooker, J.S., Beck, E.J. and Weber, J.N. (1993) Recruitment in clinical trials: the impact of HIV infection. *PO. B45-2563. IXth International Conference on AIDS*, Berlin.
6. Penman, D.T., Holland, J.C., Bahna, G.F. *et al.* (1984) Informed consent for investigational chemotherapy: patients' and physicians' perceptions. *J. Clin. Oncol.*, **2**, 849–855.
7. Byrne, G., Stafford, M. and Kitchen, V. (1995) Factors which motivate healthy female volunteers to participate in HIV-related clinical trials. *Volunteers in Research and Testing Conference*, Manchester.
8. Alpha International Coordinating Committee (1996) The Alpha Trial: European/Australian randomized double-blind trial of two doses of didanosine in zidovudine-intolerant patients with symptomatic HIV disease. *AIDS,* **10**, 867–880.

9. Gotay, C.C. (1991) Accrual to cancer clinical trials: directions from the research literature. *Soc. Sci. Med.*, **33**, 569–577.
10. (1993) *European Good Clinical Practice Guidelines*, Brookwood Medical Publications.
11. Anon. (1993) Early announcements. *Lancet*, **342**, 1001–1002.
12. Royal College of Physicians (1990) *Research Involving Patients*, RCP, London.
13. Herxheimer, A. (1993) Publishing the results of sponsored clinical research. *Br. Med. J.*, **307**, 1296–1297.
14. Smith, R. and Roberts, I. (1997) An amnesty for unpublished trials. *Br. Med. J.*, **315**, 622.

FURTHER READING

(1993) *European Good Clinical Practice Guidelines*, Brookwood Medical Publications, London.

Pocock, S.J. (1983) *Clinical Trials: A Practical Approach*, John Wiley, Chichester.

Spilker, B. and Cranmer, J.A. (1992) *Patient Recruitment in Clinical Trials*, Raven Press, New York

ICH Harmonised Tripartite Guidelines for Good Clinical Practice (1996). *Good Clinical Practice Journal*, Vol. 3 (4), Supplement.

5 Establishing a new genitourinary medicine clinic service

J.S. Bingham

INTRODUCTION

Most trainees in genitourinary medicine (GUM) expect to become consultants within the National Health Service (NHS) in the United Kingdom (UK) and, in that capacity, will be involved in the running and management of a service or clinic. The clinic system is well developed in the UK, but comparable models exist in Sweden and in some other parts of Europe, North America, Australasia and the developing world.

Opportunities to develop a new service arise either because an existing service is moving to new, improved or better sited premises or because a service is planned in an area of perceived need, where no such service has previously existed. In either circumstance, the physicians in charge of the service, and management, will have to address a number of issues.

1. What is the need for the service and what is the population that it will serve?
2. What is the competition in adjacent areas?
3. What sources of funding are available?
4. Where should the clinic be sited?
5. How extensive is the service going to be and, dependent upon that, what staffing is required?

This chapter will focus on these issues and will then move on to a discussion of the detailed planning of the arrangements for the service.

THE NEED FOR A SERVICE

A clinic may be established in an existing facility or a new building may be required to provide larger, more appropriate premises. However, if one is starting from scratch, definite reasons for a new service will have to be identified. The Monks Report[1] advises that every health district should provide a GUM service, even if not full time. In large metropolitan conurbations, with good transport, this may be more difficult to justify but, in suburban or rural areas, the need to fill a geographical gap may be apparent. An NHS hospital trust or directly managed unit will know the size and characteristics of the population within the district health authority boundaries. By examining national incidence figures, the expected workload can be estimated. Such calculations require knowledge of the social, ethnic and cultural mix and the presence of any groups with special needs such as political refugees or foreign students. There may also be local knowledge of high-risk groups, such as prostitutes or drug users, who could well become patrons of a new clinic.

Local general practitioners (GPs) or other specialists may be aware of which clinics the local residents attend. Information about the workload generated by the local population of the proposed new clinic could be sought from them.

It is unlikely that a genitourinary physician (GUP) will be in post when such deliberations are going on and this sort of exercise is usually undertaken by the local Department of Public Health. However, to ensure that they are properly advised, the opinion of an experienced GUP should be sought, particularly when the stage of planning the service is reached.

THE COMPETITION

When running any organization, it is always wise to know as much about the opposition as possible. With tighter management in the NHS, a clinic's throughput and workload are likely to be monitored closely and it is important to be attracting patients and perhaps to be offering services that adjacent clinics do not. You should therefore find out, as mentioned already, how many patients they are seeing from your district. If you provide a good or better service, it should be possible to win them for the home team despite the fact that some people will travel well away from their area of residence for reasons of discretion. It is useful to know what their workload is, where they get their patients from, what services they provide and their opening hours. Some of this information is available in the National Clinic Guide[2] but an enquiry, colleague to colleague, should provide you with the rest of it. It may be possible, on the grapevine, to find out if the services are well run, provide the sort of service the patients want and whether or not they are contributors to research and the general advancement of the specialty. This background information should enable you to establish the sort of service that is required.

FUNDING

The availability of funding varies from one part of the country to the other and from time to time. In the UK, following the Monks Report and the Minister of Health's executive letter of February 1989[3] on HIV and AIDS resources, ring-fenced regional HIV monies have been used to improve existing service facilities, hire additional staff and to establish new facilities. If you are in an existing clinic and are planning to move to new, improved premises then, hopefully, the present revenue will continue and what is required is the identification of capital monies to finance the new development.

If a totally new service is being developed then all the money required is new. It will be necessary to make a case to persuade local medical opinion that a GUM service is required and is desirable. Once there is agreement locally, then the purchasing authority has to be convinced of the need and must agree to purchase such a service. As this is new money it may not actually be available to the purchasers and may need to be obtained from a higher authority – originally the regional health authority but now the Department of Health.

One of the main functions of a GUM service is to prevent new infections, both traditional STDs and HIV infection, by the provision of information, health advice, supply of condoms and by partner notification. This is a reasonable call on HIV prevention monies and a number of new clinics have been established in recent years on this rationale. Funds for the treatment and care of HIV cases are now separate and purchasers like to insist that they are used specifically for this purpose. Naturally, therefore, unless there is a large anticipated workload, a call on such monies is not appropriate, although funds might be provided from this source on a pump-priming basis. Ringfencing of HIV treatment monies has been lifted but it is likely that HIV prevention money will continue to be protected for some time.

A logical rationale for the service will have to be produced and an outline business plan, so that it is clear what the extent of the proposed service is – building, salaries and equipment. There is always a tendency to underestimate the amount of money needed and it is therefore advisable to make a good case to obtain the maximum required, since it is very difficult to raise extra funding later. This all sounds quite straightforward but it can be complicated by having to identify a site where new premises can be constructed or an area of the hospital which is being vacated by another service which can be converted or modernized for your new clinic. Timing, to fit in with local arrangements, may be of the essence. Matters may be further complicated if a part-time service is envisaged, with the consultant linked to another hospital, perhaps with a different purchasing authority, and the effort to coordinate funding at either end may be considerable.

Of course, capital funds will be released first in order to take care of the building work and clearly, estimates for such work will have been obtained previously. The local works department can assist with this.

Occasionally, but not often, a trust or directly managed unit will have spare capital funds which can be bid for, so winning over local opinion is essential. Sometimes, management may suggest the use of charitable monies. It is likely that this sort of money will not be available to a totally new service but, where a unit is moving to new premises, particularly in a hospital with wealthy special trustees, it would be reasonable to approach them for assistance. Alternatively, with a well-established department on the move, a loan from the trust or a bank could facilitate work starting and a fundraiser could be hired in the hope of raising the funds to repay the loan.

The spending of revenue funds will be dealt with later in the chapter.

SITING OF THE CLINIC

The siting of a new department is important. In the past, there has been a tendency for GUM departments to be situated in basements of hospitals or perhaps not even in the main hospital at all. In more recent times, new departments have been sited within or adjacent to the main outpatient facility of a hospital and this is probably the best place for them to be. It allows the unit to be perceived by the public as just another outpatient department and this may help to reduce the anxiety that some patients have felt in the past about attending clinics that were separate and easily recognized as 'VD clinics'. Of course, it may not be possible to identify space for this arrangement but GUPs should try to ensure that, when a new outpatient block is being planned, plans are made for the GUM department to be within it. If this proves impossible, then the clinic should be on the main hospital site, as failure to have this arrangement can result in the professional isolation of the staff and inconvenience in terms of the accessibility of support services, such as imaging. This has become more important in the HIV/AIDS era when, with multisystem disease, ready access to support services is essential.

It is important to make sure that the new clinic is accessible to the disabled and to the very ill. This has become more apparent with the management of sick AIDS patients who are unable to climb stairs and so it is obviously best if the clinic is on the ground floor, with wheelchair access and with efficient lifts, large enough for wheelchairs or stretchers if necessary.

Most hospitals are conveniently sited for public transport and, outside the large metropolitan areas, often have reasonable car-parking facilities. If your new clinic is in a multihospital trust/district, while you would still wish to have the clinic in the outpatient area, you would probably want to site it in the hospital nearest to where you think the bulk of the patients is going to come from. If you are planning to set up a satellite or outreach clinic, then you would site it, if possible, in the area of need but also with access to public transport.

EXTENT OF THE SERVICE PROVIDED

The extent of the service will ultimately depend upon the identifiable need and on the monies available to meet it. An existing clinic moving to new premises is likely to be doing so because of the inadequacy of the existing space and therefore might be expected to expand its services. On the other hand, when starting a new service from scratch, those who planned it may have already decided the extent of the service.

In the Bromley Hospitals Trust, where I used to spend part of my time, the old district health authority was advised that it should have a GUM service. With a resident population of over 300 000, a full-time clinic could possibly have been justified but because it is in a dormitory area with many of the residents working in central London and having access to the services there, there was doubt as to whether or not a full-time clinic would be necessary. It was therefore decided to start with a part-time service. Because it was recognized that it is not ideal to have a consultant working in isolation on his or her own, with no immediate GUP colleagues, and because a full-time post is easier to fill than a part-time one, the job was linked to a central London teaching hospital and this solved both these problems.

However, such an arrangement immediately creates another problem and that is the number of clinical sessions that the new clinic can have. Given that they should be consultant led, the Genitourinary Medicine Committee at the Royal College of Physicians and the British Medical Association, now endorsed by the Association for Genitourinary Medicine, recommend that a consultant should not do more than seven clinical sessions each week. Thus, if the consultant is working for half of his or her time in the new clinic, only three and a half sessions can be carried out each week. This is perfectly reasonable because if more time than that is spent on clinical work (and it usually is), there is not enough time to attend directorate and other medical meetings, to visit the wards, to attend grand rounds, audit meetings or even to go to the library to keep up to date, not to mention work that you might be doing at a national level, as well as attending scientific meetings outside the trust. If you have a dedicated facility, its use is therefore limited straight away.

STAFFING

General

Staffing arrangements will be governed by the extent of the service to be provided and the monies available. The first person to be appointed will always be the consultant physician. It will be his or her task to assess the local circumstances and decide upon the best means of delivering the new service. It is likely that a job plan will have been drawn up when the post was advertised but, with a new service, this may have been arranged by someone with little experience of such work. The arrangements must therefore be scrutinized closely and an informed decision taken as to the best way of organizing the service.

For example, at the new part-time clinic in Bromley, the job plan stated that the physician would initially carry out three clinical sessions per week. This was perfectly acceptable but it was also stated that a fourth, evening clinic should eventually be commenced. The planners had arranged for two part-time nurses to work in the clinic and both of these had already been offered the posts before I arrived. They were to be supervised by the sister in the main outpatient department.

I believe that the physician in charge of the service should be able to decide on the best arrangement and should be able to select the staff personally. In my particular case, I met with management, informed them that their proposals were unacceptable and, despite the fact that the nurses had already been hired, made my counterproposals. A sister/charge nurse in charge of the clinic was the only sensible way forward, rather than having the clinic casually overseen by the outpatient sister. The point is that, after the physician, the nurse in charge is the most important appointment. She or he should preferably have previous experience of the subject and have demonstrable leadership qualities and an ability to work independently. Unfortunately, if the planners have got it wrong, there will not be enough money to employ such an individual. The physician should be able to spend the staff budget as he or she sees fit and can therefore rearrange the salaries within that.

In my little department the planners had originally made provision for two part-time nurses, a part-time health adviser, clerk and secretary. In order to provide the salary to attract an experienced full-time sister, I decided to forego the health adviser's post in the first instance and to carry out those duties myself, together with the sister who had previous experience of health advising.

Nursing staff

Ideally, any additional nursing staff should have previous experience but this may not always be possible. It is advantageous, since nurses in GUM clinics carry out specialized tasks such as Gram staining, microscopy, cryotreatment of warts and other duties, such as assisting at colposcopy. It is particularly important to have experienced nurses if the nurse in charge is on leave or is ill.

Nurses have traditionally carried out duties such as setting up consulting and examination rooms and chaperoning male doctors when examining female patients. These tasks do not require a nursing qualification and could be performed by a care assistant. This is cheaper than employing a nurse, which may allow for greater flexibility in running a clinic. In a small part-time clinic it may be helpful and cost-effective to use an experienced nurse in an extended role. If you decide to do this, the nurse must work to an agreed protocol and agreement must be obtained from nursing management.

Additional medical staff

Most new clinics are in district general hospitals and not in teaching hospitals. It is unlikely, therefore, that the staffing arrangements will include junior medical staff. There may be provision for clinical assistant or staff grade posts. It should be remembered that the recent Royal College of Physicians Report on staff grade doctors[4] has recognized that such an appointee might eventually expect to progress to an associate specialist post and these aspirations need to be borne in mind. A clinic with only one consultant is not ideal. Finding a suitable locum to cover annual or study leave can be difficult. If the workload of the new clinic builds up, the appointment of a second consultant or even a part-time consultant should be considered. The government has recently had a programme to encourage these developments. Some clinical assistant sessions may need to be sacrificed to pay for such a post.

Clerical staff

In a small clinic one receptionist will be enough. If part-time staff are employed, it may be necessary to have more than one. It is best to hire an individual with previous outpatient clerical experience. She or he should be capable of answering the telephone, booking appointments, dealing with queries and maintaining a medical record filing system. Additional tasks include operating a VDU and simple software package to register patient details, recording of KC60 return data and keeping a record of attendances. In a new part-time clinic the clerk could be employed, initially, only when the clinic is opened. As work builds up additional time is required for administrative duties, dealing with enquiries and making appointments when the clinic is not actually open for patients. Thus the clerk (or clerks) hired should be agreeable to this arrangement.

Secretarial support

In a new small clinic it may not be possible to justify a full-time secretary, but the secretary needs to be there whenever the consultant is. Communication with GPs is very important when setting up a clinic and a secretary with the correct temperament and manner is essential. She or he needs to have excellent telephone skills, be a competent typist and be able to use a word-processor. It is wise to check these attributes before appointment. In a small clinic it can be very helpful if the secretary is prepared to cover clerical duties during lunch breaks and when the clinic is extremely busy.

The people who planned the department may have fixed the staffing arrangements before your appointment, but you should adjust these as you see fit. You will, of course, have to justify such changes to management. Once agreement has been reached, job descriptions, including person specifications, need to be drawn up. Personnel (human resources) staff can help with this, but it is helpful to scrutinize the documents in use in other established departments and adapt these to your own needs. Advice on delineation of work responsibilities is available.[5]

Finally, salaries have to be agreed. The personnel department can advise on this but, particularly with the senior nurse's post, you will have to be prepared to pay whatever it takes to get the right person for the job. Advice on salary scales is easily obtainable from other departments. In a new clinic, with a small staff, holiday and sick leave causes problems as there will not be enough staff to provide cross cover. It is vital, therefore, that you build into your budget funds to pay for locum cover for these periods. Consultants sometimes have to attend meetings, within the hospital and elsewhere, at short notice and these can coincide with previously arranged clinical sessions. It is often important that the consultant attends these meetings and yet the session must take place. If a clinical assistant is going to be doing the session anyway, you can arrange for a colleague in a nearby clinic to be available to provide advice over the telephone, if necessary. I think that it is wise to have such an arrangement in any case, so that the service can be maintained in the event of illness.

It takes time to plan the clinic, purchase equipment, make the necessary local links and advertise the service, so it is sensible to employ staff in stages. The consultant, senior nurse and secretary (perhaps only for limited hours) are all that are initially necessary. This may permit saving on the revenue side which you may transfer to the capital side to help with the purchase of equipment. It is not always possible to hire staff who have had previous GUM experience and, after employment, there needs to be a commitment to training for these individuals, both within the clinic and by attachment to larger established departments.

THE DESIGN OF THE NEW CLINIC

This section will not go into detail about the detailed design of a clinic, which depends on the amount and the configuration of space available. The Department of Health Building Notes on STD Clinics[6] give a rough outline of the requirements for a clinic. Architects can misinterpret these and, without being properly advised by doctors and nurses, may produce a design which is unrealistic. Without my supervision, a clinic was designed in a proposed new hospital with a minute waiting room, no safe play area for children, an inadequate number of examination rooms, no proper 'hot lab' facility, no staff rest area nor locker area, no staff lavatories, a huge phlebotomy room but no designated storage area and no consultant's office. I think that they had provided soundproofing between rooms and all the rooms had doors; it is accepted now that curtains do not provide the auditory privacy required. The need for confidentiality had been stressed to the architects whose solution was to provide no windows! So they do need guidance.

The clinic should be designed so that patients flow through it smoothly. Unless the clinic is large with male and female services on different floors, it is usually adequate to have a mixed reception area. Most clinics have separate male and female sides, with separate waiting areas for each sex. In the small part-time clinic which I ran, there was not enough space so a single waiting room was shared by all. This produced few problems although we had to move quickly on occasions, such as when our sister spotted a patient's wife arriving, whilst he was already attending with his girlfriend. In a new large central London clinic a few years ago, the communal waiting room had to be divided into two after the first week, in order to stop the fighting between patients.

In a large clinic with a sizeable number of HIV-infected patients, it is best to have a separate area for this group if possible. They take longer to deal with, have particular needs, may prefer particular literature, may use the clinic as a day centre and day-care facilities may be necessary.

FURNISHING THE CLINIC

Decent furnishings can provide the ambience that you desire in a clinic. Colour coordination of the walls, doors, floor coverings and curtains and blinds can be beneficial. Carpeting lends warmth and it should be used in waiting areas, consulting rooms and offices. Selection of decent-looking furniture, for example desks, chairs and bookcases, improves the appearance of the clinic. If it can all be of the same type or style, so much the better. Comfortable chairs, coffee tables and pot plants improve the appearance of the waiting area and decent ceiling and wall lighting helps too. A huge selection of all of these is available from government suppliers.

Some departments provide televisions and video machines which can be used for health promotional videos in the waiting area and, of course, for watching important events, such as the test match and Wimbledon – staff included!

CLINIC HOURS

Clinic hours depend on the sessions when the consultant physician is available. It is important that he or she fulfils fixed commitments, but enough time must be available for administrative and liaison duties. In any clinic, but particularly in a part-time clinic, it is best to spread the opening hours evenly over the week. Many patients prefer to attend clinics outside working hours. In a small clinic with only a basic staff, it will not be possible to open over the lunch hour as staff are entitled to a lunch break. It should be possible to provide services early in the morning, say from 8.00 am, and in the evening without incurring overtime payments, by adjusting the sessions in the clinics. For instance, in Bromley, I had a clinic starting at 8.00 am one day and on another day we commenced at 12.30 pm, thereby covering the lunch hour and running until 7.30 pm. Saturday morning clinics are popular with many patients, but some departments have found such sessions to be poorly attended, which does not justify bringing in staff at such a time. If a clinic is to open on a Saturday morning, it will have to be closed for an equivalent session during the week, unless additional funding is forthcoming. If attendances are low on Saturdays, it may be best to terminate such sessions. However, in order to comply with the Patient's Charter, consideration needs to be given to clinics outside the hours of 9.00 am to 5.00 pm on Mondays to Fridays.

SERVICES TO BE OFFERED

Appointments

When starting from scratch, it is wise to focus on the basic function of the clinic, ensuring that those needs are met. Thus it is best to start with a walk-in service available to all comers. It will soon become clear whether or not the population attending is the sort that requires an appointment system or is likely to keep appointments. If that seems to be the case, an appointment system should be introduced but with enough flexibility to allow walk-in cases to be seen fairly promptly – remember the timings promoted in the Patient's Charter.

Family planning services

Some patients attending GUM services may not be accessing other services, such as family planning clinics. Doctors trained in GUM can provide such a service and many nurses have these skills too. Provision of contraceptive advice can be done on an *ad hoc* basis but as numbers build up, it may be sensible to provide most of this in a particular designated session. It should be seen as introductory advice and GUM clinics should have a complementary role in this rather than a primary one, which should remain with general practitioners and family planning clinics.

Problems clinic

As the service builds up and additional medical help is available, you might find it convenient to run a consultant session for problem cases and this is most easily accomplished with appointments. In a new clinic it will probably take time to accumulate a significant number of people with HIV disease. These can be accommodated within the ordinary arrangements of the clinic, although with symptomatic patients, an adequate appointment time needs to be set aside. Such cases can conveniently be seen within a 'problems clinic'. Difficult wart cases could be seen too, but you might wish to set up a separate wart clinic, possibly run by the nursing staff.

Colposcopy

Most GUM clinics nowadays provide colposcopy. In a new clinic there will not be a need for that initially. If capital money is available at the time of setting up the service, purchase all the equipment necessary so that it is available later on. In large inner-city clinics, where there are groups of patients not accessing primary health-care services, opportunistic cervical cytological screening is useful. I have found, however, that in a suburban area the majority of women have regular cytology performed at appropriate intervals by their GPs or family planning clinics, so the need for opportunistic screening is less. Nevertheless, even in small clinics, there is a requirement to perform cervical smears and when the results indicate that colposcopy is required, it is convenient to provide that service yourself. It is usually quicker than referring on to a gynaecologist, as in some areas the waiting list for colposcopy can be very long. Thus, establishing another colposcopy service in the trust or district may be very helpful to patients. GPs may even decide to refer direct, but care should be taken because fundholding GPs might perceive your service as a way of having colposcopy carried out free on their patients. If you link your service to that of the gynaecologists, you may be able to receive payment for the service. If you decide to offer treatment with your colposcopic service, make suitable arrangements with your gynaecological colleagues for assistance in case haemorrhage or other complications occur.

Psychosexual services

Patients with psychosexual problems sometimes present to GUM clinics. They come to us because, often, they do not know where else to go and they should be received sympathetically. Some GUM physicians have the expertise to deal with these problems, but it can be very time consuming and in a small, new service, it would not be appropriate to devote significant time to this. Rather, you should refer on to the local psychosexual services. Sometimes these clinics may not provide services for male impotence and if you have the training and skills to run such a clinic, then that is a rather more circumscribed service to offer.

Vulval clinic

Most of us have noticed an increase in the number of women presenting with vestibulitis and other vulval problems. If there is an interested dermatologist or gynaecologist in your hospital, you might wish to establish a joint vulval clinic. Such an enterprise will help to make you better known in your hospital and will also provide a useful service to patients.

Finally, you may be pressurized to set up outreach clinics of various sorts. These could range from special clinics, off your premises, for prostitutes, drug users or maybe an alternative HIV testing centre. With the latter it is important that you try to link this to your service. GUM staff are properly trained in this area, whereas not all others are.

The services which you end up offering will depend upon your own skills, interests and the time and monies available. You will probably start off with a very basic clinic but if it builds up into a busy unit, most of those mentioned above could be incorporated and might be expected in the provision of a sexual health service.[7]

EQUIPMENT

It would be superfluous in this chapter to list all the equipment required in a GUM clinic but some features are worth mentioning. Appropriate examination couches are important. For male patients, an ordinary couch is adequate, but you might prefer to order one with inbuilt drawers and pull-out flaps that are convenient to hold equipment commonly used for the taking of standard genital tests. For female patients, there is a range of couches, some with steps, which can tip so that women can easily get into the lithotomy position ready for examination. Some of these couches have an electric tipping mechanism and this can be very helpful. Some also have retractable receivers for instruments underneath or arms for the attachment of a colposcope.

When carrying out clinical examinations, especially of the genitalia, it is important to have adequate lighting. A large range of lights is available. Flexible-arm, fibre-optic lights are most convenient, if a little expensive, but they do provide the best illumination, usually from a halogen light source.

You will have to talk to your local CSSD department to decide which instruments to purchase. Some hospitals strike very good deals with particular companies. You will need to calculate with the CSSD whether or not you would be better to use, for instance, disposable plastic rather than reusable metal Cusco speculae. There are merits to both types. Most departments now use disposable proctoscopes.

You will probably want to have a cryotherapy device for the treatment of warts. It is wise to ask the various manufacturers to demonstrate their wares and, comparing performance, convenience and price, make your decision as to which one(s) to purchase. Cautery instruments are still used by some but most have moved on to the hyfrecator for heat treatment and you may wish to purchase one of these.

You will need to purchase all the equipment for your laboratory ('hot lab'). Your microbiology colleagues can be helpful here. A large range of microscopes is available at varying prices. Most people prefer separate microscopes for light- and dark-ground microscopy. Choice is usually dictated by price but if, for instance, a teaching arm is required, that may further reduce the range available to you. You will need an incubator and storage racks/jars and, if you can afford it and a supply of CO_2 is available, then a CO_2 incubator is preferable. You should also purchase the necessary 'boxes' or containers for the transportation of your specimens, particularly to the microbiology department.

Virtually all clinics now have a computer system. It is very helpful with Körner, KC60, AIDS Control Act and other *ad hoc* statistics. It can also hold the patient details database with the precise address and postcode recorded and can be used for appointments for various clinics. Most of the software packages have modules for health advisers and for many features relating to HIV, e.g. all the individual diagnoses, attendances and a means of recording those patients with CD4 counts below 200. The more sophisticated systems have pharmacy modules where all drugs prescribed are recorded and, for the annual AIDS Control Act requirements, all of this is very helpful. Most of these systems have a colposcopy package too and a cytology recall mechanism with the generation of recall letters for that and indeed for other diagnoses. A number of companies are competing for this business and it may be helpful to look at all of them and make your decision with the advice of your own local computer department. Some will actually provide a 'front-sheet' for your notes, complete with soundex code. Some will provide specimen labels but, more and more, computers are going to be necessary for audit, not just locally but also across the region. There may be a regional policy as to which software package to use and perhaps a special deal may be available through this. Some companies will sell you the hardware too, but it is worth checking with your local computer department to ensure that the hardware being advised is appropriate as they may be able to supply to you more cheaply and they are likely to want to do the cabling within your department themselves.

Patients' notes are important as medical records and legal documents. In a new clinic you have the opportunity to design them as you see fit. Of course, they have to be stored and a large number of storage systems are available from which to select. Space is always at a premium in clinics and as the files accumulate, it may run out. Your hospital may be prepared to pay for microfiching and you would require a microfiche reader in the department in this event. Transfer on to CD-Rom is now also available.

ARRANGEMENTS WITH OTHER DEPARTMENTS

To avoid professional isolation, liaison with colleagues in other disciplines is essential in the practice of medicine. Certainly, upon appointment to a new post, it is wise to make efforts to meet other colleagues. This can be accomplished at audit meetings, grand rounds, committee meetings and at lunch. Try to attend social events within the hospital as well.

At the level of establishing your department, however, the microbiology department has to be the first on your list for liaison. Most of the specimens taken in a GUM department go to microbiology. You will need to find out what services are on offer and, since GUM is generally a high throughput specialty, what additional resources and facilities the laboratory[1] will need to provide the service for you. It may need additional technicians, clerks and equipment. These cost money and you should arrange with management that the costs are met which, in negotiation, could eat into your own monies. A small microbiology department may not have a virology section and may have to subcontract that type of work out to other larger laboratories; this could result in delays in receiving results. You might wish to consider negotiating with a reputable private laboratory or an adjacent large hospital laboratory which can provide a collection service to have the work done elsewhere. Whatever arrangements you make, you will need to obtain a specific commitment to the turn-around times of the various tests so that you can accurately inform patients when their results will be available. It is just as well to get this in writing, in the form of a contract, otherwise you have no come-back with the laboratory. In addition, you should ensure that the pathology transport arrangements are convenient for your service and, if not, request changes to suit the hours the clinic intends to work.

Opportunistic cervical screening is performed in most GUM clinics and good relations should be established with the cytology department. Again, you will want to know what the turn-around time for results will be. While it is best for a copy of the cytology report to go to the patient's GP, not all patients will want their GP to know of their attendance at a GUM clinic, so you need to ensure that the cytology department observes the confidentiality expected of a GUM unit.

GUM departments are different from most other outpatient departments in that they usually dispense prepacked supplies of the commonly used medicines to their patients. You need to agree this arrangement with the hospital pharmacy and establish a mechanism for restocking. The pharmacy should also be reminded that patients at GUM departments do not pay prescription charges.

Whilst any branch of medicine will require the services of departments of haematology and biochemistry, GUM departments probably use them less than most, but it is wise to meet the consultants involved and introduce yourself. The same applies to the other branches of medicine and surgery with whom we often share patients, e.g. gynaecology, dermatology, general medicine and urology. A good working relationship with these departments may result in fast-track appointment arrangements for your patients. If you do not have allocated beds or do not intend to provide inpatient care for HIV patients, then you should come to an arrangement with one of the physicians to share care.

ADVERTISING THE NEW DEPARTMENT

As the time approaches for the clinic to open, the proposed services should be advertised. During the period of planning and hiring of staff there will have been time to give thought to this. Within the hospital, consultants in other specialties will have been aware of the appointment of a GUP but it is a good idea to write formally to all consultants to inform them of the opening date. Through the FHSA, it is possible to write personally to all GPs, in your own and adjacent districts, who might wish to refer patients to you. If you have time, you might find it helpful to visit some of the larger practices and to publicize the services. It is worth arranging with the local postgraduate medical centre to address GPs on topics relating to the specialty, thereby raising your profile.

General practice managers will usually be pleased to display posters advertising your services in their waiting rooms and these should include a number of 'bullet points' clearly describing the services available and inviting those reading them to request leaflets which can give further details of what is available and what to expect when visiting a clinic. These can be placed, with agreement, in youth clubs, sports clubs, gay clubs and drug units. Management may be able to direct you to a company which can design these items, but you should supervise the work closely so that the finished product is clear and tasteful.

Fig. 5.2 Free advertising for the new GUM clinic set up in Beckenham.

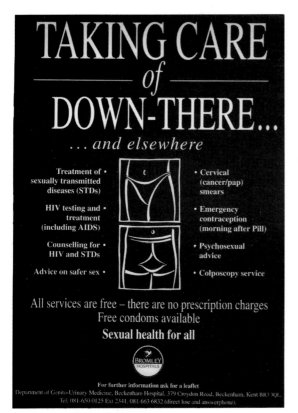

Fig 5.1 Part of a patient information leaflet.

To advertise your new unit further and to help draw its services to the attention of the public, it may be helpful to arrange interviews with the local papers, although you need to be careful in choosing your words so that the copy produced by the paper is accurate. Production of a fact sheet for the journalists will assist with this. The same applies to interviews on local radio stations. Of course, advertisements for the clinic can be placed in appropriate newspapers and in the gay press. This can be expensive, so an interview with a local journalist can achieve the same effect at no cost.

You should ensure that the clinic is listed in local voluntary and statutory directories, with a description of what is on offer, and also in the telephone directory. Finally, an official opening could be staged when the clinic has been operational for a few months. If this is performed by a well-known local or even national figure, it will give further publicity to the new department.

Most departments of genitourinary medicine are now simply known by that name, but some are called departments of sexual health. However, many are not attracted by either of these titles and there has been a patient-led drive to return to the use of eponymous names which they find more discreet, e.g. the name of the clinic at King's College Hospital in London was changed from the Department of Genitourinary Medicine to the Caldecot Centre, the name of the street in which it is sited, and the clinic at St Thomas's Hospital has always been known as the Lydia Department.

LOOSE ENDS

There are many other aspects of establishing a service which this chapter has not covered. I shall mention some of them here.

Before starting to see patients it is wise to have a clinic protocol covering the medical management of the common conditions and also other outlines and procedures as you see fit. This can provide the basis for audit. Apart from surveys of waiting times and other aspects of patient satisfaction, a small clinic can have difficulty auditing on its own. You should participate in the regional audit but you might be wise to run your own audit jointly with an adjacent small clinic.

Good communication with staff is helpful in ensuring the smooth running of the clinic. Regular staff meetings will allow teething problems to be dealt with rapidly and adjustments to the arrangements to be made. It is also good for staff morale.

It is wise to establish good relations with the purchaser and, to begin a dialogue, you will need to produce a service specification. This will form the basis of a business plan which you can then draw up with management.

It is always best to control your own budget but a new clinic is unlikely to be a directorate on its own. It is important to ensure that the monies allocated to you are actually spent on the service and, to this end, regular meetings with the business manager are vital.

Most GUM clinics are presently in the acute sector of the hospital service. Some have seen advantages in transferring to the community side because they perceive that aspects of health promotion, the concept of sexual health and indeed the provision of a primary care service are more appropriately positioned there. Additionally, the acute sector is presently subject to greater financial constraints than the community side. Some GUPs have become clinical directors of sexual health services, including the local family planning services, but others, firmly based in the hospital, have become head of a directorate of communicable diseases, including GUM and infectious diseases. There are pros and cons to both arrangements but the local circumstances will often dictate which arrangement is best.

REFERENCES

1. Monks, A. (1988) *Report of the Working Group to Examine Workloads in Genitourinary Medicine Clinics*, Department of Health, London.
2. Department of Health (1993) *List of Genitourinary Clinics in the United Kingdom and the Republic of Ireland*, Department of Health, London.
3. Mellor, D. (1989) *Letter to Chairmen of Regional Health Authorities: HIV and AIDS Resources*, Department of Health, London.
4. Royal College of Physicians (1993) *Staff Grade Doctors – Towards a Better Future*, RCP, London.
5. Allen, I. and Hogg, D. (1993) *Work Roles and Responsibilities in Genitourinary Medicine Clinics*, Policy Studies Institute, London.
6. Department of Health, Welsh Office and the Department of Health and Social Security (1990) *Health Building Note 12, Supplement 1. Genitourinary Medicine Clinics*, HMSO, London.
7. Secretary of State for Health (1992) *The Health of the Nation*, HMSO, London.

6 Setting up genitourinary medicine clinic services in a developing country

Y. Adu-Sarkodie

INTRODUCTION

In developing countries, sexually transmitted diseases are one of the most common reasons for individuals to seek medical care.[1] Their importance stems from their adverse effects on reproductive health and, more recently, the transmission of HIV/AIDS.[2] Unfortunately, in most parts of the developing world, STDs have not been treated as a priority despite the high incidence of complications such as chronic pelvic pain, pelvic inflammatory disease with subsequent infertility, ophthalmia neonatorum and complications of syphilis. With current knowledge of the interrelationship between STDs and AIDS,[2] however, this attitude seems to be changing slightly. Increased donor support of AIDS control activities in such countries is helping to address the issue. Opportunities abound for developing countries to set up services not only to control the HIV/AIDS problem but also in the long run to contribute effectively to addressing the reproductive health needs of the population. The 1993 World Bank Report[3] *Investing in Health* stresses the importance of adequate treatment and control of STDs as an intervention for containing the spread of HIV infection.

The setting up of GU medicine clinics (better known as STD clinics in many developing countries) is different from the Western perspective due to differences in the structure of the health systems, health-care delivery, funding and the presence of adequate and qualified staff to deliver the services. In the UK, the costs patients incur in attending these clinics, including consultations, laboratory investigations and medication, are paid via the NHS. In many developing countries, especially those going through economic structural adjustment programmes, there is much emphasis on cost recovery within the health sector. Thus, patients have to pay substantially for services provided. Some countries have tried to reduce the cost burden on patients with HIV/AIDS by passing legislation granting free medical care for them. In practice, however, with the increasing number of hospital beds being occupied by patients with AIDS-related diseases, the costs can be phenomenal and patients have had to contribute. The service discussed below is one that we have found useful in Ghana.

GU MEDICINE SERVICES AND HEALTH SYSTEMS IN DEVELOPING COUNTRIES

The health service in many developing countries is based on the primary health care (PHC) approach. PHC centres are supported by district hospitals, regional and teaching hospitals with appropriate referral systems. Outside teaching and some regional hospitals, specialist staff – GU physicians/venereologists, microbiologists, trained nurses and counsellors (health advisers in the UK) – are not available. Thus, the majority of the population are managed without specialist care. In this situation, the WHO algorithms for the syndromic management of STDs become useful, as proper facilities such as microbiology laboratories and basic devices such as specula are often not available. In resource-poor settings, the approach should be to have an integrated service within the existing PHC programmes. The clinicians providing services, be they physicians, medical assistants, etc., should have training in the use of the WHO algorithms for the syndromic management of STDs. However, at the teaching and regional hospitals which are tertiary referral centres, stand-alone STD clinics are needed, with qualified staff. The microbiology laboratory will take up all STD referrals and provide essential information on the antimicrobial susceptibility profile of STD pathogens to fashion effective drug-prescribing practices in GU medicine.

LABORATORY SUPPORT

Even though it is essential to have 'mini-labs' within GU clinics where Gram stains and wet film microscopy can be done to facilitate early treatment of patients, resources, including staff, may not allow this in developing countries. To this end, a close collaboration between the clinic and microbiology laboratory in teaching/regional hospitals needs to be forged in order to process Gram-stained and wet films of urethral and vaginal discharges for patients immediately so that they can receive prompt care.

THE CLINIC AND STAFFING

STD/AIDS are still stigmatized in developing countries and this should not be compounded by siting clinics away from the main outpatients area of the hospital. In parts of the developing world where STDs are seen and managed by dermatovenereologists, this problem has been contained as patients being seen by the physician are not easily identified as having a STD. The hours of operation of such specialized clinics should be the same as other clinics of the hospitals. Patients should not be kept waiting unduly as this may exacerbate the stigma. Medical staff running such clinics need to have training in genitourinary medicine/venereology. Two such postgraduate courses are offered in the UK (University of Liverpool and the British Postgraduate Medical Federation). We in Ghana have found both courses useful for our trainees.

In teaching/regional hospitals, a senior nurse with some training and skills in running a GU service is necessary. This training may be acquired from short-term attachments in established clinics locally or abroad. The responsibilities of this senior nurse will include supervising the consulting and examination rooms and acting as chaperone to doctors during examinations. As head nurse, she or he will liaise between the clinic and the hospital nursing administration. The nurse must have skills in counselling about HIV, STDs and family planning.

An examination room must be available as even in the syndromic approach,[4] the starting point of management is the visualization of genital discharges and/or ulcers, etc. This examination area can be within the consulting room but cordoned off by a screen.

In many developing countries, only clinical services are provided to GU patients. However, their needs often include psychosocial and contraceptive services. There is a trend for STD/family planning services to be integrated.[5] If this is not the case, clinic staff will need to refer patients to the appropriate place.

EQUIPMENT

In PHC settings, the most important equipment will be an examination couch and a good source of light. A good carpenter can make the couch. In referral centres, examination couches which can tilt to assist examination of female patients in the lithotomy position are mandatory, as well as a good light source. Adequate quantities of specula, microscope slides and transport media should be provided in the examination area. If cryotherapy or cautery is done for wart treatment, the necessary equipment should be provided.

LIAISON WITH OTHER DEPARTMENTS/SERVICES

Staff running PHC clinics should have clear instructions on the lines of referrals. There should also be appropriate feedback from referral institutions to the PHC clinics. At the teaching/regional hospital clinics, clinicians should liaise well with other departments of the hospital, especially microbiology, gynaecology, urology and family planning.

REFERENCES

1. World Health Organization (1991) *Technical Report Series 810. Management of Sexually Transmitted Diseases*, WHO, Geneva.

2. Laga, M., Nzila, N. and Geoman, J. (1991) The interrelationship of sexually transmitted diseases and HIV infection: implications for the control of both epidemics in Africa. *AIDS*, **5** (suppl 1), S55–56.
3. World Bank (1993) *World Development Report: Investing in Health*, Oxford University Press/World Bank, New York, p. 115.
4. Vuylsteke, B. and Meheus, A. (1996) Sexually transmitted diseases: syndromic management, in *Control of Sexually Transmitted Diseases – A Handbook for the Design and Management of Programs*, (eds G. Dallabetta N. Laga and P. Lamptey), AIDSCAP/Family Health International.
5. Cates, W. (1993) Sexually transmitted diseases and family Planning. *Sexually Trans. Dis.*, **20**, 41–44.

7 The diagnosis and management of genital dermatological conditions

B.T Goh

INTRODUCTION

Non-sexually transmitted genital dermatoses may cause anxiety to patients as they may be embarrassed to divulge the problem or be concerned that it may be sexually transmitted or malignant. Normal dermatological variation may also be noticed for the first time, especially after a new sexual relationship, and may cause anxiety. However, as sexually transmitted diseases do cause genital rashes and a few lesions will be diagnosed as premalignant or carcinomatous, it is essential that these are excluded. Diagnostic procedures, including a skin biopsy, may be necessary to confirm the diagnosis.

HISTORY, EXAMINATION AND INVESTIGATIONS

The diagnosis of many genital dermatoses may be simple. In typical cases of scabies, molluscum contagiosum or genital warts where sexual acquisition is most likely, other STDs should be excluded. The sexual history is important to estimate the risk of sexual transmission or sexual trauma. Ascertain the sites affected, the duration, whether it is itchy or recurrent in nature, any blistering or ulceration and if there is involvement of any extragenital sites. Systemic symptoms, medications or allergies should be enquired into as well as whether there is any family history of a similar problem.

Involvement of the mouth may suggest Stevens–Johnson syndrome, Behçet's disease, pemphigus, benign mucous membrane pemphigoid, lichen planus, orogenital herpes simplex infection or secondary syphilis, the last two being sexually transmitted. The regional lymph nodes may be enlarged secondary to infection or spread from genital carcinomas. Genital herpes, fixed drug eruptions, erythema multiforme, Behçet's disease, pemphigus and pemphigoid are commonly recurrent. Malignancy should be excluded in all cases of chronic lesions or ulcers.

The presentation or characteristic of the rash or lesion may help in the differential diagnoses (Table 7.1). If the diagnosis remains uncertain after the history and examination, biopsy of the lesion is indicated for histology and appropriate immunofluorescent staining. Sexually transmitted diseases should always be excluded in ulcerative lesions, particularly syphilis and genital herpes.

Following the exclusion of an STD, the history, examination and investigations will point to one of the causes of non-sexually transmitted genital dermatoses. In most non-infective or non-malignant conditions, a topical steroid is the mainstay of treatment (Table 7.2).

Table 7.1 Common presentations of genital dermatosis

Itchy lesion	Lichen planus, eczema, infection, lichen simplex, psoriasis, intraepithelial neoplasia, Fox–Fordyce disease
Burning, pain, soreness	Herpes zoster, eczema, lichen planus, lichen simplex, lichen sclerosus, vulvodynia
Erythematous lesion	Eczema, bacterial infection, erythrasma, tinea, extramammary Paget's disease, hidradenitis, lichen sclerosus, lichen simplex, pyogenic granuloma, vestibulitis, psoriasis, pemphigus, pemphigoid, chronic inflammatory bowel disease, necrolytic migratory erythema (glucagonoma), intertrigo
Lumps and bumps	Pearly penile papules, vestibular papillae, angiokeratoma, haemangioma, caruncle, cyst, lymphocele, cystocele, rectocele, skin tag, sebaceous gland, fibroma, lipoma, Fox–Fordyce disease, seborrhoeic wart, Fordyce spots, carcinoma, endometriosis, scabetic nodule
Yellow, white or pale lesion	Fordyce spots, vitiligo, postinflammatory hypopigmentation, lichen planus, lichen simplex, lichen sclerosus, intraepithelial neoplasia
Pigmented, dark lesion	Lentigo, naevus, melanoma, basal cell carcinoma, bowenoid papulosis, intraepithelial neoplasia, seborrhoeic warts, postinflammatory hyperpigmentation, acanthosis nigricans, lichen sclerosus, vestibulitis, endometriosis, angiokeratoma, caruncle
Blister, erosion, ulcer	Fixed drug eruption, Stevens–Johnson syndrome, herpes zoster, chickenpox, accidental vaccinia, pemphigus, pemphigoid, circinate balanitis, circinate vulvitis, scabies, erosive lichen planus, erosive lichen sclerosus, extramammary Paget's disease, intraepithelial neoplasia, carcinoma, idiopathic aphthoid ulcers, Behçet's disease, infective (amoebic, tuberculous), chronic inflammatory bowel disease, pyoderma gangrenosum, iatrogenic (traumatic – physical, chemical, postradiotherapy)
Oedema	Contact dermatitis, urticaria, lymphoedema (primary, secondary, e.g. filariasis, postsurgery, postradiotherapy, chronic inflammatory bowel disease), hidradenitis, cellulitis
Balanitis	Plasma cell (Zoon's), erythroplasia of Queyrat, lichen planus, psoriasis, circinate balanitis, fixed drug eruption, pyogenic granuloma, erythema multiforme, lichen sclerosus (balanitis xerotica obliterans)

Table 7.2 Topical steroid potency

Potency	Types
Mild	1% Hydrocortisone
Moderately potent	Alclomethasone dipropionate 0.05% Clobethasone butyrate 0.05% Flurandenolone 0.0125% Hydrocortisone 1% with 10% urea
Potent	Beclomethasone dipropionate 0.025% Betamethasone valerate 0.1% Desoxymethasone 0.25% Diflucortolone valerate 0.1% Flucinolone acetonide 0.05% Fluclorolone acetonide 0.025% Hydrocortisone butyrate 0.1% Mometasone furoate 0.01% Triamcinolone acetonide 0.1%
Very potent	Clobetasol priopionate 0.05% Diflucortolone valerate 0.3% Halcinonide 0.1%

SKIN BIOPSY

This is one of the most important diagnostic procedures and can be performed using either a punch biopsy or skin snip technique. The lesion is first infiltrated with 1–2% lignocaine with 1:200 000 adrenaline, ensuring that it is not given in excess as the adrenaline may cause ischaemic necrosis, particularly if used on the penis.

A punch biopsy ranging from 2 mm to 8 mm in size is pressed into the skin with a back-and-forth twisting motion; the material is then gently lifted and excised at the subcutaneous level.

For the skin snip technique, the skin is gently lifted with forceps and the elliptical base snipped off with a pair of scissors. Haemostasis is secured using a silver nitrate stick for small biopsies or with a suture for larger skin defects. Early-maturing lesions should be biopsied, preferably at the edge, and sufficient depth and breadth of the tissue should be obtained. A 3/0 or 4/0 synthetic monofilament suture such as Prolene,

Ethilon or Dermalon gives good cosmetic results. The suture is removed in 7–10 days.

NORMAL VARIATION, DYSPLASIA, HETEROTOPIA

Penile papillae

These are congenital hypertrophic papillae arranged in rows around the coronal sulcus. They may have a filiform arrangement and can be mistaken for acuminate warts.

Fordyce spots

These are ectopic sebaceous glands presenting as multiple small yellow or white spots in the submucosa of the prepuce (see Plate 5) or vulva.

Phimosis

This can be congenital or acquired. Lichen sclerosus et atrophicus in men, also known as balanitis xerotica obliterans (see Plate 6), and various infections can lead to tightness of the prepuce, preventing it from being drawn back from the glans. Any underlying cause should be treated. Circumcision should be considered if it is causing problems such as recurrent balanitis.

Paraphimosis

The retracted phimotic prepuce which cannot be reduced causes a constriction around the shaft of the penis. This can lead to painful oedema of the prepuce and, in severe cases, linear ulcers along the constriction. The constricting pressure may be relieved by compressing the end of the penis with moist cold pads and gently pulling the swollen prepuce with the constricting band over the glans penis while simultaneously pressing the glans against the band. Otherwise, surgical release of the constriction and a circumcision for the phimosis may be required.

Vitiligo

This presents with depigmented areas and may be associated with autoimmune diseases (see Plate 7).

ERYTHEMATOUS, SQUAMOUS AND PAPULAR DERMATOSES

Lichen planus

This appears as small, flat-topped violaceous polygonal lesions with white Wickham's striae on the glans penis, prepuce, foreskin, anal or vulval region. In the moist subpreputial area the lesions may be white but sometimes they may be annular or erosive. These appearances are markedly different in pigmented skin (see Plate 8). The lesions are usually itchy. The wrists, forearms, legs, mouth and nails may be involved. The lesions may resolve spontaneously but moderately potent to potent local steroid cream may speed the process and relieve the itch.

Psoriasis

This presents as dusky red plaques with or without silvery scales on the penis, vulva and perineum (see Plate 9). On flexural or moist surfaces, it may be red and shiny without any scaling. Although the genitalia may be the only site involved, it commonly affects the knees, elbows, palms, soles, scalp, hair margin and the nails where there may be pitting and onycholysis. Severe generalized or pustular psoriasis may occur. Psoriatic arthropathy may develop in less than 5% of cases. Supervening HIV infection may cause a recrudescence of the psoriasis or make it worse. The histology of the rash is similar to keratoderma of Reiter's disease. Treatment of the genital area is with local mild steroid cream.

ECZEMA, DRUG ERUPTIONS AND REACTIVE DISORDERS OF SKIN

Contact dermatitis

Irritating chemicals or medication, such as strong antiseptics applied by patients to clean the genitalia, may result in a rash or, if used to treat a genital rash, may cause it to worsen. Dequalinium, a quaternary ammonium antibacterial agent, can cause genital ulcers directly when applied to the genitals and indirectly by the use of a speculum or needle sterilized in the agent. Foscarnet, an antiherpes virus agent which is used mostly for treating CMV infection in AIDS patients, can cause genital, oral and oesophageal ulcers and may mimic a fixed drug eruption. The genital ulceration is thought to result from repeated exposure to high urinary concentrations of foscarnet or deposition of crystallized foscarnet in arterioles and capillaries. Injection of substances into the dorsal vein of the penis in drug abuse may lead to local bruising and, if extravasation occurs, a necrotic ulcer may result. Treatment of genital warts using topical podophyllin, podophyllotoxin or trichloroacetic acid as well as cryo- or electrocautery may cause iatrogenic traumatic ulcers.

Allergic contact dermatitis is due to a delayed hypersensitivity reaction. Common allergens in the genital area include rubber (condoms), perfumes, dyes, washing powder and medicaments such as topical antibiotics and anaesthetics. Once sensitized, a rash occurs within 48–72 h following exposure to the allergen. The rash tends to be acute with weeping lesions, vesicles and blisters. A patch test is required to identify the allergen.

Lichen simplex chronicus

This affects older people and presents with an itchy localized area of thickened coarse lichenified eczema secondary to chronic scratching and rubbing. In dark-skinned patients this can lead to postinflammatory hypo- or hyperpigmentation. Treatment is with potent local steroids, oral antihistamines and management of any underlying precipitating factor as the condition can be triggered by a stressful event.

Fixed drug eruption

A fixed drug eruption is more common on the male genitalia and other mucosal surfaces and recurs on the same site. A few hours after ingestion of the drug, a well-defined round or oval lesion appears, which is dusky red and oedematous, sometimes with a purplish hue. It is preceded by burning or itching. In some cases, there may be severe erythema with bulla formation and ulceration (see Plate 10) It heals with desquamation and hyperpigmentation. Drugs that are implicated include tetracycline, sulphonamides, dapsone, phenolphthalein, chlordiazepoxide, barbiturates, salicylates, oxyphenbutazone and quinine. Tetracycline, which is commonly used in the treatment of certain STDs, may cause confusion if it also triggers a fixed drug eruption, because it may then be mistaken for worsening of the original problem or for another STD. Readministration of common offending drugs may be required if the patient denies taking any medication despite the clinical diagnosis. Treatment is by local mild to moderately potent steroid application and removing the offending drug.

Erythema multiforme and Stevens–Johnson syndrome

Erythema multiforme may result in target or iris lesions and maculopapular or bullous rash, particularly on the extremities; i.e. dorsa and palms of hands, wrists, forearms, elbows, feet and knees. Stevens–Johnson syndrome is a severe form of erythema multiforme affecting the mucous membranes of the mouth, eyes and genitalia, particularly the glans penis, inner prepuce, urethra, vulva and vagina (see Plate 11). The conjunctivitis may be associated with corneal lesions.

The gastrointestinal, respiratory and renal systems may be involved. The condition may be limited to the mucous membranes and may be recurrent. Precipitating agents include viruses, especially herpes simplex virus, bacteria such as mycoplasma, fungi such as histoplasma, drugs such as sulphonamides and phenytoin, autoimmune diseases and malignancy, although in many cases no cause can be identified.

Symptomatic treatment is sufficient in mild cases. In severe cases, high doses of a systemic steroid such as prednisolone and antibiotics are required to prevent secondary infection. Precipitating causes should be avoided and recurrent herpes can be suppressed with continuous oral acyclovir.

PHYSICAL AND SEXUAL TRAUMA

The cause of the lesions may not be obvious if the patient does not give a history of trauma. This may be intentional, as in cases of dermatitis artefacta, or through embarrassment, as in excessive trauma during intercourse from genital bites, fisting (putting a fist into the rectum) or sadomasochism. Dermatitis artefacta lesions are unusual, bizarre and do not fit the history or known causes and may be seen in psychiatric patients or malingerers. Genital bites may lead to symmetrical marks or ulcers around the penis and fisting or the use of 'sex toys' in the rectum can cause anorectal tears. In women, the first penetrative intercourse may cause the hymen to tear and tampon use may be associated with recurrent vaginal and cervical ulcers. Accidental zip injury can lead to a linear ulcer in men who do not wear underwear. Condom 'catheters' tightly applied to the shaft can result in necrosis and ulceration of the penis and glans. Tight-fitting jeans may lead to friction dermatitis with excoriations in the genital area.

Traumatic lesions are usually treated with saline washes and antibiotics if there is secondary infection.

INFECTION

Bacterial infection

Erythrasma due to *Corynebacterium minutissimum* can cause a rash on the groins which has to be differentiated from tinea cruris. Wood's light will show a coral pink appearance and treatment is with a course of erythromycin.

The moist and hairy genital region, obesity and warm climate predispose to pyogenic infection with streptococci or staphylococci. Folliculitis starts as perifollicular pustules while furuncles are dark red, painful, perifollicular nodules leading to fluctuant abscesses which may break down, resulting in ulceration. Underlying scabies or pediculosis with secondary infection should be excluded. Recurrent or severe pyogenic infection should alert one to predisposing causes such as diabetes mellitus, malignancies or immunodeficiency.

Erysipelas due to streptococci or, rarely, staphylococci may lead to oedematous erythematous areas with vesicles and bullae which may ulcerate.

Systemic antibiotics such as flucloxacillin and erythromycin or topical antibiotics such as fucidic acid and mupirocin, depending on severity, will usually control the infection.

Vincent's fusiform organisms (*Leptotrichia buccalis*) and Gram-negative bacteria may cause erosive balanitis in patients with phimosis. This may lead to offensive purulent discharge and tender inguinal lymphadenopathy. Retraction of the prepuce may reveal erythematous areas with shallow erosions or irregular ulcers.

Genital tuberculous infection is usually secondary to that of other sites, in particular the lungs, although primary genital tuberculosis, possibly transmitted through intercourse, may occur.

Primary tuberculosis of the penis may present as a nodule which may ulcerate with marked lymphadenopathy. In patients highly sensitized to the bacillus, a papulonecrotic tuberculid of the genitalia may occur. This results in an intense tissue response with necrosis, leading to scarring and a 'wormeaten' appearance of the glans penis. In men, tuberculous epididymitis may lead to a discharging sinus in the overlying scrotum and spread from the bladder may lead to multiple shaggy meatal ulcers. In women, urinary or endometrial tuberculosis may spread to the vulva and cervix. The cervical lesion may appear as a granulomatous or papillomatous growth or as an ulcer. This may lead to a smelly vaginal discharge, abnormal vaginal bleeding, menstrual disturbance, dyspareunia, pelvic pain and infertility. Intestinal tuberculosis may spread to the anal region. In addition to biopsy of the lesion for histology and culture for mycobacterium, a chest X-ray is essential. A tuberculin skin test may also be helpful. Urological examination may be required to exclude urological involvement. Choice of antituberculous therapy will depend on the sensitivity of the organism. It is important to remember that tuberculosis may be the first manifestation of HIV infection.

Primary cutaneous diphtheria can rarely affect the external genitalia, resulting in a tender pustule which then ulcerates with a punched-out base and greyish membrane. In children, genital lesions may be seen in pharyngeal and nasal diphtheria.

Fungal infection

Balanitis or vulvitis may be due to candidiasis. Severe candidiasis may cause superficial ulceration and fissuring of the prepuce, glans, fourchette and vagina.

Tinea cruris is more common in men and presents with a scaly rash on the groins with a well-defined, spreading margin. This is commonly due to *Epidermophyton* or *Trichophyton*. A scrape from the margin mounted on a slide with potassium hydroxide will show the mycelium. It may be associated with tinea pedis. Treatment is with local imidazole cream but sometimes griseofulvin or an oral imidazole may be required. Loose-fitting underwear should also be advised.

Seborrhoeic dermatitis associated with pityrosporum may affect the genitalia and perineum, particularly the flexural areas. The face, scalp and chest may also be affected, presenting with a scaly erythematous rash with ill-defined edges. Treatment is with a combined steroid and imidazole cream such as Daktacort or Canesten HC cream.

Viral infection

Molluscum contagiosum, caused by a member of the pox virus group, is commonly seen in children and may affect the genital region. The lesions appear as pearly or erythematous umbilicated papules. If treatment is required, cryotherapy or pricking the lesions with phenol or trichloroacetic acid may be used.

Generalized chickenpox from varicella-zoster virus infection may occasionally affect the genitalia. Reactivation of the same virus in the third sacral nerve root results in genital herpes zoster or shingles (Plate 12). The lesions in shingles are unilateral and may be preceded by burning pain, malaise, headache and low-grade fever. Erythematous areas appear followed by large grouped vesicles over one side of the buttocks, vulva or penis and scrotum. The vesicles then ulcerate and crust. There may be localized lymphadenopathy which may be tender if there is secondary bacterial infection. Postherpetic neuralgia may occur. Treatment is with acyclovir 800 mg five times a day, famciclovir 250 mg three times a day or 750 mg once daily or valaciclovir 1 g three times a day for a week. Antibiotics may be required for secondary infection. Symptomatic relief is provided by systemic analgesia and topical calamine lotion. Postherpetic neuralgia may respond to carbamazepine or amitryptiline. Underlying malignancy or immunodeficiency must be sought, especially in young persons with multidermatomal lesions.

Parasitic infection, infestation and spider bite

Scabies may affect children and in males commonly involves the genitalia, causing excoriated papules and nodules. Treatment is with malathion or permethrin application.

Genital amoebiasis, which is rare, may cause serpiginous ulcers with greyish slough in the perianal region, cervix, vagina and, rarely, the vulva and penis. Women may present with vaginal discharge and dyspareunia. There may be associated diarrhoea. Amoebiasis may occasionally be transmitted by anal or vaginal intercourse. Diagnosis is by cytology, histology or direct microscopy of the vaginal discharge or ulcer scrapings. Treatment is with oral metronidazole.

Genital schistosomiasis usually results from *S. haematobium* infection. Papillomatous lesions or nodules are seen in the external genitalia while ulcerative lesions may be found in the cervix. Treatment is with metrifonate, niridazole, hycanthone or praziquantel. Papillomatous lesions may be removed surgically.

Cutaneous leishmaniasis may rarely affect the external genitalia, resulting in erythematous nodules which may ulcerate.

Filariasis may affect the lymphatics of the external genitalia, leading to elephantiasis with thickened dry skin. Treatment is with diethylcarbamazine but surgery may be required for the elephantiasis.

Loxosceles reclusa spider bite can result in a chronic vulval ulcer.

BEHÇET'S DISEASE

This disease is more common in the Mediterranean, Middle East and the Far East. The diagnosis is based on recurrent painful oral ulcerations (> three episodes per year) and two of the following criteria.

- Recurrent genital ulcers or scarring (Plate 13).
- Eye lesions (anterior or posterior uveitis, cells in vitreous on slit-lamp examination, retinal vasculitis).
- Skin lesions (erythema nodosum, pseudofolliculitis or papulopustular lesions, acneiform nodules in postadolescent patients).
- positive pathergy test where a papule or pustule develops at the site of a hypodermic needle puncture.

The genital ulcer is preceded by a tender nodule and may heal with scarring and loss of tissue. Characteristic splash fibrosis may be seen in the mouth. The pathergy test is performed with a 20 gauge or smaller needle inserted obliquely and read at 24–48 h.

The lesions can be treated with potent topical steroids such as triamcinolone acetonide oral paste.

ZOON'S BALANITIS (PLASMA CELL BALANITIS)

This benign chronic balanitis appears as moist, shiny, red patches on the glans or inner prepuce with 'cayenne pepper' stippling due to haemosiderin deposition (see Plate 14). The biopsy will show intense plasma cell infiltration of the epidermis. Treatment is with a mild to moderately potent topical steroid.

BENIGN TUMOURS

Sebaceous cysts

These are retention cysts of the sebaceous glands appearing as round yellow swellings commonly on the labia majora or scrotum (see Plate 15). They may calcify, in which case idiopathic calcinosis of the scrotum should be considered in the differential diagnosis.

Fibroma

This appears as a firm painless nodule on the scrotum or labia majora and may be pedunculated. Neurofibroma may cause pain on pressure and may be associated with *cafe au lait* spots elsewhere.

Lipoma

This is a soft, semifluctuant, lobular lump which can occur on the vulva or scrotum (see Plate 16).

Haemangioma

This may occur on the penis, scrotum, vulva, vagina or rectum. Angiokeratoma occur typically as cherryred spots on the scrotum which do not require treatment. Pyogenic granuloma, a proliferating capillary haemangioma, occurs as a small red papule which enlarges. Minor trauma may cause it to bleed. The lesion may be excised or cauterized.

Seborrhoeic wart

This occurs usually as a solitary brown, well-defined papule or plaque with a rough stippled or fissured surface. It results from accumulation of immature keratinocytes. It is important to differentiate this from genital warts, as the latter require screening for sexually transmitted infections in the patient and sexual partners. The lesion can be removed using cryotherapy or curettage.

Pigmented naevus

This appears as a brown or black, circumscribed lesion with uniform distribution of pigment.

PREMALIGNANT CONDITIONS

Lichen sclerosus et atrophicus

Vulval involvement leads to shiny white sclerotic, smooth papules which become atrophic with telangiectasia and purpura. In men, the glans penis and prepuce are commonly affected, leading to meatal stenosis and phimosis (see Plate 17). Occasionally sexual intercourse may lead to the formation of a haemorrhagic bulla at the meatus which may lead to haematuria. Treatment is with emollients and a potent or very potent steroid in the acute phase or if symptoms are severe. Meatoplasty and circumcision may be required for complications in men.

Bowen's disease

This presents as a raised, irregular, reddish brown, scaly plaque on the penis, vulva or perianal area. It may be present for many years and may result in a squamous cell carcinoma. It is associated with human papillomavirus infection.

Erythroplasia of Queyrat

This presents as a well-defined, velvety, slightly raised red plaque on the glans penis, prepuce and occasionally vulva which may later ulcerate and crust. This is premalignant and usually affects elderly men. Treatment is by excision, radiotherapy or local application of 5% 5-fluorouracil.

Extramammary Paget's disease

This is seen mainly in postmenopausal women and may give rise to erythematous weepy plaques with crusting in the anogenital region.

MALIGNANT TUMOURS

Anogenital carcinoma

Carcinoma should be excluded in any chronic non-healing anogenital ulcer. Predisposing factors include being uncircumcised, lichen sclerosus et atrophicus, erythroplasia of Queyrat, Bowen's disease, Paget's disease, leucoplakia of the penis and vulva, psoriatic patients exposed to PUVA therapy and anogenital warts. Squamous cell carcinoma of the scrotum is associated with exposure to mineral oil, tar and soot, as in chimney sweeps of the past.

Genital carcinoma usually presents as a single persistent asymmetrical ulcer in an elderly patient although young people are not exempt. The ulcer is often asymptomatic and may be ignored or go unnoticed. The lesion may develop under a tight prepuce as a warty growth or ulcer with an indurated edge which may be palpable as a hard mass. It may lead to a purulent or bloodstained subpreputial discharge. The lesion may ulcerate through the prepuce or spread to the glans penis, the shaft or regional lymph nodes. Vulval carcinoma commonly presents as an ulcerated nodule on the labia majora or minora and sometimes on the clitoris. Surgery and radiotherapy are the mainstays of treatment depending on the stage of invasion.

Basal cell epithelioma

This is a malignant or semimalignant condition and may occasionally occur in the anogenital region. The slow-growing lesion may appear as a hard brown or grey nodule, the centre of which may become depressed and may later ulcerate. Telangiectatic vessels are present across the surface. The lesion may also appear as a well-defined superficial erythematous plaque which may become pigmented. Treatment is usually surgical.

Melanoma

This may be a superficial spreading, nodular or circumscribed lesion occurring in the glans penis, coronal sulcus, shaft of the penis, meatus, vulva, vagina and cervix. The colour ranges from black to brown or red with varying pigmentation which may spread irregularly into the surrounding skin. If untreated, the lesion rapidly metastasizes to the regional lymph nodes. Surgical excision with a margin of 1–3 cm, depending on the depth of lesion, is the treatment of choice, preferably by a specialist or plastic surgeon. Radical surgery with regional lymphadenectomy for more invasive lesion may be necessary.

Sarcoma

Kaposi's sarcoma, particularly those associated with AIDS, may affect the genitalia. This presents as purplish or dark erythematous lesions. Treatment is not necessary if lesions are asymptomatic. Otherwise radiotherapy or chemotherapy may be indicated (see Chapter 18).

BULLOUS ERUPTIONS

Pemphigus vulgaris

This severe chronic blistering disease, occurring in the middle aged, commonly involves the mucous membranes of the mouth and genitalia. The cutaneous bullae rupture, leaving raw eroded areas which are painful. In women the vulva, cervix and vagina may be affected and may cause superficial dyspareunia. The diagnosis is confirmed by biopsy and immunological staining. Treatment is with corticosteroids and immunosuppressives and other agents.

Benign familial pemphigus (Hailey–Hailey disease)

This causes localized pustules and erosions in the intertriginous areas and may affect the scrotum, penile shaft, vulva, vagina and perianal region. It is inherited as an autosomal dominant form and there is a family history in about two-thirds of patients.

Benign mucous membrane pemphigoid

This occurs in elderly patients and has mucosal and cutaneous involvement. The glans and the prepuce may be affected, leading to adhesions between both surfaces and phimosis. In women, the vulva and vagina may be affected, leading to ulceration, stenosis and urinary obstruction. The eyes may be affected, leading to corneal scarring and blindness.

Other bullous eruptions

Epidermolysis bullosa and herpes gestationis may also affect the genitalia.

CHRONIC INFLAMMATORY BOWEL DISEASE

This may present in the anorectal region with granulomatous lesions with ulcers, fistulae or oedematous skin tags. Islands of ulceration may be seen. The penis or vulva may also be affected and surgery of an anal lesion may lead to a spread to the perineum, groins and genitalia. The ulcers generally respond to metronidazole rather than conventional treatment with steroids, sulphasalazine and other antibiotics. Chronic inflammatory disease and other conditions such as rheumatoid arthritis and multiple myeloma may also be associated with pyoderma gangrenosum which may cause deep painful necrotic genital ulcers with undermined overhanging edges.

HYDRADENITIS SUPPURATIVA

This is a chronic inflammatory apocrine disorder affecting the anogenital region, axilla, breast, periumbilicus and neck, presenting with recurrent acute painful erythematous swellings and abscesses. These may lead to sinus tracts, fistulae, scarring and oedema with fibrosis and restriction of movement. Comedones may also be seen in other sites, including the retroauricular area. Other complications include amyloidosis, interstitial keratitis and squamous cell carcinoma. Predisposing factors include obesity and the condition may worsen premenstrually in women. Medical treatment includes weight reduction, antiseptic wash, antibiotics with tetracycline or minocycline, intralesional or very potent topical steroids and antiandrogen, with varying success. Radical surgery to remove the affected apocrine area may be required in severe disease.

FOX–FORDYCE DISEASE

This is a chronic apocrine disorder characterized by severe itch and conical follicular papules in the anogenital region and axillae. Apocrine anhidrosis may occur. The condition occurs mainly in women, appears after puberty and may regress after menopause. Potent to very potent topical steroids, antihistamines, oral contraceptives and Retin-A have been used with variable success.

OTHER GENITAL DERMATOSES

Lymphocele (benign transient lymphangiectasis)

This is a firm translucent cord-like swelling resulting from the blockage of lymphatics in the coronal sulcus or adjacent to it and resolves spontaneously within a few weeks (see Plate 18). It may be associated with hectic sexual intercourse or a genital lesion.

Reiter's disease

Sexually transmitted non-gonococcal urethritis or gastrointestinal infections such as shigellosis may precipitate Reiter's disease. This is characterized by arthritis, conjunctivitis, psoriasiform skin lesions and mucosal involvement of the mouth and genitalia. The latter results in painless superficial erythematous erosions with a slightly raised circinate edge on the glans penis and rarely on the vulva, also known as circinate balanitis or vulvitis. The genital and skin lesions respond to mild to moderately potent topical steroids.

Eosinophilic granuloma

This may cause a granulomatous vegetating lesion or ulcer in the genital region or perineum.

Darier's disease

This occurs in the seborrhoeic and intertriginous areas. The lesions are symmetrical, reddish brown, greasy crusted papules. In the genitocrural area and around the scrotum and vulva, there may be maceration and warty vegetations with secondary infection, leading to a foul odour, or the area may be white or moist with papulovesicles. Other sites include the face and chest and there may be wedge-shaped notches on the distal nail edge and pitting of the palms. The lesions may be aggravated by ultraviolet light. Treatment is with potent local steroids or retinoids with antibiotics for secondary infection.

Idiopathic aphthosis

This is characterized by recurrent painful ulcers in the mouth and rarely on the scrotum or vulva. It may be associated with chronic cyclic neutropenia.

Acanthosis nigricans

This condition may be associated with a malignancy such as adenocarcinoma of the stomach or breast. Marked hyperpigmentation and thickening with papillomatoses are seen in the flexures of the groin and axillae. Treatment is by removing the underlying cause.

Necrolytic migratory erythema

This is a manifestation of glucagonoma, a rare pancreatic tumour causing characteristic erythematous or bullous lesions with an extending circinate edge, with healing behind, in the genital and other areas. Treatment is by removing the tumour.

CONCLUSION

A genital dermatosis can provoke anxiety because of embarrassment and the possibility that it might be sexually acquired, malignant or associated with malignancy. A full history, including a sexual history, clinical examination and investigations are required to exclude sexually transmitted genital dermatoses before considering other causes. Malignancy should be considered in chronic lesions and non-healing ulcers. A biopsy may be required to establish the definitive diagnosis, allowing the appropriate treatment or reassurance. Joint clinics such as dermatovenereological or vulval clinics involving dermatologists are an excellent way to provide integrated care as well as being educational. This aspect can be further enhanced by joint clinicopathological meetings to review histological findings.

FURTHER READING

Cerio, R. (ed.) (1995) *Genital Skin Disorders: A Guide to Non-Sexually Transmitted Conditions*, Chapman & Hall, London.

Champion, R.H., Burton J.L. and Ebling F.J.G. (1992) *Textbook of Dermatology*, 5th edn, Blackwell Scientific, Oxford.

Dorkenoo, E. and Elworthy, S. (1996) *Female Genital Mutilation: Proposals for Change*, Minority Rights Group International, London.

Korting, G.W. (1981) *Practical Dermatology of the Genital Region*, W.B. Saunders, Philadelphia.

Kaufman, R.H. (ed.) Vulvovaginal Disease in Clinical Obstetrics and Gynecology, Vol. 34, No. 3, Sept. 1991: 581–681. Lippincott-Raven Publishers, Philadelphia.

Lynch, P.J. and Edwards, L. (1994) *Genital Dermatology*, Churchill Livingstone, New York.

Leibowitch, M., Staughton, R., Neill, S., Barton, S. and Marwood, R. (1997) *An Atlas of Vulval Disease: A Combined Dermatological, Gynaecological and Venereological Approach*, 2nd edn, Martin Dunitz, London.

Ridley, C.M. and Neill, S. (1998) *The Vulva*, 2nd edn, Blackwell Scientific, Oxford.

Ridley, C.M., Oriel, J.D. and Robinson, A.J. (1992) *A Colour Atlas of Diseases of the Vulva*, Chapman & Hall, London.

8 The diagnosis and management of vaginal discharge

P.E. Hay

INTRODUCTION

Many women experience abnormal vaginal discharge at some time in their lives. At least 25% of women attending GUM clinics receive treatment for one of the three common causes of abnormal vaginal discharge: bacterial vaginosis, candidiasis and trichomoniasis (Fig. 8.1).

Trichomonas vaginalis was described by Donné in 1936 and vaginal candidiasis by Wilkinson in 1849. Other abnormal discharges were grouped together as non-specific vaginitis. In the 1920s Schroeder described three grades of vaginal flora, the most abnormal containing anaerobes, but this concept was not taken up widely. In 1955, Gardner and Dukes described the criteria for recognizing what we now call bacterial vaginosis. An abnormal discharge may originate from anywhere in the lower genital tract, although abnormal bleeding from the uterus is usually considered as a separate entity. Discharge can arise from the uterus or cervix due to endometritis or cervicitis caused by chlamydia, gonorrhoea, herpes simplex virus or non-specific infection. Excess mucus may be produced by the cervix in association with a benign ectropion. As these conditions will be covered elsewhere, this chapter will concentrate on abnormal discharge arising from the vagina itself.

Abnormal discharge is a subjective symptom. It may not be reported by a woman due to reluctance for self-examination or discussing such symptoms. In some cases a woman may perceive a physiological discharge as abnormal, particularly if she worries that she may have acquired a sexually transmitted infec-

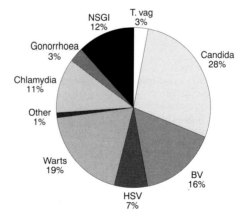

Fig. 8.1 This pie chart shows the proportions of diagnoses reported for women in the 1992 KC60 annual return from England and Wales. Episodes in which no diagnosis was recorded have been excluded. Abbreviations: Other, includes syphilis and infestations; HSV, genital herpes; BV, bacterial vaginosis; T. Vag, *Trichomonas vaginitis* infection; NSGI, pelvic inflammatory disease and contacts of men with non-gonococcal urethritis.

tion. Two misconceptions about vaginal discharge are common amongst clinicians, although they are mutually exclusive. One is that all abnormal vaginal discharge is due to thrush (candidiasis). This avoids any need to discuss the possibility of a sexually transmitted infection. The other, that it is a symptom of a sexually transmitted disease. Of the three common causes of abnormal discharge arising from the vagina itself, only *Trichomonas vaginalis* is regarded as a sexually transmitted infection.

Vaginal physiology

At birth the vagina is lined with stratified squamous epithelium, produced in response to maternal oestrogen. Infant girls may have a withdrawal bleed in the first month of life. With the waning of this hormonal influence after the first month of life, the vagina is lined with cuboidal epithelium until puberty. For this reason, prepubescent girls infected with chlamydia or gonorrhoea develop a generalized vaginal infection. At puberty, the vagina once again becomes lined with stratified squamous epithelium which regresses slowly after the menopause, to become atrophic.

Normal vaginal discharge is white, becoming yellowish on contact with air, due to oxidation. It consists of desquamated epithelial cells from the vagina and cervix, mucus originating mainly from the cervical glands, bacteria and fluid which is formed as a transudate from the vaginal wall. More than 95% of the bacteria present are lactobacilli.

Vaginal fluid has a pH between 3.5 and 4.5. This is principally due to the production of lactic acid by the vaginal epithelium metabolizing glycogen and by the lactobacilli.

It has proved very difficult to define a normal vaginal flora. Variations in swabbing technique and differences between study populations have lead to discrepancies between studies. One review summarized the normal flora as shown in Table 8.1.[1] This list of organisms is not exhaustive. The predominant organisms are lactobacilli and many women are colonized with two or three different species.[2] These

Table 8.1 Bacterial vaginal flora among asymptomatic women without vaginitis (from ref[1])

Organism	Proportion of women colonized (%)
Facultative organism	
Gram-positive rods	
Lactobacilli	50–75
Diphtheroids	40
Gram-positive cocci	
Staphylococcus epidermidis	40–55
Staphylococcus aureus	0–5
β-*haemolytic streptococci*	20
Group D streptococci	25–35
Other streptococci	35–55
Gram-negative organisms	
Escherichia coli	10–30
Klebsiella spp.	10
Other	2–10
Anaerobic organism	
Peptococcus	5–65
Peptostreptococcus spp.	25–35
Bacteroides spp.	20–40
Bacteroides fragilis	5–15
Fusobacterium spp.	5–25
Clostridium spp.	5–20
Eubacterium spp.	5–35
Veillonella spp.	10–30

Table 8.2 Management of a woman presenting to a GUM clinic with abnormal vaginal discharge

History and examination
Screen for gonorrhoea, chlamydia, syphilis and vaginal swabs
Microscopy of urethral, cervical, vaginal secretions

Outcome
1. Organism or condition found which is compatible with symptoms: prescribe appropriate treatment and review one week after completion of treatment, with results of other tests.
2. Clinical features are suggestive of candidiasis, but no spores or hyphae are seen: treat for candida and review with culture results.
3. Other clinical diagnosis, e.g. genital herpes, warts, mucopurulent cervicitis: treat appropriately and review.
4. No clinical diagnosis
   - Discuss with patient and review after 1–2 weeks
   - Repeat all swab tests and clinical examination
   - If no diagnosis on second visit consider:
     – physiological discharge and the reasons for the woman being concerned about her discharge
     – increased mucus from cervical ectropion
     – dermatological condition or exposure to irritants (these usually present with soreness and itching rather than discharge as the principal symptom)

Avoidance of douching, bubble bath, etc. should be discussed with all patients.

are anaerobic or facultative anaerobic organisms appearing as large Gram-positive rods on microscopy. They contribute to the pool of lactic acid in the vagina and exert inhibitory effects on the growth of other microorganisms through the production of acid, hydrogen peroxide and lactocins. Lactobacilli may have a crucial role in maintaining the health of the vaginal ecosystem. In a prospective study, hydrogen peroxide-producing lactobacilli were present in 96% of 100 women with normal vaginal flora.[3] Their presence was associated with a significantly reduced risk of subsequent episodes of bacterial vaginosis, trichomoniasis, chlamydia and gonorrhoea.

TRICHOMONIASIS

Trichomonas vaginalis can cause an intensely irritant vaginitis and a profuse, purulent discharge. Before metronidazole became available in the 1960s, treatment was often lengthy and unsuccessful. This was summarized well by Keighley in 1971. She remarked that since the introduction of:

> ... 'Flagyl' a whole generation has no knowledge of the sufferings of women with trichomoniasis; the indignities and discomfort of the perpetual local treatments, douches, paintings, insufflations and insertions of pessaries, etc. All these things women suffered for months and sometimes years on end, only to relapse when the treatment was discontinued.[4,5]

Microbiology

Trichomonas vaginalis is a pleomorphic organism usually 10–20 μm in diameter. When viewed on a wet mount, live organisms show amoeboid movement in addition to the characteristic jerky motion produced by the flagellae. There are four flagellae located anteriorly. A fifth flagellum is attached to the undulating membrane which extends two-thirds of the length of the organism and there is a terminal spike. No cyst form has been described.

Two other species are found in humans. *Trichomonas tenax* is found in the mouth, often in association with gingivitis. *Pentatrichomonas hominis* has been found in the large intestine, sometimes from patients with diarrhoea. It is not certain whether either of these species is pathogenic.

Epidemiology

Trichomoniasis has become less common over the last 15 years. The number of cases reported from England and Wales in the KC60 returns has dropped from over 20 000/year in 1982 to below 6000/year in the early 1990s. Published series from the mid 1970s gave a prevalence of 5% in women attending a family planning clinic and 7–32% among women attending a GUM clinic.[6] Currently, approximately 1% of women attending our clinic, at St George's Hospital, are diagnosed as having trichomoniasis. There are few data on the prevalence in men, as infection is frequently asymptomatic and the organism is detected rarely unless there has been recent intercourse with an infected woman. In one study, *T. vaginalis* could be found in 70% of men within 48 h of intercourse with an infected woman and in only 33% after two weeks.[7] *T. vaginalis* was found in 12% men attending an STD clinic in the USA, when cultured from a centrifuged deposit of urine.[8]

No controlled trials have been performed to investigate the value of treating male partners in reducing the incidence of relapse in women. This is, however, standard practice. The strongest evidence to support such an approach is that the long-term success rate of treatment with metronidazole in a women's prison was 98.3%.[4] This is higher than in any other cohort and has been attributed to the low likelihood of reinfection from a male partner.

Vertical transmission occurs in about 5% of pregnancies. The infection usually remits in the neonate as the influence of maternal sex hormones diminishes, but treatment is appropriate if infection lasts more than one month. The organism can survive in moist places and many patients suggest they might have acquired it from swimming pools or sharing a towel, but there is no documented evidence of non-venereal transmission. The organism only colonizes the urogenital tract in women, preferring the squamous epithelium of the vagina, although it has been isolated from aspirated bladder urine, the rectum and fallopian tubes. Asymptomatic carriage occurs, but it is estimated that one-third of women will develop symptoms within six months and ultimately 50–90% will become symptomatic if not treated.

Pathology

Trichomoniasis is characterized by an acute inflammatory response. Histological examination of vaginal biopsies shows a neutrophil infiltrate. The organism does not invade through the mucosa. Vascular proliferation and double cresting of mucosal capillaries is seen, with microhaemorrhages. This is responsible for the so-called strawberry appearance of the cervix in women with trichomoniasis. Protective immunity does not appear to develop and reinfection is common. Serum antibodies can be detected at low titre but are not sufficiently reliable for diagnostic use. Secretory

IgA is present in vaginal fluid from up to 75% of women. A delayed hypersensitivity response can be demonstrated in some patients by injecting trichomonal antigens. Protective immunity has been produced following systemic vaccination with organisms in a murine model[9] and a bovine model which used *Trichomonas foetus*.[10]

Symptoms and signs

Three-quarters of the women with trichomoniasis present with an abnormal vaginal discharge. About half of the infected women report vaginal soreness, pruritus or dyspareunia, whilst 10% experience dysuria and urinary frequency. A similar proportion report an offensive smell which may be due to associated bacterial vaginosis, as 40% of women with trichomoniasis have a bacterial flora characteristic of bacterial vaginosis. On examination, an obvious vulvitis is present in approximately one-third of patients with trichomoniasis and in severe cases, the discharge may be profuse and frothy, with a yellow or green colour. Vaginal erythema is present in the majority of patients. A strawberry cervix can be seen with the naked eye in 1–2% of patients. On colposcopy, punctate haemorrhages and dilated vessels can be detected in almost 50% of infected women.

Diagnostic tests

The vaginal pH is usually raised to a level between 4.5 and 7.0 in women with trichomoniasis. Wet-mount examination has a sensitivity of 40–80% when compared to culture. Experienced observers can recognize non-motile organisms, but it is usually the movement of the flagellae which leads to identification of the organism. In addition, there is usually a high concentration of polymorphs in the vaginal fluid. *T. vaginalis* grows in normal saline at room temperature, but grows optimally in specialized media such as Fineberg–Whittington or Bushby. *T. vaginalis* is not detected easily on a Gram-stained smear of vaginal fluid. Often there is a high concentration of polymorphs and bacterial vaginosis or an intermediate vaginal flora. Cervical cytology, using a Papanicolaou stain, has a sensitivity of 60–70% compared to culture.

Treatment

Most strains of *Trichomonas vaginalis* are sensitive to metronidazole and other imidazole antibiotics. More than 90% of women are cured by a single 2 g stat dose of metronidazole or a five-day course of 400 mg twice daily. Tinidazole 2 g can be used as an alternative. Many physicians are reluctant to prescribe metronidazole during the first trimester of pregnancy because high doses were associated with mutagenesis in animal studies. There is, however, no evidence of such complications when it has been used in pregnancy in humans. As clotrimazole has an inhibitory effect on *T. vaginalis*, some clinicians prefer to use clotrimazole pessaries during the first trimester of pregnancy before giving definitive treatment with metronidazole at a later date.

When treatment is ineffective, the history should be reviewed to exclude the possibility of reinfection from a new or untreated sex partner. Antibiotic sensitivity testing is not available as a routine and it is therefore not easy to distinguish poor drug absorption from antibiotic resistance. Retreatment with a higher dose of metronidazole is usually effective.[11] A dose of 400 mg three times a day orally for seven days can be prescribed, working up to a 2 g suppository twice a day for 5–10 days, if lower doses are unsuccessful. Neurological toxicity manifesting as peripheral neuropathy or fits has been reported at high doses. Metronidazole suppositories have been used intravaginally; however, this route of administration may be less successful than systemic treatment as the 3-hydroxy metabolite of metronidazole, produced by hepatic metabolism, is more active. Nonoxynol-9 was apparently successful in one woman with highly resistant infection who experienced severe neurological toxicity with high doses of metronidazole.[12] Paromomycin has been successful in another case.[13] An *in vitro* study suggests that a combination of clotrimazole with metronidazole may be synergistic.[14]

Sexual partners of women with trichomoniasis should be examined and screened for sexually transmitted pathogens, including *T. vaginalis*. Even if the organism is not detected, contacts should receive treatment with metronidazole 2 g stat dose or 400 mg twice daily for five days.

Complications

Complications of trichomoniasis are rare. *In vitro*, attachment of potentially pathogenic bacteria to *T. vaginalis* has been demonstrated. It is conceivable, therefore, that this motile organism might carry other pathogens into the upper genital tract. On occasions, it has been detected in the fallopian tubes of women with salpingitis. It is, however, rarely thought to be an important, direct aetiologic factor in pelvic inflammatory disease. In pregnancy, several studies have reported that women with *T. vaginalis* infection are up to twice as likely to have a preterm labour as women with normal vaginal flora. It is not known whether this is a direct effect of the organism or due to an association with other pathogens.

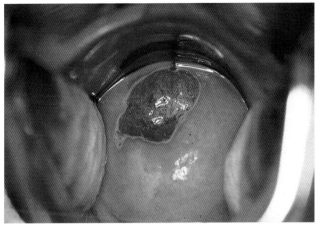

Plate 1 Cervical ectopy and the squamocolumnar junction.

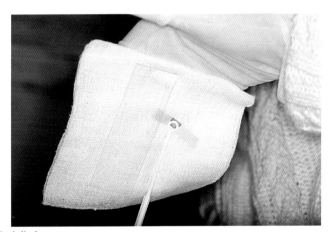

Plate 2 pH/KOH test on vaginal discharge.

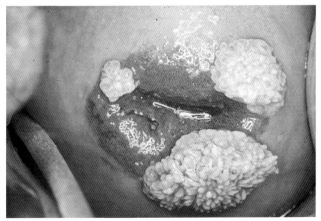

Plate 3 Wart infection of the cervix.

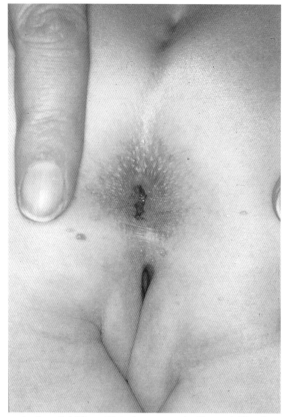

Plate 4 Knee/chest position for examination of children. Note the warts and reflex anal dilatation.

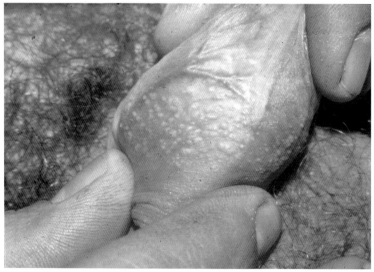

Plate 5 Fordyce spots may be mistaken for warts but are merely ectopic sebaceous glands.

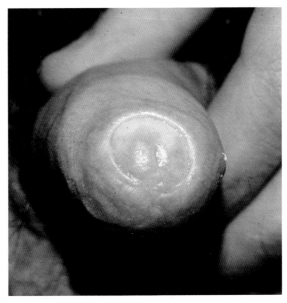

Plate 6 Balanitis xerotica obliterans (BXO) with urethral stenosis.

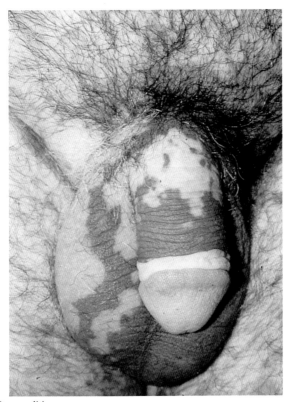

Plate 7 Genital vitiligo is a benign condition.

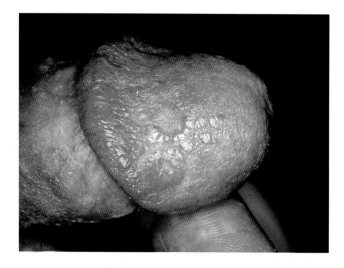

Plate 8 (a) Penile hypertrophic lichen planus in a Caucasian. (b) The same condition in pigmented skin.

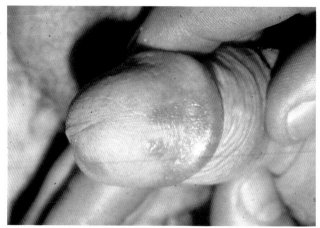

Plate 9 Psoriasis on the penis.

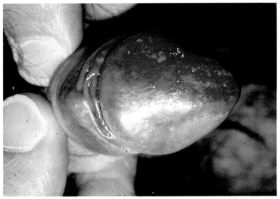

Plate 10 A case of a fixed drug eruption secondary to tetracycline therapy.

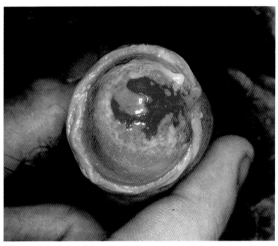

Plate 11 Stevens–Johnson syndrome, secondary to mycoplasma infection.

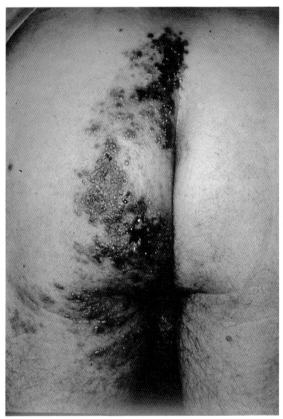

Plate 12 Recurrent varicella-zoster affecting the buttocks; be especially wary of underlying immunosuppression in multidermatomal presentation.

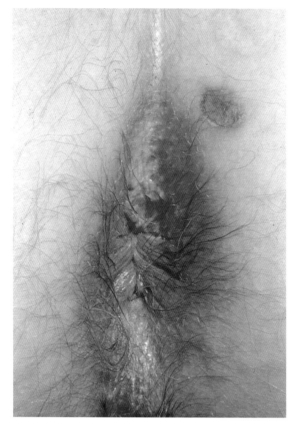

Plate 13 A perianal ulcer: the differential diagnosis includes infective or non-infective dermatoses and malignancy.

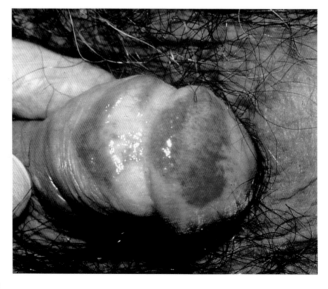

Plate 14 Zoon's balanitis.

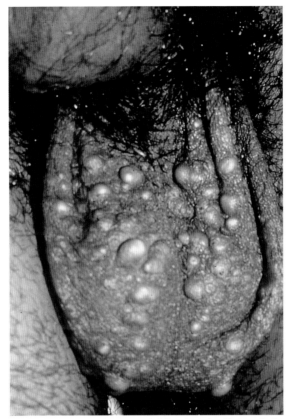

Plate 15 Scrotal sebaceous or epidermoid cysts can become painful and tender if secondarily infected.

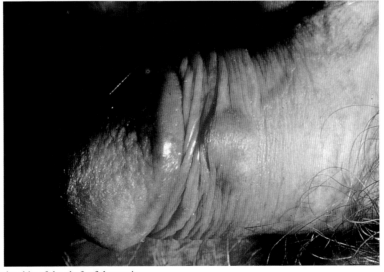

Plate 16 A lipoma in the skin of the shaft of the penis.

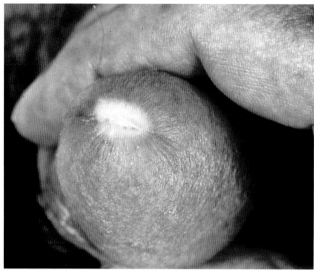

Plate 17 Balanitis xerotica obliterans is a consequence of lichen sclerosus of the glans penis which involves the urethra.

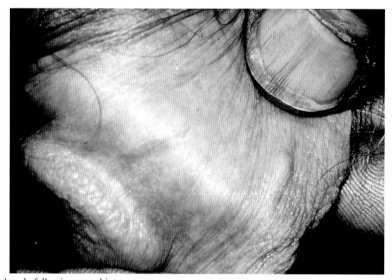

Plate 18 Benign lymphocele following sexual intercourse.

CANDIDIASIS

Introduction

Vaginal candidiasis is often considered a relatively trivial condition. A woman with severe candidiasis, however, can experience severe discomfort which can totally dominate her life until it is relieved. It is estimated that 80% of women will experience at least one episode of vaginal candidiasis in their lifetime. Fortunately, for most women, it responds to simple treatments and is a transient experience.

Microbiology

Candida albicans, a yeast-like fungus, is responsible for approximately 80% of cases of vaginal candidiasis. *Candida glabrata* accounts for approximately 10% of cases and a variety of other Candida species for the rest. Candida grows as a non-capsulated oval blastophore. It reproduces by budding and long pseudohyphae can develop. It has a speckled Gram-positive appearance on Gram staining.

Epidemiology

Vaginal candidiasis is common worldwide. Over half the women who experience at least one episode of vaginal candidiasis will have a subsequent recurrence. The number of cases reported from GUM clinics has remained stable over the last 10 years. The fungus is commonly carried as a commensal. It can be isolated from the mouth, anorectum, nails and vagina in asymptomatic women. Male-to-female transmission is rarely an important aetiologic factor but transmission from a woman to a man, causing balanoposthitis, does occur. Sometimes male partners can develop a hypersensitivity response to candida, developing irritation of the glans penis shortly after intercourse. Trauma following sexual activity may lead to the development of symptomatic candidiasis in a previously asymptomatic woman.

Pathology and immunology

Candidal pseudohyphae invade into the mucosa and its presence is associated with a lymphocyte and neutrophil infiltrate. Vaginal candida is common in women with defective cell-mediated immunity, such as those with AIDS or those taking steroids. One study has shown low levels of anticandidal IgA in women with recurrent candidiasis.[15] Women with immunoglobulin deficiencies do not, however, suffer from recurrent candidiasis. It has been suggested that women with recurrent candidiasis exhibit macrophage dysfunction related to reduced T-cell reactivity to candidal antigens.[16] Increased levels of anticandidal IgE may contribute to producing symptoms in some women.[17,18]

Vaginal candida is also common in pregnancy and in women with diabetes mellitus. It was reputedly common in women using a combined oral contraceptive pill but is not associated with use of current low-dose oestrogen preparations.[19] A course of broad-spectrum antibiotics can produce an attack of symptomatic candidiasis.[20]

The interaction between lactobacilli and candida is not elucidated fully as, for instance, in pregnancy there is an increased concentration of vaginal lactobacilli and an increased incidence of symptomatic candidiasis. Altered hormone levels may be important in pregnancy. There is *in vitro* evidence that progesterone affects the growth of candida directly, but also inhibits neutrophil activity.[21] Tight-fitting clothing or nylon tights induce a moist, warm environment, which is thought to encourage growth of the yeast. Deodorants, perfumed soaps and antiseptic douches may irritate the vulva and decrease the number of lactobacilli, contributing to the severity of symptoms. In the majority of women with recurrent vaginal candidiasis, none of these predisposing factors can be identified.

Clinical features

Itching of the vulva or vagina is the most common symptom of women with candidiasis. This may be accompanied by soreness in the vagina extending to the vulva or into the groin. A burning sensation, external dysuria, frequency and superficial dyspareunia are common. Examination shows erythema and inflammation of the vagina and vulva. Marked oedema occurs in severe cases. The typical discharge is white or yellow, thick and with a curdy consistency, but in some cases the discharge is thin and watery or looks normal. There may be small erosions and fissuring of the skin at the introitus.

Differential diagnosis

The discharge associated with trichomoniasis is usually profuse, yellow, thin and frothy. Bacterial vaginosis usually has an offensive smell and is rarely associated with inflammation. Herpes simplex infection may cause localized itching, soreness and fissuring. Dermatological conditions such as eczema, lichen planus or lichen sclerosus may produce similar symptoms. In some cases of eczema there is a history of atopy but it is not always easy to diagnose eczema on the first presentation. It must also be remembered that vaginal infections can coexist with these dermatological conditions, causing diagnostic difficulty.

Laboratory diagnosis

The vaginal pH is usually normal, between 3.5 and 4.7 in women with vaginal candidiasis. Microscopy of a wet mount or Gram stain for the identification of spores and pseudohyphae has a sensitivity of approximately 50% when compared to culture. Direct inoculation of vaginal material onto Sabouraud's medium produces the best results. Sometimes severe vaginitis is associated with relatively low concentrations of organisms.

Treatment

Many preparations are available for the treatment of vaginal candidiasis. The main decision is whether to use topical or oral preparations. In uncomplicated cases both are effective in at least 80–90% of cases and the choice should be determined by the patient's preference. A study from the UK showed that approximately 50% of women preferred oral treatment, 25% intravaginal cream, 14% hard pessaries and 10% soft pessaries.[22]

Topical treatments

Relief of vulval symptoms is obtained quickly by applying an antifungal cream. It is essential, however, that an intravaginal application is also prescribed. Varying doses and durations of treatment are available, ranging from single applications of, for example, a 500 mg clotrimazole pessary to a 14-day course of nystatin pessaries. In general, the single-dose treatments are suitable for uncomplicated candidiasis. The longer duration treatments are associated with lower relapse rates and are preferred for women with recurrent episodes. Creams containing a combination of an antifungal agent and 1% hydrocortisone are useful when there is a lot of inflammation.

Oral treatment

Short-course treatments are available, such as a single-dose 150 mg fluconazole capsule or itraconazole 100 mg bd for one day. It was hoped that by also acting to eliminate candida from reservoirs such as the bowel, these treatments would be associated with a lower relapse rate than topical treatments, but this does not seem to be the case.

Complications

There are a few direct complications from vaginal candidiasis. *Candida albicans* has been detected in amniotic fluid from women in preterm labour but only infrequently. There is often psychological morbidity associated with recurrent candidiasis.

Recurrent candidiasis

There are some women who suffer from recurrent episodes of vaginal candidiasis who do not have any identifiable predisposing factors for candidiasis. The management of these women can be difficult. It is important to exclude an underlying dermatosis or vulval condition such as eczema or vestibulitis. In some cases a biopsy may be required. In women with erythema but no increase in discharge, candida may be difficult to detect on microscopy. If this is the case, it is useful to confirm that candida has been isolated on previous occasions before reinforcing a possibly wrong diagnosis by prescribing further treatment for candida.

The first approach to treatment is to give a longer course, such as 100 mg clotrimazole pessaries daily for 14 days or fluconazole 50 mg daily for 14 days. Some women experience symptoms around the time of menstruation and can be prescribed a single-dose treatment, either topical or systemic, to take at the appropriate stage of the cycle for a six-month period. Occasionally a low dose of an oral antifungal agent taken daily for up to six months is required to overcome persistent symptoms.

There are few good data about diet. One crossover study, in which compliance with the protocol was poor, demonstrated a significant reduction in the number of attacks of candidiasis in six months, from 2.54 to 0.38, when 8 oz of yoghurt containing *Lactobacillus acidophilus* was consumed daily.[23] Candidiasis has been associated with a diet high in refined carbohydrate but we do not have a controlled study demonstrating benefit from altering diet.[24]

It is important that women do not wash excessively or use soap on inflamed mucosa as this will exacerbate rather than assist their condition. Vaginal deodorants, bubble bath and other additives, such as Dettol, should be avoided. In some studies candidiasis has been associated with frequent sexual intercourse[25] but there is no evidence that sexual transmission plays an important role, even in recurrent vulvovaginal candidiasis. It is usual to examine the partner, who may develop a balanitis after intercourse. If nothing else, it may be helpful to explain the nature of the condition to the partner.

BACTERIAL VAGINOSIS

Introduction

Bacterial vaginosis (BV) is the commonest cause of abnormal vaginal discharge in women of childbearing age. It was found in approximately 10–15% of women attending a gynaecology clinic.[26] A higher prevalence

Table 8.3 Terms used for bacterial vaginosis

Non-specific vaginitis
Haemophilus vaginalis vaginitis (1955)
Gardnerella vaginitis (1980)
Anaerobic vaginitis (1983)
Bacterial vaginosis (1983)
Vaginal bacteriosis

has been reported from other populations, for instance 28% of women undergoing termination of pregnancy[27] and up to 40% of women attending clinics for sexually transmitted diseases.[28] Bacterial vaginosis is characterized by a replacement of the usual lactobacillus-dominated flora with an overgrowth of many anaerobic organisms. This pattern was recognized by Curtis and Schroeder early in the 20th century but it was not until 1955 that Gardner and Dukes described the clue cell and specific criteria for diagnosis.[28,29]

The terms that have been used for bacterial vaginosis are shown in Table 8.3. The organism which Gardner and Dukes termed *Haemophilus vaginalis* has subsequently been renamed *Corynebacterium vaginalis* and now *Gardnerella vaginalis*. Despite a better understanding of the microbiological changes associated with BV, the primary aetiology remains poorly understood.

The principal symptom of BV is an offensive vaginal malodour which is characteristically described as 'fishy'. This may be accompanied by an increased vaginal discharge. It is a recurrent relapsing condition which may have a spontaneous onset and resolution.

Microbiology

Healthy women are colonized by several species of lactobacillus, which form the dominant vaginal organism. In women with BV, lactobacilli may be absent or present in low or even normal concentrations, in which case they are swamped by an overgrowth of anaerobic bacteria. The vaginal pH is increased and the concentration of many anaerobic species is increased up to a thousandfold greater than seen in normal women. The change is quantitative rather than qualitative. For instance, *Gardnerella vaginalis* can be cultured in lower concentrations from many women with normal flora. *Ureaplasma urealyticum* may be found in high concentrations in women who have and who do not have bacterial vaginosis and its presence is not significantly associated with BV.

The effect of pH on the growth of lactobacilli and anaerobes in the vaginal flora has been examined. Lactobacilli which produce hydrogen peroxide inhibit the growth of anaerobic organisms at low pH, *in vitro*.[3] The degree of inhibition is reduced by adding catalase, an enzyme which destroys hydrogen peroxide, or by raising the pH to 7.0. In a two-year prospective study, women colonized by hydrogen peroxide-producing lactobacilli were less likely than women lacking such organisms to develop BV, trichomoniasis, gonorrhoea and chlamydial infections.[30]

Bacterial vaginosis is not usually associated with an inflammatory response. Bacteroides species have been shown to secrete substances which inhibit leucocyte chemotaxis.[31] It is possible, therefore, that bacteria secrete substances which inhibit an inflammatory response to their presence. *In vitro*, mobiluncus produce cytotoxins. Supernatant from a culture of mobiluncus was highly toxic to bovine fallopian tube cultures, producing complete deciliation.[32] It is possible that cytotoxins released by anaerobes might cause desquamation of vaginal epithelial cells, leading to the production of clue cells and increased discharge.

Epidemiology

Bacterial vaginosis is not a sexually transmitted disease. Clue cells were found in 12% of virgin adolescent girls in a study in the United States.[33] However, the prevalence of BV increases with increasing numbers of sex partners and increasing frequency of sexual intercourse. In some studies it has been associated with the presence of sexually transmitted pathogens such as chlamydia or gonorrhoea, but this may be due to the influence of sexual lifestyle. The prevalence is high in women with pelvic inflammatory disease.[34] It is more common in women with an intrauterine contraceptive device and has been associated with the practice of vaginal douching and cunnilingus.[30] It is less common in women using oral contraceptives than in those having unprotected intercourse and also less common in those using barrier methods. It is commoner in women of Afro-Caribbean race. One study reported that the vaginal pH of adolescent black American women was greater than that of white women, 5.4 compared to 4.5.[35] It was hypothesized that this might predispose the black women to developing BV.

The prevalence of bacterial vaginosis is high among lesbian women, a group with a low incidence of traditional bacterial STDs.[36,37] The high concordance rate of lesbian couples to have or not have BV can be interpreted as evidence of sexual transmission between women.[36] BV may develop and resolve spontaneously within a few days. Prospective studies employing daily Gram-stained vaginal smears have shown that in some women the vaginal flora is in a very dynamic state. BV develops most often around the time of menstruation and resolves mid-cycle.[38–40] Hormonal influences may therefore be critical in the control of the vaginal flora. In one study unprotected sexual intercourse with a regular partner was associated with resolution of BV and BV developed after the onset of vaginal candidiasis.

There is uncertainty about the prevalence of bacterial vaginosis in postmenopausal women. Confusion can arise because the normal reduction in lactobacilli following the menopause and colonization by skin flora may be interpreted as a lactobacillus-deficient flora of bacterial vaginosis microscopically. I have only diagnosed bacterial vaginosis in a postmenopausal woman on one occasion, when it was associated with trichomoniasis.[40]

Clinical features

An offensive vaginal smell is the principal symptom of bacterial vaginosis. This is usually associated with an increased watery discharge which may be white or yellow. Itching and soreness, which indicate inflammation, are present rarely and suggest the coexistence of candidiasis or trichomoniasis. The malodour is thought to be due to increased levels of polyamines, such as putrescine and cadaverine, as well as trimethylamine which is the principal constituent of rotting fish.[41,42] These amines are released by the addition of alkali so that the malodour may be worse following intercourse or around the time of menstruation.

On examination, the typical discharge is described as thin, homogeneous and adherent to the walls of the vagina. It may be frothy on occasions. Signs of inflammation are usually absent. The smell is frequently obvious whilst examining the patient.

DIAGNOSIS

Conventionally, the diagnosis is made by fulfilling at least three of four composite criteria, as shown in Table 8.4. Clue cells are vaginal epithelial cells so covered with adherent bacteria that the border of the cell is obscured. These are visible in the wet mount of vaginal fluid and their presence alone is almost diagnostic of bacterial vaginosis. Occasionally cells with adherent lactobacilli are seen and cellular debris can also be mistaken for clue cells. With experience, the typical vaginal discharge can be recognized reasonably accurately but it is not always detected in women with BV. It appears to be the least reliable diagnostic sign.

Vaginal pH is nearly always raised in women with BV. However, it is not a specific indicator and, in one series, was also present in nearly 50% of women attending a STD clinic.[43] In the absence of infection, a

Table 8.4 The composite (Amsel's) criteria for diagnosis of bacterial vaginosis. At least three of the four criteria must be present for the diagnosis to be made

---

pH of vaginal fluid >4.5
Typical thin, homogeneous vaginal discharge
Release of strong fishy smell on adding alkali (10% KOH) to a sample of vaginal fluid
Clue cells present on microscopic examination of a wet mount of vaginal fluid

---

falsely raised pH may be found if cervical mucus, which has a pH of 7, has inadvertently been included in the vaginal swab. It also occurs around the time of menstruation and following unprotected intercourse for up to 48 h. The amine test or 'whiff' test is performed by placing a sample of vaginal fluid on a glass microscope slide and adding a drop of 10% potassium hydroxide to it. A pungent fishy odour is released if the woman has bacterial vaginosis.

There are characteristic changes on the Gram-stained vaginal smear from women with bacterial vaginosis. Various scoring systems have been developed which rely on assessing the concentration of lactobacilli in relation to other organisms. It is likely that the Nugent criteria (Table 8.5) will be adopted most widely for studies of bacterial vaginosis. When performing research studies, the Gram stain has the advantage that the samples can be stored and reviewed subsequently by an independent observer. It also allows recognition of an intermediate flora which is neither normal nor full-blown bacterial vaginosis.[44,45]

Table 8.4 The Nugent scoring system for reading Gram-stained vaginal smears. A score of 0–3 is normal, 4–6 intermediate and 7–10 corresponds to bacterial vaginosis. Lactobacillus morphotypes are large Gram-positive rods. Gardnerella are Gram-variable coccobacilli. Mobiluncus are curved rods. A woman with normal flora may have 4+ lactobacilli, no gardnerella and no mobiluncus, giving a score of 0. A woman with BV may have 1+ lactobacilli, 4+ gardnerella, 4+ mobiluncus giving a score of 9. A woman with 2+ lactobacilli, 3+ gardnerella and no mobiluncus would have a score of 5 corresponding to an intermediate flora

	Number of morphotypes				
	None	1+	2+	3+	4+
Lactobacillus	4	3	2	1	0
Gardnerella	0	1	2	3	4
Mobiluncus	0	1	1	2	2

There are other tests being developed to assist in the diagnosis of bacterial vaginosis without requiring microscopy. An assay which relies on detecting the presence of aminopeptidase, an enzyme present in 80% of women with bacterial vaginosis, has been developed. It has a sensitivity of approximately 80%.[46]

Treatment

Up to 50% of women with bacterial vaginosis are asymptomatic at the time of diagnosis. Symptoms can often be elicited after discussing the diagnosis but there is no absolute indication to treat healthy asymptomatic women. It is not infrequent, however, for women to appreciate the difference between a normal discharge and that of bacterial vaginosis after they have accepted treatment.

Oral metronidazole has been the mainstay of treatment for BV. A common regimen is 400 mg twice daily for five days. In 1992 Larsson reviewed all the published trials of treatment for BV and summarized the weakness of design of many of them.[47] He concluded that the cure rate immediately after treatment with metronidazole was up to 95%, but after four weeks this fell to 80% in open label studies and less than 70% in blinded studies. Topical treatment with clindamycin 2% cream once daily for seven days is now available, although more expensive than metronidazole. Topical metronidazole gel, applied once or twice a day for five days, has similar efficacy. All these treatments produce early cure rates of more than 80% and have similar long-term relapse rates. Other oral antibiotics which are effective against anaerobes are also helpful, although not commonly prescribed. In women unable to tolerate metronidazole or for whom topical treatment is inappropriate, oral clindamycin 300 mg twice daily for five days or co-amoxiclav one tablet three times daily for five days can be used, although neither preparation is licensed for this indication.

In women with frequently relapsing bacterial vaginosis it is logical to try different agents, although there are no published studies demonstrating improved efficacy from this. Many strains of gardnerella and mobiluncus are in fact resistant to metronidazole at the levels achieved with oral treatment. Higher intravaginal levels are achieved with metronidazole gel. Even after treatment with topical clindamycin, the vagina often becomes recolonized with BV-associated organisms within a week.[48] This suggests that relapse is likely if conditions favouring anaerobic overgrowth recur, regardless of the potency of the antibiotic used.

The relationship between candidiasis and bacterial vaginosis needs further exploration. One study reported that a combination of metronidazole and nystatin pessaries prevented relapse in 32/32 women, compared to 26/34 given placebo pessaries. All the women had an IUCD.

Published placebo-controlled studies have failed to demonstrate any benefit from treating male partners with metronidazole or tinidazole.[50] There is therefore no indication for 'epidemiological treatment' of male partners.

Complications

Bacterial vaginosis is a benign condition for the vast majority of women. Pelvic inflammatory disease (PID) is an important cause of gynaecological morbidity, with the subsequent increased risk for ectopic pregnancy and tubal factor infertility. Women with PID have a high incidence of BV, but is the BV the cause or a consequence of other infections? A study in the USA concluded that although BV or intermediate flora was present in 62% of women with PID, the majority of cases (77%) were associated with chlamydia and/or gonorrhoea.[34] In only two of 84 subjects were anaerobic organisms isolated in the absence of an STD pathogen. The authors concluded that BV was not an important primary cause of PID. In contrast, another study reported that plasma cell endometritis was associated with BV in women in whom gonorrhoea and chlamydia were not detected.[51] Histological evidence of plasma cell endometritis was present in endometrial biopsies from 10 of 22 women with bacterial vaginosis compared with one of 19 controls. Despite these studies, most women with bacterial vaginosis do not have pelvic inflammatory disease.

There are several studies which show that women with BV are at increased risk of developing endometritis or pelvic inflammatory disease following termination of pregnancy. A double-blind, placebo-controlled trial from Sweden demonstrated that the risk of endometritis was reduced from 12.2% in placebo-treated women to 3.8% of women prescribed oral metronidazole before termination.[52]

BV has also been associated with the development of postpartum endometritis and vaginal cuff infections and abscesses following vaginal hysterectomy.[54]

Pregnant women with bacterial vaginosis have a significantly increased incidence of intrauterine death, late miscarriage and preterm birth.[45,55,56] It is thought that infection ascends into the uterine cavity. Preterm birth then occurs as a consequence of deciduitis or chorioamnionitis. If preterm labour does not ensue, chorioamnionitis may progress to amniotic fluid infection with foetal sepsis and intrauterine death.[57]

The women at greatest risk for preterm birth are those who have had a previous preterm birth. In two double-blind, placebo-controlled studies, BV was treated during the second trimester of pregnancy with metronidazole[58] or a combination of metronidazole and erythromycin.[59] The incidence of preterm birth in women with a history of previous 'idiopathic' preterm birth was reduced by 50% and 32% respectively.

What about widespread screening? A large cohort study of 1260 women was performed in Denver, Colorado.[60] Pregnant women were screened and treated for chlamydia and gonorrhoea routinely. For seven months women were additionally screened for BV and trichomonas infection, but not treated. In the second eight months women with BV were treated with clindamycin 300 mg twice a day for seven days. Their rate of preterm birth was reduced by 50% compared to the control period.

Two published studies of the use of intravaginal clindamycin cream have failed to show a reduction in the incidence of preterm delivery in women with BV.[61,62] The treatment was administered up to 24 weeks of gestation and it may be that this was too late to counter infection already present in the upper genital tract. A small open label study found that the incidence of preterm birth in women whose BV was treated with clindamycin cream at 12 weeks gestation was the same as that of women who did not have BV.

In conclusion, screening and treating pregnant women for bacterial vaginosis offers the possibility of reducing the incidence of preterm birth significantly. We have data from two double-blind, placebo-controlled trials that in 'high-risk' pregnancies, i.e. women with a previous late miscarriage or idiopathic preterm birth, screening and treatment for BV improve the chance of a successful outcome.[63,64] There is not sufficient evidence at present to recommend generalized screening in pregnancy. Double-blind, placebo-controlled trials to assess this approach are continuing. Treatment in pregnancy should be before 16 weeks gestation, preferably at 12 weeks. We need trials to find out whether a topical treatment given this early can prevent preterm birth as effectively as an oral treatment. Blanket screening and treatment would mean treating up to 20% pregnant women with antibiotics. It would be better if we could identify other markers of risk to allow targeting of the most high-risk pregnancies.

OTHER CAUSES OF ABNORMAL DISCHARGE

Abnormal discharge may be associated with the presence of a foreign body, such as retained tampons or condoms. Such women are at risk of the rare toxic shock syndrome. Vaginitis may occur due to a dermatological condition such as eczema, which may be the first site at which the eczema presents. There are cases recorded where a woman has been allergic to her partner's sperm. IgE reactive to either antigens of the female genital tract or components of semen can be transferred from the male partner during coitus, producing an allergic inflammatory response in the woman.[65]

Inflammatory conditions have also been attributed to a cytolytic vaginitis and vaginal lactobacillosis. Cytolytic vaginitis is described in women with persistent vaginal soreness. A Gram-stained vaginal smear shows nuclei and cellular debris, a cytolytic pattern associated with the presence of high levels of lactobacilli and an acidic pH. Treatment with bicarbonate of soda is recommended.[66] Vaginal lactobacillosis has been described recently in women with a history of vaginal candida.[67] The lactobacilli are said to line up and fuse, giving the appearance of what was previously thought to be *Leptothrix vaginalis*. Treatment with a broad-spectrum antibiotic such as co-amoxiclav has been helpful in some cases. Neither of these conditions is recognized universally and their importance is yet to be established.

Women with abnormal discharge who do not fall into any obvious diagnostic category may have evidence of non-specific genital infection in the form of urethritis or cervicitis. In these cases, appropriate treatment with a tetracycline or erythromycin may be beneficial, as well as screening and treating sexual partners. A high vaginal swab can be useful to detect carriage of streptococci which occasionally cause vaginitis. The urge to administer blind treatment without making a diagnosis should be resisted.

REFERENCES

1. Eschenbach, D.A. (1983) Vaginal infection. *Clin. Obstet. Gynecol.*, **26**, 186–202.
2. Redondo-Lopez, V., Cook, R.L. and Sobel, J.D. (1990) Emerging role of Lactobacilli in the control and maintenance of the vaginal bacterial microflora. *Rev. Infect. Dis.*, **26**, 856–872.
3. Klebanoff, S.J., Hillier, S.L., Eschenbach, D.A. and Waltersdorf, A.M. (1991) Control of the microbial flora of the vagina by H2O2-generating lactobacilli. *J. Infect. Dis.*, **164**, 94–100.
4. Keighley, E.E. (1971) Trichomoniasis in a closed community: efficacy of metronidazole. *Br. Med. J.*, **1**, 207–209.
5. Forgan, R. (1972) History of treatment of trichomoniasis. *Br. J. Vener. Dis.*, **48**, 522–524.

6. Rein, M.F. and Chapel, T.A. (1975) Trichomoniasis, candidiasis and the minor venereal diseases. *Clin. Obstet. Gynecol.*, **18**, 73.
7. Weston, T.E.T. and Nicol, C.S. (1963) Natural history of trichomonal infection in males. *Br. J. Vener. Dis.*, **39**, 251.
8. Borchardt, K.A., al-Haraci, S. and Maida, N. (1995) Prevalence of *Trichomonas vaginalis* in a male sexually transmitted disease clinic population by interview, wet mount microscopy, and the InPouch TV test. *Genitourinary Med.*, **71**, 405–406.
9. Abraham, M.C., Desjardins, M., Filion, L.G. and Garber, G.E. (1996) Inducible immunity to *Trichomonas vaginalis* in a mouse model of vaginal infection. *Infect. Immun.*, **64**, 3571–3575.
10. Corbeil, L.B. (1995) Use of an animal model of trichomoniasis as a basis for understanding this disease in women. *Clin. Infect. Dis.*, **21** (s2), S158–161.
11. Lossick, J.G. (1990) Treatment of sexually transmitted vaginosis/vaginitis. *Rev. Infect. Dis.*, **12(s)**, S665–681.
12. Livengood, C.H. and Lossick, J.G. (1991) Resolution of resistant trichomoniasis with the use of intravaginal nonoxynol-9. *Obstet. Gynecol.*, **5**, 954–956.
13. Nyirjesy, P., Weitz, M.V., Gelone, S.P. and Fekete, T. (1995) Paromomycin for nitroimidazole-resistant trichomonosis. *Lancet*, **346**, 1110.
14. Debbia, E.A., Campora, U., Massaro, S., Boldrini, E. and Schito, G.C. (1996) In vitro activity of metronidazole alone and in combination with clotrimazole against clinical isolates of *Trichomonas vaginalis*. *J. Chemother.*, **8**, 96–101.
15. Romero-Piffiguer, M.D., Vucovich, P.R. and Riera, C.M. (1985). Secretory IgA and secretory component in women affected by recidivant vaginal candidiasis. *Mycopathologia*, **91**, 165–170.
16. Witkin, S.S., Yu, I.R., Ledger, W.J. (1983) Inhibition of *Candida albicans*-induced lymphocyte proliferation by lymphocytes and sera from women with recurrent vaginitis. *Am.. J. Obstet. Gynecol.*, **147**, 809–811.
17. Witkin, S.S., Jeremias, J. and Ledger, W.J. (1989) Vaginal eosinophils and IgE antibodies to *Candida albicans* in women with recurrent vaginitis. *J. Med. Vet. Myco.*, **27**, 57–58.
18. Regulez, P., Garcia Fernandez, J.F., Moragues, M.D. *et al.* (1994) Detection of anti-*Candida albicans* IgE antibodies in vaginal washes from patients with acute vulvovaginal candidiasis. *Gynecol. Obstet. Invest.*, **37**, 110–114.
19. Davidson, F. and Oates, J.K. (1985) The pill does not cause 'thrush'. *Br. J. Obstet. Gynaecol.*, **92**, 1265–1266.
20. MacDonald, T.M., Beardon, P.H., McGilchrist, M.M. *et al.* (1993) The risks of symptomatic vaginal candidiasis after oral antibiotic therapy. *Quart. J. Med.*, **86**, 419–424.
21. Nohmi, T., Abe, S., Dobashi, K., Tansho, S. and Yamaguchi H. (1995) Suppression of anti-Candida activity of murine neutrophils by progesterone in vitro: a possible mechanism in pregnant women's vulnerability to vaginal candidiasis. *Microbiol. Immunol.*, **39**, 405–409.
22. Tooley, P.J. (1985) Patients' and doctors' preferences in the treatment of vaginal candidiasis. *Practitioner*, **229**, 655.
23. Hilton E., Isenberg., H.D., Alperstein, P. *et al.* (1992) Ingestion of yoghurt containing-*Lactobacillus acidophilus* as prophylaxis for candidal vaginitis. *Ann. Intern. Med.*, **116**, 353–357.
24. White, E.J., Emens, M. and Shahmanesh, M. (1991) Recurrent vulvovaginal candidiasis. *Int. J. STD AIDS*, **2**, 235–239.
25. Foxman, B. (1990) The epidemiology of vulvovaginal candidiasis: risk factors. *Am. J. Public Health*, **80**, 329–331.
26. Hay, P.E., Taylor-Robinson, D. and Lamont, R.F. (1992) Diagnosis of bacterial vaginosis in a gynaecology clinic. *Br. J. Obstet. Gynaecol.*, **99**, 63–66.
27. Blackwell, A.L., Thomas, P.D., Wareham, K. and Emery, S.J. (1993) Health gains from screening for infection of the lower genital tract in women attending for termination of pregnancy. *Lancet*, **342**, 206–210.
28. Hillier, S. and Holmes, K.K. (1990) Bacterial vaginosis, in: *Sexually Transmitted Diseases*, 2nd edn, (eds K.K. Holmes, P.A. Mardh, P.F. Sparks and P.J. Wiener), McGraw-Hill, New York, pp. 547–559.
29. Gardner, H.L. and Dukes, C.D. (1955) *Haemophilus vaginalis* vaginitis. A newly defined specific infection previously classified 'nonspecific' vaginitis. *Am. J. Obstet. Gynecol.*, **69**, 962–976.
30. Hawes, S.E., Hillier, S.L., Benedetti, J. *et al.* (1996) Hydrogen peroxide-producing lactobacilli and acquisition of vaginal infections. *J. Infect. Dis.*, **174**, 1058–1063
31. Sturm, A.W. (1989) Chemotaxis inhibition by *Gardnerella vaginalis* and succinate producing vaginal anaerobes: composition of vaginal discharge associated with G vaginalis. *Genitourinary Med.*, **65**, 109–112.
32. Taylor-Robinson, A.W., Borriello, S.P. and Taylor Robinson, D. (1993) Identification and preliminary characterisation of a cytotoxin isolated from *Mobiluncus spp. Int. J. Exper. Pathol.*, **74**, 357–366.

33. Bump, R.C. and Buesching, W.J. (1988) Bacterial vaginosis in virginal and sexually active adolescent females. Evidence against exclusive sexual transmission. *Am. J. Obstet. Gynecol.*, **158**, 935–939.
34. Soper, D.E., Brockwell, N.J., Dalton, H.P. and Johnson, D. (1994) Observations concerning the microbial etiology of acute salpingitis. *Am. J. Obstet. Gynecol.*, **170**, 1008–1014.
35. Stevens-Simon, S., Jamison, J., McGregor, J.A. and Douglas, J.M. (1994) Racial variation of vaginal pH among healthy sexually active adolescents. *Sexually Trans. Dis.*, **21**, 168–172.
36. Berger, B.J., Kolton, S., Zenilman, J.M. *et al.* (1995) Bacterial vaginosis in lesbians: a sexually transmitted disease. *Clin. Infect. Dis.*, **21**, 1402–1405.
37. Skinner, C.J., Stokes, J., Kirlew, Y., Kavanagh. J. and Forster, G.E. (1996) A case-controlled study of the sexual health needs of lesbians. *Genitourinary Med.*, **72**, 277–280.
38. Priestly, C.J.F., Jones, B.M., Dhar, J. and Goodwin, L. (1997) What is normal vaginal flora? *Genitourinary Med.*, **73**, 23–28.
39. Keane, F.E.A., Ison, C.A. and Taylor-Robinson, D. (1997) A longitudinal study of the vaginal flora over a menstrual cycle. *Int. J. STD AIDS*, **8**, 489–494.
40. Hay, P.E., Ugwumadu, A. and Chowns, J. (1997) Sex, thrush and bacterial vaginosis. *Int. J. STD AIDS*, **8**, 603–608.
41. Pheifer, T.A., Forsyth, P.S., Durfee, M.A. *et al.* (1978) Nonspecific vaginitis: role of *Haemophilus vaginalis* and treatment with metronidazole. *N. Engl. J. Med.*, **298**, 1429–1434.
42. Brand, J.M. and Galask, R.P. (1986) Trimethylamine: the substance mainly responsible for the fishy odour often associated with bacterial vaginosis. *Obstet. Gynecol.*, **68**, 682–685.
43. Eschenbach, D.A., Hillier, S.H., Critchlow, C. *et al.* (1988) Diagnosis and clinical manifestations of bacterial vaginosis. *Am. J. Obstet. Gynecol.*, **158**, 819–828.
44. Hillier, S.L., Krohn, M.A., Nugent, R.P. and Gibbs, R.S. (1992) Characteristics of three vaginal flora patterns assessed by Gram stain among pregnant women. Vaginal Infections and Prematurity Study Group. *Am. J. Obstet. Gynecol.*, **166**, 938–944.
45. Hay, P.E., Lamont, R.F., Taylor-Robinson, D.J. *et al.* (1994) Abnormal bacterial colonisation of the genital tract and subsequent preterm delivery and late miscarriage. *Br. Med. J.*, **308**, 295–298.
46. Schoonmaker, J.N., Lunt, B.D., Lawellin, D.W. *et al.* (1991) A new proline aminopeptidase assay for diagnosis of bacterial vaginosis. *Am. J. Obstet. Gynecol.*, **165**, 737–742.
47. Larsson, P.G. (1992) Treatment of bacterial vaginosis. *Int. J. STD AIDS*, **3**, 239–247.
48. Hillier, S., Krohn, M.A. and Watts, D.H. (1990) Microbiologic efficacy of intravaginal clindamycin cream for the treatment of bacterial vaginosis. *Obstet. Gynecol.*, **76**, 407–413.
49. Pulkkinen, P., Saranen, M. and Kaaja, R. (1993) Metronidazole combined with nystatin vagitories in the prevention of bacterial vaginosis after initial treatment with oral metronidazole. *Gynecol. Obstet. Invest.*, **36**, 181–184.
50. Moi, J., Erkkola, R., Jerve, F. *et al.* (1989) Should male consorts of women with bacterial vaginosis be treated? *Genitourinary Med.*, **65**, 263–268.
51. Korn, A.P., Bolan, G., Padian, N. *et al.* (1995) Plasma cell endometritis in women with symptomatic bacterial vaginosis. *Obstet. Gynecol.*, **85**, 387–390.
52. Larsson, P.G., Platz-Christensen, J.J., Thejls, H., Forsum, U. and Pahlson C. (1992) Incidence of pelvic inflammatory disease after first-trimester legal abortion in women with bacterial vaginosis after treatment with metronidazole: a double-blind, randomized study. *Am. J. Obstet. Gynecol.*, **166**, 100–103.
53. Watts, D.H., Krohn, M.A., Hillier, S.L. and Eschenbach, D.A. (1990) Bacterial vaginosis as a risk factor for post-caesarean endometritis. *Obstet. Gynecol.*, **75**, 52–58.
54. Soper, D.E. (1993) Bacterial vaginosis and postoperative infections. *Am. J. Obstet. Gynecol.*, **169**, 467–469.
55. Kurki, T., Sivonen, A., Renkonen, O.V. *et al.* (1992) Bacterial vaginosis in early pregnancy and pregnancy outcome. *Obstet. Gynecol.*, **80**, 173–177.
56. Hillier, S.L., Nugent, R.P., Eschenbach, D.A. *et al.* (1995) Association between bacterial vaginosis and preterm delivery of a low-birth-weight infant. *N. Engl. J. Med.*, **333**, 1772–1774.
57. McGregor, J.A. (1988) Preterm birth and infection: pathogenic possibilities. *Am. J. Reproduct. Immunol. Microbiol.*, **16**, 123–132.
58. Morales, W.J., Schorr, S. and Albritton, J. (1994) Effect of metronidazole in patients with preterm birth in preceding pregnancy and bacterial vaginosis: a placebo-controlled, double-blind study. *Am. J. Obstet. Gynecol.*, **171**, 345–347.
59. Hauth, J.C., Goldenberg, R.L., Andrews, W.W. *et al.* (1995) Reduced incidence of preterm delivery with metronidazole and erythromycin in women with bacterial vaginosis. *N. Engl. J. Med.*, **333**, 1732–1736.
60. McGregor, J.A., French, J.I., Parker, R. *et al.* (1995) Prevention of premature birth by screening and treatment for common genital tract infections:

results of a prospective controlled evaluation. *Am. J. Obstet. Gynecol.*, **173**, 157–167.
61. McGregor, J.A., French, J.I., Jones, W. *et al.* (1994) Bacterial vaginosis is associated with prematurity and vaginal fluid mucinase and sialidase: results of a controlled trial of topical clindamycin cream. *Am. J. Obstet. Gynecol.*, **170**, 1048–1059.
62. Joesoef, M.R., Hillier, S.L., Wiknjosastro, G. *et al.* (1995) Intravaginal clindamycin treatment for bacterial vaginosis: effects on preterm delivery and low birth weight. *Am. J. Obstet. Gynecol.*, **173**, 1527–1531.
63. Dennemark, N., Meyer-Wiles, M., Schulter, R. and Gries, K. (1996) Comparison of two treatments of bacterial vaginosis versus non treatment: influence on incidence of premature births. BV 96, Croydon, UK.
64. Hay, P.E., Morgan, D.J., Ison, C.A. *et al.* (1994) A longitudinal study of bacterial vaginosis during pregnancy. *Br. J. Obstet. Gynaecol.*, **101**, 1048–1053.
65. Witkin, S.S., Jeremias, J. and Ledger, W.J. (1988) Recurrent vaginitis as a result of sexual transmission of IgE antibodies. *Am. J. Obstet. Gynecol.*, **159**, 32–36.
66. Cibley, L.J. and Cibley, L.J. (1991) Cytolytic vaginosis. *Am. J. Obstet. Gynecol.*, **165**, 1245–1249.
67. Horowitz, B.J., Mardh, P.A. and Rank, E.L. (1994) Vaginal lactobacillosis. *Am. J. Obstet. Gynecol.*, **170**, 857–861.

9 The diagnosis and management of genital ulceration

S. McCormack

INTRODUCTION

This chapter is divided in two:

1. a practical guide to the management of genital ulceration in genitourinary medicine (GUM) clinics;
2. a review of the sexually transmitted causes of genital ulceration, in particular herpes simplex virus.

PRACTICAL GUIDE TO CLINICAL MANAGEMENT

The definition of an ulcer is 'a local defect which is produced by the sloughing of inflammatory necrotic tissue'. Genital ulceration is the presenting complaint of 2–70%[1] of those attending GUM clinics. With improved laboratory techniques for the detection of herpes simplex virus,[2] it is clear that this agent is by far the commonest cause of genital ulceration, a fact that has long been suspected by genitourinary (GU) physicians.

Objectives

The objectives of this section are to enable doctors working in GUM clinics:

- to recognize the clinical presentations most likely to be genital herpes;
- to identify the cases in which genital herpes is unlikely to be the causative agent;
- to recognize the risk factors associated with genital ulceration due to other sexually transmitted agents.

History of the ulcer

Onset
Genital ulcers due to herpes simplex usually start within eight days of sexual contact with the index case. Although the genital ulcer caused by syphilis (chancre) can start within 10 days, it is more usual for symptoms to commence three weeks after exposure. The mucous patches associated with secondary syphilis present four weeks onwards following exposure and in a third of cases, the primary chancre will still be present. Ulceration in association with lichen sclerosis or intraepithelial neoplasia is commonly preceded by other symptoms such as chronic itching.

Duration
The ulcers of genital herpes last for 2–3 weeks in the 'classic' first episode, although a significant proportion of people with genital herpes never experience this (see HSV, below). Recurrent herpes lasts about one week. Primary syphilitic ulcers last about one month in the absence of treatment. The genital ulcers caused by

lichen sclerosis or intraepithelial neoplasia may have been present for several months.

Pain

The ulcers caused by herpes simplex are associated with discomfort, which can vary from mild irritation to extreme pain. When symptoms are mild they may pass unnoticed, but on examination one would expect to elicit tenderness. Syphilitic ulcers, by contrast, are painless and this is unexpected in the context of clinical signs. Genital ulcers due to the dermatoses and candida are usually painful. Neurological phenomena such as tingling, shooting pains in the buttocks and thighs are typical of herpes.

Past medical history

Previous episodes

A past history of similar episodes would strongly favour a diagnosis of genital herpes. A history of oral herpes is relevant because previous exposure to the virus may affect the clinical presentation (see HSV, below).

Skin disease

A history of skin disease is particularly important if of long standing. Specific enquiries should be made about eczema, psoriasis and seborrhoeic dermatitis. Extragenital lesions may be present, including a scaly scalp.

Drug history

Fixed drug eruptions may present as genital ulceration and so recent (three months) drug history is relevant.

Topical agents

Other agents that may cause contact dermatitis and ulceration when in contact with the genital skin include bath additives, washing powder and douches. An occupational history is of relevance here as chemical substances on the hands can come in contact with the genitals.

Sexually transmitted diseases

Chlamydia is associated with circinate balanitis and Reiter's disease. Although genital warts rarely ulcerate, the human papillomavirus, the causative agent, is linked to genital neoplasia. It is useful to know whether there is a past history of syphilis serology, which would be taken as part of a screen prior to blood donation or antenatally.

Cytology

The risk factors for vulval and vaginal intraepithelial neoplasia are similar to those for cervical intraepithelial neoplasia and therefore a history of an abnormal cytology is relevant.

Smoking

Smoking is linked to a history of sexually transmitted diseases and cervical cancer. This may be due to a direct effect of nicotine or an indirect effect due to an alteration in local immunity.[3] Alternatively, the association may be through a confounding variable such as sexual behaviour, since unprotected sex with multiple partners and smoking are both examples of 'risk-taking behaviour'. This explanation was not supported by the UK study of sexual attitudes and lifestyles.[4] Whatever the underlying reason, an enquiry into smoking habits is relevant.

Sexual history (Table 9.1)

Three months will cover the typical incubation periods for the sexually transmitted organisms that cause genital ulceration. Details of all partners during this time period must be elicited with emphasis on the following.

Table 9.1 Useful information from sexual history

When	First and last contact
Where	In which country did sex take place? From which country was the partner?
Symptoms in partner	Genital herpes
Condoms	With or without lubricants
Use of toys for sex	Whether shared with partner

When

It is important to clarify the time of the first and last contact with each partner as this will help with the diagnosis (see Onset, above) and will also have implications for partner notification.

Where

This is a crucial question as the possibility of a tropical connection will affect the choice of both investigations and medication.

Other aspects

Direct questioning may elicit a history of symptoms in the partner which suggest genital herpes.

Examination

Number and distribution
Multiple ulcers that are widely distributed suggest first-episode genital herpes or secondary syphilis. A single ulcer is more typical of syphilis, chancroid, intra-epithelial neoplasia or a fixed drug eruption. Several ulcers localized to one area suggest recurrent herpes. Two or three ulcers widely distributed may be indicative of primary syphilis.

Character
Ulcers can be described according to depth and the definition of the edges. The ulcers associated with herpes simplex are superficial and may coalesce. The primary syphilitic ulcer is well defined and deep, as is the ulcer associated with chancroid, but the ulcers of secondary syphilis are superficial. Circinate balanitis and β-haemolytic streptococcus have the very distinctive appearance of large areas of superficial ulceration with well defined edges.

Tenderness
Herpetic ulcers are extremely tender and in the case of recurrent herpes, the tenderness is frequently excessive in the context of the clinical signs. Syphilitic ulcers are said to be painless, but the vigorous swabbing required to yield a good specimen for dark-ground examination is usually uncomfortable for the patient. However, it is the discrepancy between clinical appearance and degree of discomfort that is the hallmark in both herpes and syphilis.

Lymphadenopathy
Inguinal lymphadenopathy is common and the presence of palpable nodes, particularly in a thin patient, may not be significant. Nodes larger than 2 cm are certainly significant. Tender lymph nodes are to be expected in 50% of those suffering with genital herpes and appear earlier in recurrent disease than first episode. The lymph nodes should not be tender in syphilis. The inguinal swellings (buboes) associated with LGV and chancroid may be large (> 5 cm) and fluctuant and the overlying skin is frequently abnormal. In a recent prospective study in Thailand, *Haemophilus ducreyi* was the commonest organism grown in pustular bubo aspirates and *Chlamydia trachomatis* the commonest in non-pustular aspirates.[106]

The oral cavity
The mucosa of the oral cavity is similar to the genital tract and as such is vulnerable to the same organisms and pathological processes. Oral sex is common and so the oral cavity may be the primary site of a sexually transmitted disease. In secondary syphilis the snailtrack ulcers are a result of systemic spread and in HIV the aphthous-like ulcers seen in the oral cavity are thought to be mediated by immune complexes. Non-sexually transmitted conditions like Stevens–Johnson syndrome also cause orogenital ulceration.

General examination
If syphilis or HIV is suspected, then it is advisable to do a general examination, focusing on the skin, the lymph nodes, liver and spleen, heart and central nervous system.

Investigations

Figure 9.1 is an algorithm of the recommended investigations, together with the diagnoses that the results of such investigations may suggest.

HSV
Swabs should be taken from the ulcers for viral culture or antigen testing. A cervical swab should also be taken in female patients with milder disease, in whom it is possible to pass a speculum. This is the appropriate time to take a baseline serology for HSV, but this is only available as a research tool at present.

Dark-ground examination and syphilis serology
In order to take a good specimen for dark-ground examination, the ulcer must first be cleaned with normal saline. The ideal material for dark ground is the serous exudate from the base of the ulcer and squeezing the sides of the ulcer together helps to produce this. If there is plenty of exudate, the specimen can be collected with a plastic loop or by simply pressing a slide onto the base. In the absence of an exudate, one needs something firmer, such as a curette, to scrape material from the base. Alternatively, the side of a glass slide can be used to scrape across the base of the ulcer.

Serology for syphilis is performed on a clotted specimen. If one suspects primary syphilis, it is important to let the laboratory know so that they can arrange a FTA antibody screen, as this is usually the first test to become positive in syphilis.

Haemophilus ducreyi
Microscopy of the ulcer material may reveal the presence of small Gram-negative bacilli grouped in chains but this lacks sensitivity and specificity, particularly in inexperienced hands. Culture media for *Haemophilus ducreyi* are not routinely available in the developed world and it is usually necessary to discuss this with a microbiologist. The choice of medium may depend in part on the suspected source of the infection, as growth requirements of the organism often vary

100 Handbook of Genitourinary Medicine

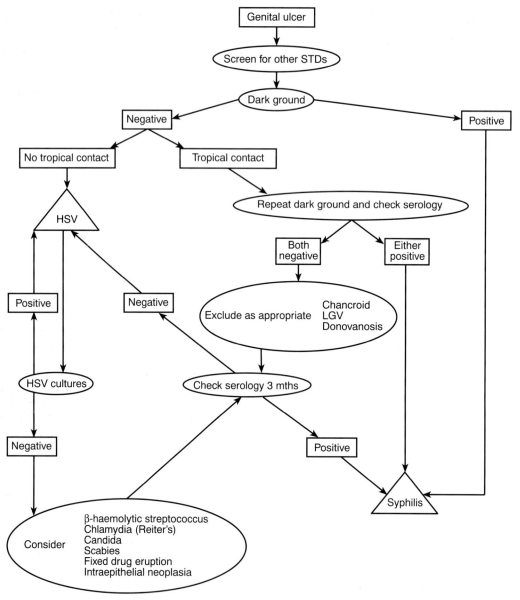

Fig. 9.1 Investigation and differential diagnosis algorithm for genital ulcers.

geographically. Cotton, rayon, cotton wool or calcium alginate swabs can be used and specimens should be obtained from the base or undermined edges of the ulcer.[5] It is advisable to inoculate the culture medium directly as a suitable transport medium is not available and organisms will only survive for 2–4 h unless refrigerated.[6] Ideal incubation is in a humid, carbon dioxide-enriched environment and so if it is not possible to return the specimen to the laboratory immediately, it should be placed into a candle extinction jar containing a moistened towel.

Lymphogranuloma venereum

LGV can be isolated by culture from bubo pus in about 50% of cases using a chlamydia culture system. Serology using the microimmunofluorescence technique is the most accurate serological method, but reagents are not commercially available. A fourfold rise in titre using the complement fixation test strongly favours the diagnosis of LGV. High titres may be indicative of LGV but can occur with other *C. trachomatis* infections.

Donovanosis
Donovanosis is even harder to confirm since the causative agent will not grow on artificial culture media and diagnosis relies on the identification of the Donovan bodies in a tissue specimen. A suitable specimen can be obtained by scraping off a piece of granulation tissue using a scalpel or curette and crushing it between two slides. Suitable sites are the base of the ulcer or a leading edge and the lesion should be cleaned with saline first. Slides should be air dried.

Biopsy is recommended for early and sclerotic lesions, in which the number of organisms is likely to be scarce.

Other investigations
If the history and examination are atypical of genital herpes, then it is advisable to take swabs for bacterial and fungal cultures from the ulcers at the outset. If the history is of long standing, especially in the older patient, then early referral for colposcopy and biopsy is recommended (Table 9.2).

Management

The principal aims in the GUM clinic are to identify, treat and control sexually transmitted diseases. The last, which is probably the most challenging of these aims, is the one that is fundamental to a good genitourinary medicine service.

Identifying the sexually transmitted causes of genital ulceration
In developed countries, the commonest cause of genital ulceration is genital herpes. Although syphilis is far less common, genitourinary physicians must also be competent in recognizing and excluding this disease.

A proportion of cases of genital ulceration seen in the clinic are not due to a sexually transmitted agent. Most clinics have staff with the expertise to deal with the non-sexually transmitted causes of genital ulceration, but in the author's opinion, this should not be done in the walk-in clinic.

Treatment
The treatment of genital herpes and early syphilis is outlined in detail below. Early treatment in genital herpes is important and this should be started whilst awaiting the results of investigations.

Control
Identifying and treating STDs contribute to their control but there are three more components to consider in the prevention of further transmission: adequate follow-up, assiduous contact tracing and health education. In general, patients should be followed up until the genital ulceration is healed and advised not to have penetrative sex during this period. Exactly how contact tracing is undertaken will depend on the diagnosis, but all patients should see a health advisor at the outset and every effort should be made to contact partners. Health education should be part of each communication between the patient and staff in the GUM clinic and it is vital that information is consistent and in keeping with any written information given to patients.

Pitfalls

Missing a STD in a patient
This may happen because of missing the risk factor, omitting part of the screen or inadequate follow-up. Direct questioning is often necessary to establish a tropical contact. The part of the screen that is most frequently omitted is the dark-ground examination and

Table 9.2 Checklist for differential diagnosis

	Likely	*Less likely*
Tropical contact	Syphilis, chancroid, LGV, donovanosis	HSV
>2 weeks between contact and onset of ulcer	Syphilis, donovanosis	HSV, chancroid, LGV
Ulcers painless	Syphilis, donovanosis, primary LGV in women	HSV, chancroid
Ulcers deep	Syphilis, chancroid, tertiary LGV	HSV, primary LGV, syphilis, donovanosis
Solitary ulcer	Syphilis, LGV	HSV, chancroid
Lymph nodes >2 cm	Chancroid, LGV, donovanosis*	HSV, syphilis

*Swelling in inguinal region, but not due to lymph nodes

a proportion of genital ulcers due to syphilis will be missed because serology is not positive for a week after the onset of the ulcer. The importance of follow-up must be stressed to the patients and it may be facilitated by offering appointments. Follow-up is helped by the patient seeing the health advisor at the first visit, as they will check details such as address and telephone number, ensure that the patient reattends and pursue them if they do not.

Missing a STD in a partner
In order to avoid this pitfall, it is essential to take a good contact history and refer to a health advisor on the first visit. It cannot be assumed that advice regarding abstinence or barrier methods has been adhered to and it is important to continue to take a sexual history at subsequent visits.

Missing an opportunity for early treatment
This is most important for cases of ulceration due to genital herpes. Experience will help but in general, this situation can be avoided by examining every patient carefully.

HERPES SIMPLEX VIRUS

Herpes simplex virus is a DNA virus belonging to the herpes family of viruses. Other viruses in this family include varicella-zoster virus, cytomegalovirus and Epstein–Barr virus. There are two distinct types of herpes simplex virus on biological grounds: HSV-1 and HSV-2.

Objectives

To provide doctors with the background information necessary for advising patients and colleagues on:

- the diagnosis of genital herpes;
- the management of genital herpes;
- the prevention/reduction of further transmission.

Epidemiology

There are several points to consider in interpreting the published literature on the epidemiology of herpes simplex virus infection.

Case ascertainment
HSV infection may be identified on clinical or laboratory grounds. In a clinic setting, cases may be 'passively' identified, when they are invariably symptomatic, or 'actively' identified using techniques such as colposcopy. In the laboratory, cases may be identified by techniques which detect the virus or a response to the virus. In the former, it has been possible to differentiate between HSV-1 and HSV-2 since the late 1960s, but with serological responses it has only been possible to do this reliably since 1985.[7]

Many factors may contribute to an increase in case ascertainment: increased awareness in the population; increased awareness amongst doctors; increased sensitivity of diagnostic techniques.

Source of data
Surveillance systems are in operation in some countries, both through clinical reporting systems (genital herpes and neonatal infection) and through laboratory reporting systems (viral isolation and serology). The common problem with such systems is that they frequently lack a denominator. Trends can still be examined by using proxy denominators such as the number of new patient visits or total diagnoses of STDs.

Community-based seroepidemiological studies have been undertaken and have yielded very useful information. They are more applicable to the general population and are not usually subject to biases in case ascertainment. However, the majority of studies have been undertaken in clinic-based populations (most commonly STD clinics, antenatal clinics and student health clinics), in which cases are identified using a combination of clinical and laboratory criteria.

Clinical disease

In the UK, there is a nationwide surveillance system for reporting STDs through the genitourinary medicine clinics. In this population, cases are defined using clinical and laboratory criteria. A fourfold increase in the proportion of visits due to genital herpes occurred between 1976 and 1991 and this trend has continued to a lesser extent (Table 9.3). Similar rising trends in genital

Table 9.3 Data from Department of Health Report (1994) on new cases seen at GUM clinics in England[8]

Condition	1990–91	1993–94
Male		
All herpes simplex	10 437	11 794
First episode	6030	6243
Females		
All herpes simplex	10 222	14 255
First episode	6861	8630

herpes have been reported from patient consultations to US doctors. Using colposcopic techniques in addition to genital examination, 11% of 779 women attending a STD clinic in Seattle had clinical genital herpes, 54% of whom were experiencing their first episode.[9]

Reports of neonatal infection have also increased over the last 20 years in the USA[10] and in 1984 prevalence was estimated to be a minimum of four per 100 000 live births.[11] This is probably higher than the UK where the estimated incidence is 2–3 per 100 000 births.[12]

Viral isolation

The majority of isolates in genital herpes are HSV-2, although recent reports have suggested that the proportion due to HSV-1 is increasing, particularly in women.[13,14,15] The clinical significance of this is that both clinical recurrences and asymptomatic shedding of virus are less frequent when the causative agent of genital herpes is HSV-1.[16,17,18]

Seroepidemiology (Table 9.4)

Reported rates of antibodies to HSV-1 in adults vary from 27% in students at the University of Washington to 97% in prostitutes from Kinshasa, Rwanda.[2] Seroprevalence rises with age and tends to be inversely associated with socioeconomic status.[19] The correlates of HSV-1 infection that remained after controlling for associated factors in a community-based study in San Francisco were Hispanic or black race, less education and age 20. Age-specific prevalence is declining in some populations, which may reflect a change in the acquisition of HSV-1 in early life. This may influence subsequent acquisition of HSV-2 infection.[2,21]

HSV-2 is less common and reported rates vary from 2% in University of Washington students to 95% in prostitutes in Kinshasa. About 30% attenders at STD clinics in developed countries have antibodies to HSV-2, with higher percentages consistently reported in female patients. Prevalence rates tend to rise with age, most notably in the age group 30–44. Correlates that remained significant after controlling for associated factors in a large population survey in the US were black race and female gender.[22] In a large UK study of STD clinic attenders and blood donors, correlates of HSV-2 antibody included female gender, years of sexual activity, number of lifetime partners and past history of a sexually transmitted disease.[23]

Transmission

Transmission usually occurs through close contact with a person who is shedding virus from a mucosal surface or peripheral site. The virus gains access through cracks in the skin or through direct contact with the

Table 9.4 Seroprevalence of antibodies to HSV-1 and HSV-2. Results of several studies from around the world adapted from Corey (1994)[2] (figures represent percentage of cohort positive)

Source of data	Seronegative	HSV-1 only	HSV-2 only	HSV-1 and 2
UK STD				
Men (n = 294)	18	55	10	17
Women (n = 347)	5	58	12	25
King County STD				
Men (n = 50)	26	42	17	15
Women (n = 776)	21	33	18	25
Seattle HIV+ve gay men (n = 171)	20	33	17	30
Uni Washington (n = 186)	70	27	2	0
Stockholm pregnant women			HSV-2 total	
1969 (n = 941)		57	17	
1983 (n = 1759)		43	32	
1989 (n = 1000)		47	33	
Kinshasa prostitutes				
HIV+ve (n = 181)	0	5	7	88
HIV–ve (n = 187)	0	25	3	72
Nairobi STD (n = 115)	0	54	3	58
Peru STD (n = 395)	1	18	10	73

cervix, urethra, oropharynx or conjunctivae. It may be possible for infection to be transmitted by autoinoculation from an infected lesion to a distant site.[19]

Transmission can occur from asymptomatic shedders and this has important implications for the control of genital herpes. Mertz *et al.* prospectively followed 144 couples in which each source partner had symptomatic, recurrent disease for a median of 334 days.[24] Transmission occurred in 14 (9.7%) couples and in nine of these cases, the source partner had no symptoms of infection. Transmission was significantly higher in couples where the susceptible partner was female and where the female partner was seronegative for HSV-1 at entry, suggesting that previous infection with HSV-1 plays a protective role. This is supported by the fact that the isolation of different strains of HSV-2 is an uncommon occurrence in symptomatic recurrent genital herpes.[25]

The incubation period is from one to 26 days with the majority presenting within the first eight days, although a retrospective study of clinical case notes suggested that the incubation period may be longer.[26]

Pathogenesis

There are several aspects of pathogenesis that are important to the clinician.

The effect of different viral strains on disease

Herpes simplex virus exhibits considerable genetic diversity and there are several strains of HSV-1 and HSV-2, but these minor variations do not appear to affect the clinical presentation. There are some generalizations that can be made about the differences in disease caused by HSV-1 and HSV-2, but essentially they are clinically indistinguishable (Table 9.5).

The effect of host response on disease

HSV can infect many different cell types but infection tends to be limited to the skin around the site of inoculation and the relevant sensory ganglia. This is probably due to the cell-mediated immune response of the host, since disseminated disease occurs in adults in whom cell-mediated immunity is compromised.[27] The absence of HSV antibodies in the exposed neonate is associated with neonatal infection, suggesting that they may play a protective role in the prevention of transmission.[28]

Latency

HSV kills the cells in which it replicates, but it is able to exist in a latent state within neurons by mechanisms that remain poorly understood, in part because there is not yet an adequate *in vitro* system in which to study latency. Elucidation of the viral and host factors that contribute to the establishment and maintenance of latency is crucial to the development of therapeutic and preventive measures.

The opportunities for therapeutic or preventive measures

Antiviral agents exert their effects by interfering or interrupting the life cycle of the target virus and this is illustrated in Figure 9.2.

Clinical presentation

Herpes simplex virus is most commonly associated with lesions of the mucous membranes and skin around the mouth (oral herpes) and genitalia (genital herpes). Genital herpes is divided into first-episode and recurrent genital herpes according to the clinical presentation. If definitive serology is available, it is possible to further classify first episode into:

- first episode of primary genital herpes;
- first episode of non-primary genital herpes.

In the studies done to date in the US using definitive serology, over 85% of first episodes that were fully

Table 9.5 Clinical differences between HSV-1 and HSV-2

More characteristic of HSV-1	*More characteristic of HSV-2*
Oral lesions	Genital lesions
Encephalitis	Meningitis
Lower risk of neonatal transmission	Higher risk of neonatal transmission
Moderate protection against HSV-2 genital herpes	Minimal protection against HSV-1 oral herpes[55]

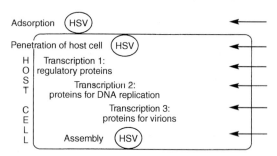

Fig. 9.2 Lifecycle of HSV and opportunities for therapeutic interventions.

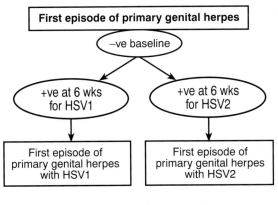

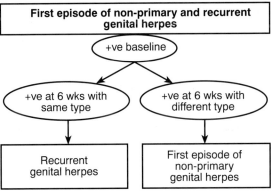

Fig. 9.3 Classification of genital herpes according to serological response.

evaluated were due to primary genital herpes and over 80% of these were in association with HSV-2.[16,18]

Amongst the 14 couples that seroconverted (nine primary, five non-primary) in Mertz et al.'s study[24], eight had a classic first-episode presentation, one had asymptomatic seroconversion followed later by a symptomatic episode, three had atypical episodes (one urethritis and two with symptoms but no signs) and in two cases, the susceptible partner remained asymptomatic and culture negative throughout follow-up. Unfortunately, it is not clear what proportion of classic first episodes were associated with primary disease.

Classic presentation of primary genital herpes

The clinical events are classically: vesicle or papule formation; ulceration, crust formation (crust formation does not occur on mucosal surfaces); and reepithelialization. The distribution is typically bilateral and widespread. New lesions develop in the majority and appear between days 4 and 10. Local pain increases during the first 6–7 days, reaches a peak between days 7 and 11 and the lesions are usually completely healed at 21 days. Scarring is uncommon and should alert the clinician to an alternative diagnosis, such as hydradenitis suppurativa. The inguinal lymph nodes become tender during the second week and may remain so after resolution of the ulcers. The lymph nodes are characteristically firm and non-fluctuant.

Dysuria, which can be internal or external, is more common in women than men and this is reflected in the urethral HSV isolation rate which is considerably higher in women.[19] In men the dysuria is accompanied by urethral discharge which is usually clear and mucoid. The dysuria is often disproportionate to the clinical signs and this is particularly noticeable when taking a urethral specimen.

Systemic symptoms are reported by over 50% of patients with this classic presentation and comprise fever, headache, malaise and myalgia. They appear early in the course of the disease and usually resolve before the lesions have healed.

Clinical presentation with recurrent or non-primary HSV

The lesions tend to be unilateral and fewer in number. The duration of pain and time to healing are also reduced, as is the time of viral shedding. Systemic symptoms are uncommon. However, pain out of proportion to the clinical signs is also a feature of recurrent disease (often noted on swabbing the ulcers).

Over half of the patients notice prodromal symptoms up to five days before an attack. These symptoms are due to sacral neuralgia and vary from mild tingling to shooting pains in the buttocks and thighs (see below).

Atypical genital herpes

Although the above descriptions are characteristic of genital herpes, the advent of definitive serology and the detection of viral DNA in clinical lesions using the polymerase chain reaction has led to the recognition of a broader range of clinical disease. It is clear that a considerable proportion of patients never have symptomatic disease and many lesions such as linear fissures, furuncles, are now recognized as herpes.[9,2] Therefore any breach of mucosal surface should be evaluated for HSV. Several studies have demonstrated that careful questioning and thorough examination, in combination with education of the patient, will enable the majority of individuals with antibodies to HSV-2 to identify clinical episodes.[2,29]

Specific sites

Herpes cervicitis

The majority of women with first-episode primary and non-primary genital herpes shed virus from the cervix and this may be symptomatic with discharge. The cervix is usually abnormal on inspection, ranging from focal erythema to necrotic cervicitis. The duration of viral shedding mirrors that from the peripheral lesions.

In contrast, recurrent genital herpes is associated with cervical shedding in 10–30% and clinical examination without colposcopy is invariably normal. Viral shedding from the cervix is shorter and similar in duration to that from peripheral lesions in recurrent disease.[9]

Herpes proctitis

The presentation of primary herpes proctitis is that of severe rectal pain of sudden onset, with tenesmus and discharge. Systemic symptoms and fever are common and there may be evidence of sacral dysfunction. Perianal lesions are present in half of the cases.[30] Proctoscopy, if possible, reveals anything from a friable inflamed mucosa to discrete ulcerations.

Complications

Because HSV is a neurotrophic virus, neurological phenomena are common and include urinary retention, meningitis and encephalitis.

Neurological phenomena

The central nervous system complications from genital herpes are:

- aseptic meningitis;
- encephalitis;
- transverse myelitis;
- autonomic dysfunction (radiculomyelopathy).

HSV accounts for approximately 3% of cases of aseptic meningitis, usually in association with HSV-2. Presenting symptoms of aseptic meningitis are fever, headache, vomiting, photophobia and neck stiffness and onset is usually 3–12 days after the clinical lesions, with resolution over one week. Herpes encephalitis is rarely associated with HSV-2.[19]

Autonomic dysfunction may be part of a transverse myelitis or may present alone. Symptoms of autonomic dysfunction include hyperaesthesia or anaesthesia of the perineal, sacral and lower back regions, difficulty passing urine and emptying bowels and impotence in men. Signs include a large bladder, altered sensation, impaired sphincter tone and loss of the bulbocavernous reflexes. Occasionally this occurs in association with a transverse myelitis, with muscle weakness and loss of deep tendon reflexes. Symptoms and signs usually resolve over eight weeks, but there are case reports of residual disability years later.[31] Significant autonomic dysfunction occurs in 1% of cases, usually in association with primary infection. Dysesthesia at the site of the developing lesion or at a remote site (for example, shooting pains in the thighs) is common in association with reactivation. It precedes the development of the lesion, but can occur in the absence of a lesion.[32] Symptoms may last for hours to days and there are usually no signs of neurological deficit on examination. Chronicity with neurological deficit has been described in an immunocompetent patient.[33]

Disseminated infection

Dissemination is rare. Predisposing conditions include immunosuppression, pregnancy, malnutrition and alcoholism.[27,32] It has also been reported in association with burns and atopic eczema.[34] Deficits in cell-mediated immunity seem most important but cases have occurred in apparently immunocompetent hosts.

Dissemination with resistant strains of HSV has been reported in HIV-infected patients.[35]

Investigations

The investigation of choice is culture, taken from clinical lesions. In centres where definitive serology is available, a baseline sample and six-week sample give additional useful information. The investigations are discussed in detail elsewhere. The sensitivity of culture varies according to the stage of the clinical lesions, with the greatest yield coming from vesicles and early ulcers (Table 9.6).

It is essential to screen for other sexually transmitted diseases. The screen may not be practical in patients with a first episode since a full examination may be too uncomfortable. Of prime importance in the differential diagnosis is the screen for syphilis, ideally by darkground examination of samples from the lesions in addition to serology, although the former may be too painful.

Secondary bacterial infection is unusual but it is worth taking a standard mcs from any ulcers that are particularly inflamed or purulent. Secondary fungal infection is common in women, particularly in the second week of infection.[15]

Table 9.6 Viral isolation and stage of disease in genital HSV infection (figures are % of lesions culture positive)[36]

Maculopapular $n = 9$	Vesicle $n = 136$	Pustule $n = 68$	Ulcer $n = 132$	Crust $n = 93$
25	94	87	70	27

Management

In the UK, an independent Advisory Panel has been formed to help establish a consensus on the management of patients with genital herpes in the light of the available evidence. The group has published reviews of therapy[107] and other relevant issues such as asymptomatic shedding.[108]

Antiviral therapy

The initial diagnosis of genital herpes is made on clinical grounds. The signs found on examination are probably most helpful in differentiating between 'primary' and 'non-primary' types of clinical presentation. However, these are dependent on the day of presentation and must be considered in conjunction with the history. The presence of systemic features favours a diagnosis of primary infection.

Treatment of acute episode

The lesions should be kept clean and dry.

Oral acyclovir is indicated for all acute episodes suggestive of primary infection that present during the period of new lesion formation, which is usually up to 10 days in a primary first episode. It is of proven benefit in reducing the duration of the attack and alleviating local and systemic symptoms.[37,38] Oral acyclovir will also reduce the duration of the acute episode in recurrent genital herpes, but the effect on local symptoms is less marked[19] and one must be guided by the clinical severity of individual attacks. Benefits are maximum with early treatment and therefore patient-initiated treatment has advantages over physician-initiated treatment.[39]

Even at high doses, oral acyclovir does not prevent recurrences.[40]

The frequency of the recurrent attacks will determine whether medication should be given daily (suppressive therapy, see below) or per attack.

Suppression of further episodes

It is hard to be dogmatic about suppressive therapy. As a rule of thumb, >6 attacks of clinical herpes a year would justify the use of continuous acyclovir. However, some patients may easily manage this clinical burden whilst others with less severe disease may require suppressive therapy in order to cope. Although the popular formulation for suppressive therapy is 400 mg twice a day, a more frequent dose may be necessary to achieve control of attacks and this can be at a lower dose, for example 200 mg four times a day. The dose should be reduced appropriately (2–3-month intervals) so that the lowest dose possible is used. It is very unusual to achieve control at a dose lower than 200 mg twice daily and the majority require 200 mg three times a day or 400 mg twice a day. After an agreed period, usually 6–12 months, acyclovir should be discontinued to see if the natural history of the disease has changed. Renal and liver function should be monitored twice a year at least.

In the immunosuppressed higher doses of acyclovir are usually needed. The potential for resistant strains is higher in this population and if patients are not responding it is important to try to confirm the presence of a resistant strain in the laboratory, but if this is not possible alternative therapy with foscarnet should be considered.

Two new antiviral agents have been evaluated in phase III randomized placebo-controlled clinical trials for episodic use in patients with recurrent genital herpes.[109,110] In both trials the cohorts were randomized at screening and given therapy to self-initiate in the event of a recurrence. Only those that returned for evaluation after starting treatment were included in the analyses (approximately two-thirds of each cohort). Results demonstrate a statistically significant reduction in the duration of the lesions and symptoms and viral shedding for both drugs when compared with placebo, with a maximum reduction in median time of 1.9 days. In addition, both the 500 mg and 1000 mg doses of valaciclovir were shown to prevent the development of vesicular ulcerative lesions in a significantly higher proportion of recipients compared to those on placebo: 31%, 28% and 21% respectively, However, these statistically significant results may not be clinically significant to an individual patient and such marginal benefits should be fully discussed before embarking on a potentially long-term commitment of treatment.

Psychological support

There is considerable psychological morbidity associated with first-episode genital herpes but this is significantly improved at three months follow-up, especially in the absence of recurrences.[41] A minority of patients are unable to adjust psychologically to recurrent herpes and these may require referral to a psychologist. Continuing morbidity in patients may manifest as negative feelings about themselves and reduced sexual activity.[42] Suppressive acyclovir may help to reduce illness concern and anxiety in recurrent genital herpes sufferers[43] but it would be difficult to design a study to establish this, because of the ethical considerations in withholding suppressive acyclovir from a control group.

It is important to involve the health advisor at the outset and it is helpful to hand on as much information about the patient as possible before this consultation. Health advisors can facilitate the transfer of information, psychological support and partner notification. However, patients will only benefit from different approaches if the same accurate information is given and this should concur with any written information.

Many larger centres have a specific clinic run by a dedicated team for patients suffering from herpes.

Issues to be covered

Below is a summary of some of the available information relating to particular topics of concern. How this information is delivered depends on the intelligence and personality of the individuals giving and receiving the advice.

Infectivity

Table 9.7 summarizes the annual transmission rates extrapolated from Mertz's *et al.* prospective study of 144 couples. This study looked at risk factors for transmission in heterosexual couples in whom the source partner had symptomatic recurrent genital herpes and the susceptible partner had no history of genital herpes.

Although these rates are only applicable to highly motivated heterosexual couples, they are invaluable to the clinician as a basis for advice to all patients. Higher transmission rates in males compared to females have been reported in other studies, as have lower transmission rates in the presence of antibodies to HSV-1.[44]

It is important to reiterate that in 9/14 cases (79%), the transmission occurred when the source partner was asymptomatic. Asymptomatic shedding has been reported in 12% (2/17) to 50% (4/8) of individuals[18,45] and although dramatically reduced with suppressive acyclovir, it is not entirely prevented (six of 1611 days on acyclovir compared to 83 of 1439 days on placebo).[111] Asymptomatic shedding is more common within three months of primary HSV-2 disease; when the causative agent is HSV-2; and in the absence of antibodies to HSV-1.[18]

Recurrence rate

The recurrence rate varies greatly between individuals, even amongst those who harbour the same strain,[46] and may in part be determined by genetic influences on host immune response. Benedetti studied 457 individuals with culture-positive, serologically confirmed symptomatic first-episode genital herpes for a median of 391 days.[16] The strongest correlate of subsequent recurrences was duration of first episode >34 days. This study confirmed that other correlates of frequent recurrences included causative type (HSV-2), but it is the first to identify male gender in association with higher recurrence rates. Younger age of acquisition was also identified as a variable influencing more frequent reactivation, although longer follow-up was achieved in the cohort who had more severe first episodes.

HSV and cervical cancer

A link between genital herpes and cervical cancer was first suggested in 1969 when dysplasia was observed in 25% of cervical biopsy specimens from patients with cytological evidence of HSV. Subsequent seroepidemiological studies carried out in the 1970s appeared to confirm an association. However, more recent studies which have been able to differentiate between HSV-1 and HSV-2 antibodies have not supported an association. In order to establish oncogenic characteristics in a virus, it is also necessary to demonstrate that the virus can transform cells. Evidence suggests that HSV can initiate transformation in rodent cell lines, but to date this has not been demonstrated in primate cells.

Current evidence suggests that HSV does not have a causal role in the development of cervical cancer, but that it cannot be excluded as a possible cofactor.[47] It is therefore especially important that female patients with genital herpes participate in the screening programme for cervical cancer.

HSV, pregnancy and neonatal infection

The minimum incidence of neonatal infection in the United States was 139 cases per 3.7 million births in

Table 9.7 Risk of acquisition of genital herpes by gender, antibody status and method of contraception

Variable	Partners at risk	Cases	Annual rate	p value
Sex of susceptible partner				
Male	79	3	4.5	0.006
Female	65	11	18.9	
Initial serology of susceptible partner				
HSV-1-ve	63	9	16	0.11
HSV-2+ve	81	5	7.2	
Initial serology of source partner				
HSV-2 alone	93	6	7.5	0.17
HSV-1 and 2	41	6	15.6	
Barrier methods				
Used	42	2	5.7	0.19
Not used	101	12	13.6	

Table 9.8 Issues for couple in whom woman/partner has genital herpes

Genital herpes in partner	Risk of catching herpes
	Risk of shedding at term and transmission
	Risk to the foetus/neonate if infected
Genital herpes in woman	Risk of recurrent disease in pregnancy
	Risk of shedding at term and transmission
	Risk to the foetus/neonate if infected

Table 9.9 Summary of transmission by category of genital herpes in 15 923 pregnant women

1/52 = primary first-episode HSV-1	2/5 transmitted
4/52 = primary first-episode HSV-2	
13/52 = non-primary first episode	4/13 transmitted
3/52 = reactivation HSV-1	1/3 transmitted
31/52 = reactivation HSV-2	0/31 transmitted
4/52 = serology not available	0/4 transmitted
15 867 = culture negative	3/15 867 transmitted

1984, the equivalent of one per 26 000 or four per 100 000 live births.[11]

Risk of catching herpes during conception and pregnancy (Table 9.8)

Kulhanjian et al. looked at 277 women and their 190 partners and found a HSV-2 seroprevalence of 32% and 25% respectively.[48] Of the 190 couples studied, 51 (27%) were serologically discordant. Eighteen women who were seronegative for HSV-2 had seropositive partners, but only eight of these partners had a history of genital herpes. Five of the 18 women who were seronegative for HSV-2 were also seronegative for HSV-1. Seven of the 18 couples practised unprotected intercourse throughout the pregnancy at a mean of 5.5 times per month. Only one of the 18 women seroconverted for HSV-2 during the pregnancy, in association with a clinical episode. She was seronegative for HSV-1 and 2 at 16 weeks and she and her partner had unprotected intercourse throughout the pregnancy. The neonate did not develop neonatal herpes.

Risk of recurrent disease in pregnancy

Pregnancy does not appear to have a dramatic effect on the clinical course of genital herpes in the woman who has acquired it prior to pregnancy.[49]

Risk of transmission including that associated with symptomatic shedding

The risk of transmission seems to be considerably higher with first-episode genital herpes as opposed to recurrent disease (33% versus 3% respectively). Many studies report especially low rates of transmission with subclinical recurrences.[50]

In a prospective study in Washington of 15 923 pregnant women, 56 (0.35%) women were shedding HSV asymptomatically in early labour, 18 (35%) of whom had serological evidence of a recently acquired infection[50] (Table 9.9).

The risk of positive cultures in women with known risk factors for genital herpes who are asymptomatic is 0.96–2.4%.[51] However, the risk of positive cultures in women with symptomatic disease including prodromal symptoms may be as high as 35% (7/20).[52]

Risk to foetus and neonate

Genital HSV infection during pregnancy has been reported to result in spontaneous abortion, prematurity and intrauterine growth retardation, as well as congenital and perinatal HSV infections. Knowledge about the clinical presentation, natural history and transmission of genital herpes suggests that the risks to the foetus/neonate are greatest when the mother acquires infection during pregnancy.[53] A prospective study of 29 women who presented with first-episode genital herpes in pregnancy has confirmed that the risks are particularly high with primary first episodes, especially when infection occurs during the third trimester[54] (Table 9.10).

In the Washington cohort of 15 923 pregnant women, 10 infants developed neonatal herpes: seven of the mothers were culture positive in labour, and three were culture negative but had histories compatible with recurrent genital herpes. One infant died, three suffered severe disability, one moderate disability and five were normal.[50]

Predictors of mortality and morbidity in neonatal infection include disseminated disease, the presence of HSV pneumonitis, late presentation and prematurity.[55] In cases of neonatal infection presenting with

Table 9.10 Characteristics of infants born to women with genital herpes (adapted from Brown 1987[54])

	Primary infection			Non-primary infection
Trimester of acquisition (n)	First (5)	Second (5)	Third (5)	– (14)
Prematurity (<36 weeks)	0	1	4	0
Spontaneous abortion <20 weeks	1	0	0	0
Growth retardation	0	0	3	0
Neonatal HSV	0	0	2*	0

*1 death

mucocutaneous disease alone, three or more recurrences are associated with neurological impairment.[56]

Summary of management

Concensus guidelines on the management of pregnant women with genital herpes infection have now been published. [Smith, J.R., Cowan, F.M., Munday et al. (1998) Management of Herpes Simplex Virus Infection in Pregnancy. *Br. J. Obs. Gynaecol.*, **105**, 255–260]. For the woman who has acquired genital herpes before pregnancy, the risks are small and the most reasonable strategy for prevention of transmission to the neonate is a thorough assessment at the onset of labour.[57]

The safety profile of acyclovir is excellent and well established in large numbers of individuals but to date is not licensed for use in pregnancy. A randomized placebo-controlled study of acyclovir taken from 36 weeks gestation by 46 pregnant women who had experienced their first episode of genital herpes in pregnancy showed a significant reduction in the proportion of women requiring caesarean section: 0/21 receiving acyclovir compared with 9/25 receiving placebo (odds ratio 0.04, 95% confidence interval 0.002–0.745).[112]

Herpes and HIV

Infection with HSV as a risk factor for acquisition of HIV

Several studies have suggested that genital ulcer disease is a risk factor for acquisition of HIV and a thorough review of the literature demonstrated that all nine criteria for causal inference as defined by Sir Austin Bradford Hill were met.[113] Herpes simplex virus was evaluated specifically in 12 of the 27 articles reviewed, six prospective and six case control studies. Three of the prospective and four of the case control studies found an association between HSV-2 and HIV[58] and none of the others found a statistically significant negative association. It is possible that the increased seroprevalence of HSV-2 is a marker for other variables more closely associated with HIV acquisition. Alternatively, HSV may exert a biological effect which increases the likelihood of HIV transmission and there are two possible explanations to support this rationale. First, HSV causes a breach in the mucosa which may facilitate transmission of HIV, and second, HSV ulceration is associated with the presence of activated CD4 cells, the target cell for HIV.

Genital herpes in HIV-infected individuals

As immunosuppression occurs with advanced HIV disease, HSV infections may become more frequent, more severe and resistant to therapy (see Chapter 15).

Follow-up

Patients should be followed up until the genital ulceration has healed and this will require a minimum of two visits. This should ensure that regular partners have been contacted and that there has been an opportunity to pass on relevant information about recurrences and infectivity. It is important to see patients after any confirmatory results become available and attempts should be made to recall those that fail to return for follow-up.

Partner notification

Patients should be advised to refrain from sex until all ulcers have healed. How to advise patients thereafter is far from clear. A balance is required between advice aimed at the complete prevention of transmission to future partners and advice that minimizes the psychological morbidity frequently associated with genital herpes.

Prospects for a vaccine

The problems that hamper HSV vaccine development are:

- no natural correlate of prevention of infection/latency;
- no natural correlate of prevention of reactivation.

It is possible that the immune response required to achieve each of these targets differs and a single vaccine product may not be able to prevent infection and ameliorate disease.

The most extensive experience in the UK is that of Dr Skinner at the University of Birmingham, who has carried out several HSV vaccine studies with inactivated subunit and virus particle vaccine products.[59,60] These products have yet to be tested in randomized double-blind, placebo-controlled trials. In fact, although more than 50 different HSV vaccine products have been tried in preventive or therapeutic clinical trials, there was no evidence of their efficacy in either context in randomized double-blind, placebo-controlled trials, until recently.[61] Now, with new developments in both the understanding of the biology of HSV and in vaccine technology, an effective HSV vaccine is no longer unrealistic and a recent clinical trial of a therapeutic vaccine showed promising results.[62]

SYPHILIS

Schaudinn and Hoffman identified *Treponema pallidum* in 1905 and a year later, Wasserman described the complement fixation reaction between serum and a crude extract of the liver of an infected infant. The first specific serological test became available in 1949.[63]

Syphilis is classified in two ways:

1. into primary, secondary, tertiary and latent on the basis of clinical symptoms and signs;
2. according to the time from infection, into early (<2 years) and late (>2 years). This is deduced from both clinical and laboratory parameters.

Objectives

To provide doctors with the background information necessary for:

- not missing early syphilis as a cause of genital ulceration;
- advising and treating patients with early syphilis;
- managing positive serology in pregnancy;
- preventing further transmission.

Epidemiology

Few countries have comprehensive data. Although not ideal, the national clinic reporting system in the UK gives some information about trends in disease. The number of new cases of syphilis fell dramatically from the late 1940s, remained stable and then increased again in the 1970s. A retrospective study of the case notes of 946 patients with primary syphilis and 854 patients with secondary syphilis attending the Middlesex GUM clinic in central London looked at changes in demographic characteristics over the period 1965–1984.[64] The changes that were detected are summarized in Table 9.11. Although they may reflect a change in consulting patterns at the Middlesex, other studies have confirmed the increased proportion of homosexually acquired primary and secondary syphilis[65] (Table 9.11).

There was a decline in the number of new cases of syphilis in men from 3228 in 1981 to 1577 in 1986[66] and this is generally thought to have been associated with a change in sexual behaviour in homosexual men. In the late 1980s, reports of an increase in cases of heterosexually acquired syphilis and congenital syphilis began to emerge,[67,68] in some studies in association with prostitution and drug use.[69]

Pathogenesis

Confusion remains about the exact pathogenesis of syphilis and, with the arrival of HIV, basic scientists have been somewhat diverted.

Dose of the inoculum

Treponema pallidum gains access through a breach in the mucosa and evidence suggests that the dose of inoculum may determine both the production of a clinical lesion and the incubation period.[70]

Host response

The most fascinating question for the clinician is: what protects people from developing clinical disease? Large doses of antibodies may attenuate, but do not protect rabbits from disease. Animal experiments have suggested that cell-mediated immunity may be more effective in offering protection from disease[71] and this has been reflected by the rapid progression of syphilis that has been reported in patients coinfected with HIV.[72]

Transmission

Infectious particles are present in the lesion from the time that the first breach in the mucosa appears. The body fluids become infectious about a week later. The risk of transmission occurring after exposure to someone with active mucocutaneous syphilis is 30–62%. The mean incubation period is 21 days with

Table 9.11 Demographic changes in patients with primary and secondary syphilis, 1965–1984 (from ref[64])

	1965–69	1970–74	1975–79	1980–84
Homosexual/bisexual	62%	70%	88%	93%
British	59	64	71	76
Social class I,II,III	62	71	87	91
Homosexuals with three or more partners	27	32	38	54
Homosexual with previous GC	20	21	28	33

a range of 10–90 days.[73] How previous infection influences this is not clear, but it has been shown to shorten the time to clinical lesion in rabbits.[74]

The risk of transmission is minimal after two years, based on the premise that active lesions only occur in 1.4% of untreated patients with symptomatic primary or secondary syphilis.[75]

Clinical presentation of early syphilis

Natural history in the preantibiotic era

The natural history of syphilis is well documented by the Oslo Study, which started in 1890 and continued until 1910 and which followed the fate of 1978 patients who were diagnosed on the basis of clinical early syphilis. Fifty years later, 1404 of these patients were followed up again and data were available on 80% of this cohort, 259 of whom were still alive.[75] All the patients entered in the study had symptomatic early syphilis (Tables 9.12 and 9.13).

Primary syphilis

The first lesion appears at the site of entry of the organism and this is invariably a site associated with trauma such as the coronal sulcus or frenulum in males

Table 9.12

	Males	Females
Secondary syphilis only at presentation	50%	75%
Relapses of secondary syphilis: sore throat, oral and anogenital lesions	24% of whom: 90% occurred in first year 94% occurred in first two years All within four years	
Gumma: skin (70%), bone (10%), mucosa (10%)	14%	17%
Cardiovascular syphilis: uncomplicated aortitis (5%, males = females) complicated aortitis (17%, males > females)	14% of those age >15*	8% of those age >15*
Neurosyphilis: meningovascular (4%) general paresis (5%) tabes dorsalis (4%) gumma (1%) males > females in all categories	9%	5%

*No cases in those age <15 at acquisition of infection

Table 9.13

Clinical	Laboratory	Diagnosis
Genital ulcer, no systemic symptoms, no past history of syphilis	Positive dark ground and/or FTA-Abs alone or FTA-Abs, VDRL or VDRL, TPHA	Primary syphilis First episode
Genital ulcer, no systemic symptoms, past history of syphilis	VDRL (×4 rise in titre), TPHA	Primary syphilis Subsequent episode
Genital ulceration or systemic symptoms, no past history of syphilis	VDRL (high/rising titre), TPHA	Secondary syphilis
Genital ulceration or systemic symptoms, past history of syphilis	VDRL (×4 rise in titre), TPHA	

and the fourchette in females. Rectal and cervical lesions are usually asymptomatic. Primary syphilitic ulcers may be multiple. Ulcers can occur at other sites and have been reported from the lips, tongue, tonsils, breasts and fingers.

The lesion starts as a red macule which rapidly progresses through a papular stage to become ulcerative. The classic ulcer, known as a chancre, has well-defined margins and a thickened indurated base which, if squeezed, produces more serum than expected when compared to ulcers due to other causes. It is painless.

Regional lymphadenopathy, which is apparent about a week later, is present in 50–83% of cases, more frequently in association with genital chancres than those at other sites and more frequently in men than women.[76] Lymph nodes feel rubbery and are painless. It is of note that secondary infection of the ulcer may make the ulcer and the regional lymph nodes painful.

Without treatment, the lesion will heal within six weeks.

Secondary syphilis

Secondary syphilis is a systemic disease. The clinical presentation is highly variable but the principal sites involved are the skin and mucous membranes and the lymph nodes.

Symptoms may start from one week after the appearance of the primary ulcer. The primary ulcer is still present in significantly more men (64%) than women (49%) presenting with secondary syphilis.[76]

There may be systemic features of general malaise including sore throat, fevers and myalgia. Other manifestations include arthritis/arthralgia (2%), headache (2%), meningitis (1%) and iridocyclitis (1%).[76]

The rash of secondary syphilis, which is present in over 85% of cases, is usually widespread, the commonest site being the trunk. Palms and soles are frequently involved and the Middlesex study reported this in 43% of men and 63% of women, whereas more men than women had rashes on the genitalia (36% compared to 16%). The appearance of individual lesions depends on the thickness of the skin in the area, the chronicity of the lesion and the underlying pathological process which in turn is influenced by host response. Thus lesions which have been present for several months may be scaly and very similar to psoriasis in appearance. Lesions in the moist area of the genitals, referred to as condylomata lata, may be mistaken for genital warts. Other mucosal lesions associated with secondary syphilis include mucous patches which are more commonly found in the oral cavity than on the genitalia. These lesions are very superficial with a dull red base or a greyish slough.

Lymphadenopathy is present in 49–78% of cases of secondary syphilis and the characteristics of the nodes resemble those of the enlarged lymph nodes in primary syphilis, namely firm, rubbery, moderately enlarged and non-tender. Splenomegaly is present in 1%.[76]

Other less common manifestations include hepatitis, hepatomegaly, nephrotic syndrome, periostitis, cranial nerve palsies and peripheral neuropathy.

Investigations

The investigations of choice are dark-ground examination of exudate from primary or secondary lesions and standard serology (VDRL and TPHA). If primary syphilis is suspected, a FTA-abs should be requested as this is the first test to become positive and usually does so a week after the onset of the primary chancre, but in the majority of cases (70%) the standard serology will be positive. In 14.5% of primary syphilis and 0.9% of secondary syphilis, serology will be negative at presentation and will remain so in just over half of the primary cases during follow-up.[77]

Infection with syphilis should always be confirmed by repeating serology.

Management

Antimicrobial therapy
Penicillin remains the treatment of choice but there is some debate about the ideal regimen.

Adequate levels in the CSF
Abnormalities in the CSF may be found in primary and secondary syphilis in the absence of any clinical features and consequently the ideal regimen should achieve adequate penicillin levels in the CSF. Dunlop et al. demonstrated that treponemicidal levels could be achieved using an intramuscular regimen in conjunction with probenecid[78] but other studies have not confirmed this.[79]

Treatment failures with two injections of IM benzathine penicillin 2.4 million units a week apart have occurred in immunocompetent patients.[80] Although treatment failures have not been reported with previous regimes used in the UK (daily IM procaine penicillin 600 000 units for 10 days), such regimes may not produce adequate CSF levels.[81]

Achieving compliance
An attraction of intramuscular regimes is that it is clear whether or not the patient has completed the recommended treatment. The disadvantage is that daily attendance may be required, which may be impractical for both the patient and the clinic. Patients may find it

Table 9.14 Standard treatment regimes for primary and secondary syphilis

Standard therapies	Procaine penicillin G 600 000 IU IM daily for 10 days Benzathine penicillin 2.4 MU IM weekly for 2 weeks
Alternative therapies	Ampicillin 500 mg qds orally with probenecid 500 mg qds for 15 days Tetracycline 500 mg qds orally for 15 days Doxycycline 100 mg bd orally for 15 days Erythromycin 500 mg qds orally for 15 days

difficult to remember oral antibiotics which have to be taken four times a day.

Choosing the regimen
Most clinics have a standard policy for the treatment of primary, secondary and latent syphilis and these are outlined in Table 9.14. Alternative regimes which may be needed to suit the patient's circumstances or in the case of allergy to penicillin are also listed.

In the immunosuppressed patient
There are two aspects to be considered.

Interpretation of serology
Serological responses may be altered in HIV-infected individuals; in particular, they may be suppressed and patients coinfected with syphilis may have low or negative VDRL titres in the presence of secondary syphilis.[82] Repeated dark-ground examinations may be necessary to confirm the diagnosis. Treatment should be considered in the presence of suspicious clinical features alone, because of the consequences of missing syphilitic infection in HIV-infected individuals.

Choice of treatment regimen
A regimen that achieves adequate levels in the CSF is recommended in HIV-infected individuals, because of their increased risk of rapid progression to neurosyphilis. The only regimens that are known to do this are intravenously administered.

In pregnancy
Is antenatal screening justified?
With the apparent decline in syphilis, controversy arose about the need to continue to screen antenatal patients. The reemergence of congenital syphilis in the US in the 1980s[83] removed any sense of complacency. The cost-effectiveness of screening has been established in Norway, with a maternal prevalence rate of 0.02%.[84]

It is disappointing that antenatal screening does not prevent congenital syphilis and reasons for this include omission of syphilis serology during antenatal screen and failure of contact tracing.[85]

Risk to the neonate
Congenital syphilis can result from untreated syphilis in pregnancy, regardless of stage, but is much commoner in early disease (41%) than late disease (2%).[86] Stillbirths and neonatal death are also more common in untreated early disease.

Treatment in pregnancy
Because of the consequences of syphilis in pregnancy, overtreatment is preferable to undertreatment. All contacts of syphilis who are pregnant should be treated regardless of serology. Women with a past history of syphilis need only be treated in the absence of documented treatment or if there is any suggestion of reinfection. Although positive serology may be attributed to yaws, it is advisable to treat for syphilis, as both infections may coexist.

Recommended regimes are similar in pregnant women, except for the tetracyclines. Treatment failure has been reported in women treated in the last trimester of pregnancy[87] and in women who delivered within two weeks of starting treatment.[88]

Treating women who are allergic to penicillin poses a problem as treatment failure with erythromycin in pregnancy is unacceptably high.[89] However, this remains the most popular alternative regime, but close follow-up of the neonate is essential.

Partner notification

It is essential that all partners of patients with symptomatic early syphilis are contacted and screened, because of the infectious nature of the disease and the effectiveness of treatment. Explanation of the possible consequences will persuade the majority of patients of the importance of contact tracing, which in difficult circumstances can be facilitated by the health advisor contacting the partner directly. The whereabouts of all partners may not be known, but repeated attempts should be made to contact partners by telephone, letter and through the clinic network. It may be necessary to visit an address given, but this is much less commonly carried out now.

Partners that are contacted should be screened for features of primary and secondary syphilis, including serology. They should be informed about the symptoms and signs of early syphilis and the routes of transmission. All contacts should be asked to come back in a week for their results. Default from follow-up is common and this is why it is essential that all contacts see a health advisor at first visit. Contacts may need to attend again to ensure that all tests are negative three months after the last possible occasion that they could have caught syphilis and they must be advised to use barrier methods in the meantime. Partners in whom transmission is suspected should be seen earlier.

Follow-up

Follow-up depends on confirmation of the diagnosis, the nature of the clinical lesions, the effectiveness of contact tracing and serological results.

Early follow-up
Follow-up on consecutive days or at least once in the ensuing week is recommended whilst confirming the diagnosis (serology may take 2–3 days) and establishing the contacts. Thereafter, patients should be seen once a week if they have continuing clinical lesions or if difficulties are experienced in partner notification.

Long-term follow-up
Standard practice is to follow the patient's serological response (falling to negative VDRL titre) to ensure that treatment failure has not occurred. However, in the Middlesex cohort, 15% of patients with primary and 29% of patients with secondary disease had positive VDRL at 13–18 months, without any other evidence to support treatment failure.[77] Similar results were reported at six months follow-up in a cohort of patients in the armed forces in India, but at 30 months follow-up only 7% of primary and 8% of secondary cases still had positive VDRL.[90] Clearly, it is useful to document a negative VDRL, if possible, because this acts as a baseline for subsequent infections but once this is established there seems little point in bringing patients back to repeat serology.

CHANCROID

The causative organism is *Haemophilus ducreyi* and this was first identified in 1889 by Ducrey. The organism is a Gram-negative facultative anaerobe requiring haemin for growth.

Objectives

To help doctors:

- recognize when to consider chancroid in the differential diagnosis;
- be aware that early advice will be needed from a microbiologist;
- prevent further transmission through appropriate contact tracing.

Epidemiology

Chancroid is the commonest cause of genital ulceration in some parts of the world[114] and is endemic in Africa, Asia and the seaports of Western Europe. Cases of chancroid have increased in the United States in conjunction with several outbreaks, the first of which was reported from California in 1981,[91] and endemic foci are now established in parts of Florida, California, Texas, Boston and New York City.[92] Moreover, the Centre for Disease Control in Atlanta surveyed 115 STD clinics in all states and their findings suggest that there was considerable under reporting of chancroid. In part, this may reflect the lack of laboratory facilities, as only 16 of 115 clinics had access to culture media for *Haemophilus ducreyi*.

Chancroid is still a rare disease in the United Kingdom with 30–50 cases a year notified through the clinic reporting system, the majority due to imported infection.

The infections in the US outbreaks occurred predominantly in black and Hispanic heterosexuals, with a high male-to-female ratio (3:1 to 25:1). Contact with prostitutes was also an associated factor in most studies.[5]

Transmission

Chancroid is thought to be at least as infectious as primary syphilis, but there are few recent studies on transmission. The incubation period is rarely less than three or more than 10 days.[6]

Clinical presentation

The lesion begins as a tender papule which evolves into an ulcer within 48 h, sometimes passing through a pustular phase. The ulcer is solitary in about half of men but multiple ulcers are more common in women and they may merge to form serpiginous ulcers. They tend to be deep with a non-indurated base and have a sharply demarcated but irregular edge. Painful ulceration is more common in men and this may be related to location since the ulcers are found on the foreskin. Ulcers in women are frequently subclinical and located at the introitus. The ulcers may bleed on swabbing. A purulent urethritis may be present in men.[93]

The regional lymphadenopathy, which is present in 50%, is usually unilateral and painful. There is frequently erythema of the overlying skin. The lymph nodes may develop into abscesses, the so-called 'bubo formation'.

Without treatment, ulcers may persist for a considerable time (>30 days) and have been reported to persist for years.

Investigations

Even experienced clinicians can only expect to make an accurate diagnosis in half the cases,[5] although this can be improved in settings where chancroid is more common.[114] It is therefore important to have a high index of suspicion and to consult a microbiologist at the outset, as laboratory facilities for the diagnosis of chancroid are not routinely available.

Microscopy lacks sensitivity and specificity even in areas where the disease is endemic. There remains some controversy over the optimal culture medium and some investigators recommend the standard use of two culture media to achieve a sensitivity of 90%.[94]

Research to further improve the diagnosis of chancroid has been directed at techniques such as monoclonal antibodies and DNA probes.

Management

Antimicrobials

There is no established method for assessing antimicrobial susceptibility, which varies geographically. Resistance has been described with sulphonamides, ampicillin, tetracycline and trimethoprim through plasmids[6] and sulfonamides additionally through chromosomal mechanisms.[95]

H. ducreyi has remained sensitive to erythromycin and third-generation cephalosporins. Erythromycin 500 mg qds for seven days is probably the regime of choice, the only disadvantage being that a concurrent infection with syphilis would be masked and only partially treated. Other acceptable regimes include ceftriaxone 250 mg IM stat. Ciprofloxacin in single-dose and three-day regimes has shown cure rates of 90–100% in Africa and Asia. Spectinomycin 2 g IM stat has shown a cure rate of 94%.[115]

Drainage of buboes

Buboes >5 cm diameter will require aspiration if they do not drain spontaneously. Warm moist compresses applied to the enlarged tender nodes help.

Partner notification

The patient should be advised not to have sex until complete resolution has taken place. All sexual partners who could have been the index case or subsequent susceptibles should be traced and treated whether symptomatic or not. Because of the short incubation period, this should be relatively easy to achieve although in practice, contacts are often untraceable or abroad.

Follow-up

It is essential that patients are carefully followed up to ensure cure and optimal partner notification. Positive culture after three days treatment is a good indicator of treatment failure.[96]

They should be seen three months following the last risk exposure, as it is possible that they have also been exposed to syphilis and HIV.

LYMPHOGRANULOMA VENEREUM

LGV is one of several diseases caused by *Chlamydia trachomatis*. The serotypes L1–3 are usually responsible and differ from the more familiar serotypes D–K that cause urethritis and cervicitis in that they are associated with more invasive disease.

Objectives

To help doctors:

- know when to include lymphogranuloma venereum in the differential diagnosis of genital ulceration;
- follow up suspected cases appropriately.

Epidemiology

LGV is endemic in East and West Africa, India, parts of South East Asia, South America and the Caribbean. In the UK, the annual number of cases rarely exceeds 20 and these are almost exclusively imported.

Acute LGV is reported more frequently in men than women (5:1 and greater)[97] but this is probably because women are often asymptomatic at this stage.

Transmission

The risk of transmission is unknown, as is the length of time that the index case is infectious. Transmission probably occurs through a breach in mucosa, following sexual contact with someone with urethritis, cervicitis, proctocolitis or a genital lesion. The material in the buboes is also infectious and rupture may pose a risk to health-care workers.[97]

The incubation period is usually 3–12 days, but may be longer.

Pathogenesis

The target cell is the macrophage as opposed to the squamocolumnar cell that the other strains prefer and consequently, systemic features are common.

Lymph nodes draining the affected area enlarge and necrotic abscesses develop. Lymphangitis follows with inflammation of the tissues surrounding the lymph nodes, including adjoining nodes. Sinus formation and fistulae are common. The granulomatous inflammation is associated with fibrosis and subsequent to this, ischaemia, ulceration, stricture of the lymphatics and elephantiasis.

Humoral and cell-mediated host responses occur. The antibody response is mainly directed against the outer membrane protein. Cell-mediated hypersensitivity to chlamydial antigens is probably responsible for the extensive tissue damage that can be seen in LGV.

Clinical manifestations
Primary lesion
The primary lesion is often asymptomatic, particularly in women, where the favoured sites are posterior vaginal wall, cervix and vulva. The commonest site in men is the coronal sulcus, but lesions may occur anywhere in the genital area, including within the urethra and on the scrotum. Non-specific urethritis may be the presenting complaint. The lesion itself may be a small papule, a superficial ulcer or a herpetiform lesion. In men, penile lymphangitis may be observed, with subsequent fistula formation.

Secondary stage (inguinal syndrome)
The lymphatic stage, which is the more frequent presentation of the disease, follows the primary lesion by 2–6 weeks (range 10 days to six months). Patients present with painful, swollen, inflamed lymph nodes that rupture in one-third. The favoured site in men is the inguinal region, because of the location of the primary lesion. If the primary site is in the rectum, the vagina, cervix or posterior urethra, then the deep iliac and perirectal nodes are involved. Enlargement of the inguinal nodes above and the femoral nodes below Poupart's ligament results in the 'groove sign', which is considered pathognomonic for this infection and occurs in about one-fifth of patients.[98]

LGV organisms have been recovered from the blood and CSF at this stage, from patients with and without systemic features.[97]

Tertiary stage (anogenitorectal syndrome)
In the third stage of the disease, symptoms may become chronic and disabling. The anogenitorectal syndrome is more frequently reported in women and symptoms and signs are initially those of proctitis and proctocolitis. The disorder progresses to the development of abscesses, rectal and anal fistulae. Chronic lymphangitis and ensuing lymphoedema can lead to gross enlargement of the genitals, with increasing ulceration, and sinus and fistula formation which result in scarring and further tissue breakdown.

Systemic complications may include cardiac and pulmonary infection, aseptic meningitis and chronic inflammation of the eye.

Investigations
The diagnosis depends mainly on serological tests like the complement fixation test, which is positive within two weeks of onset, although early intervention with antibiotics may prevent this. Antibodies may remain for years. Unfortunately response to treatment and new infections are not consistently associated with changes in titre and there is crossreaction between these antibodies and those of other chlamydia serotypes. In general, a titre of 1:64 or a fourfold rise supports the diagnosis.[5]

The definitive test is the isolation of the organisms in cell culture. The best material for this purpose is an aspirate from a lymph node, although studies have reported good results from cutaneous biopsy material.

Management
Medication
The recommended treatment is doxycyline 100 mg bd for 21 days. Alternative regimes include erythromycin or sulphisoxazole. There are no clear criteria for judging efficacy of treatment, but patients with acute LGV (primary and secondary) should notice improvement within two days of starting antibiotics.[97]

Surgical intervention
Aspiration of nodes or incision for drainage may be necessary. Late complications usually require surgical repair, but the patient should always receive antibiotics first.

Partner notification
The principles for contact tracing of LGV cases are similar to those for chancroid, but the period that needs to be covered may be considerably longer (up to six months) and will depend on the clinical stage in which the patient presents.

In an endemic area treatment of suspected contacts is doubly important as repeated infections may play a part in the development of tertiary disease.[99]

Follow-up
It is essential that patients are carefully followed up to assess cure and optimal partner notification.

They should be seen again three months following the last risk exposure, as it is possible that they have also been exposed to syphilis and HIV.

DONOVANOSIS (GRANULOMA INGUINALE)

The causative organism is the bacteria *Calymmatobacterium granulomatis*, a Gram-negative bacterium.

Objectives

To help doctors:

- know when to include donovanosis in the differential diagnosis;
- follow up suspected cases appropriately.

Epidemiology

This is a rare sexually transmitted infection in the UK and cases are invariably imported. Worldwide, numbers have appeared to be declining, possibly in association with the widespread use of antibiotics. A recent report from South Africa suggested that numbers of cases may be increasing, but this may have reflected increased case ascertainment as laboratory parameters were used in addition to clinical criteria.[100]

Areas where donovanosis remains prevalent include South America, particularly Southern and Eastern parts, the Caribbean, South Africa, Zambia, Zimbabwe, Eastern Transvaal, South East India, Northern Territories of Australia (aboriginal populations), New Guinea, Vietnam and Japan.[101]

Transmission

The length of the incubation period is unclear but is probably 3–80 days.[101,102] Donovanosis is probably less contagious than the other sexually transmitted organisms that cause genital ulceration.

Clinical manifestations

The primary lesions can be single or multiple and are usually sited on the distal penis in men and the fourchette in women.[103] The lesions start as papules and then these erode to form painless ulcers with clean bases and sharp borders. As the ulcers enlarge, friable granulation tissue appears, giving the lesions a characteristic velvety appearance (traditionally described as raw beef). Lesions are frequently elevated above the skin, with a smooth rolled edge.

Inguinal lymphadenopathy does not usually occur but subcutaneous involvement of this area following inflammation around the lymph nodes can take place with subsequent ulceration or abscess formation.[101]

Investigations

This is by identifying the bacteria within the histiocyte in the crush preparation of granulation tissue stained with Wright or Giemsa stain. Routinely processed skin biopsies are also diagnostic in up to 95% of cases.[104]

Management

Medication

The recommended treatment is tetracycline although treatment failures have been reported.[101] Alternative regimes include erythromycin minocycline, doxycycline, co-trimoxazole, norfloxacin and chloramphenicol. Treatment must be continued until healing has occurred, which will depend in part on the size of the original lesion. Relapses may occur and may require a change in therapy.

Partner notification

The patient should be advised not to have sex until complete resolution has taken place. All sexual partners who could have been the index case or subsequent susceptibles should be traced and the period that needs to be covered is up to 80 days from the appearance of the primary lesion. Donovanosis is probably not as contagious as other sexually transmitted agents that cause genital ulceration, but it is important to identify those at risk for two reasons: first, because early treatment prevents progression of disease to the sclerotic stage when complications such as urethral occlusion may occur, and second, because donovanosis may facilitate HIV infection.[105]

Follow-up

Patients should be followed up on a weekly basis as treatment may need to be extended to achieve a cure. Relapses have been reported up to 18 months and so follow-up at one year and 18 months is advisable and patients should be advised to attend early if any new lesions appear.

They should be seen again three months following the last risk exposure, as it is possible that they have also been exposed to syphilis and HIV.

REFERENCES

1. Holmes, K.K. (1990) *Sexually Transmitted Diseases*, 2nd edn, McGraw-Hill, New York, p. 711.
2. Corey, L. (1994) The current trend in genital herpes. *Sexually Transmitted Diseases*, **21** (suppl 2), S38–S44.

3. Barton, S.E., Hollingworth, A., Maddox, P.H. et al. (1989) Possible cofactors in the etiology of cervical intraepithelial neoplasia. An immunopathologic study. *J. Reprod. Med.*, **34**(9), 613–16.
4. Johnson, A.M. et al. (1994) *Sexual Attitudes and Lifestyles*. Blackwell Scientific Publishers.
5. Committee on Sexually Transmitted Diseases of the American Academy of Dermatology (1991) Sexually transmitted diseases: bacterial infections. *J. Am. Acad. Derm.*, **25** (2 part 1), 287–299.
6. Holmes, K.K. (1990) *Sexually Transmitted Diseases*, 2nd edn, Mcgraw-Hill, New York p. 267.
7. Lee, F.K., Coleman, R.M., Pereira, L. et al. (1985) Detection of herpes simplex virus type 2 specific antibody with glycoprotein GL. *Clin. Microbiol.*, **22** (4), 641–643.
8. KC60 Report to Public Health Laboratory Service, CDSC, Colindale. Published in *Commun. Dis. Rep.* annually
9. Koutsky, L.A., Stevens, C.E., Holmes, K.K. et al. (1992) Underdiagnosis of genital herpes by current clinical and viral isolation procedures. *N. Engl. J. Med.*, **326**, 1533–1539.
10. Sullivan-Bolyai, J. et al. (1983) Neonatal herpes simplex virus infection in King County, Washington. Increasing incidence and epidemiological correlates. *J. Am. Med. Assoc.*, **250**, 3059–62.
11. Stone, K.M. et al. (1989) National surveillance for neonatal herpes simplex virus infections. *Sexually Trans. Dis.*, **16** (3), 152–156.
12. Ades, A.E. et al. (1989) Prevalence of antibodies to herpes simplex virus types 1 and 2 in pregnant women, and estimated rates of infection. *J. Epidemiol. and Commun. Health*, **43**, 53–60.
13. Ross, J.D., Smith, I.W. and Elton, R.A. (1993) The epidemiology of herpes simplex types 1 and 2 infection of the genital tract in Edinburgh, 1978–1991. *Genitourinary Med.*, **69**, 381–383.
14. Scoular, A., Leask, B.G. and Carrington, D. (1990) Changing trends in genital herpes due to Herpes simplex virus type 1 in Glasgow, 1985–88. *Genitourinary Med.*, **66**, 226.
15. Woolley, P.D. and Kudesia, G. (1990) Incidence of herpes simplex virus type-1 and type-2 from patients with primary (first attack) genital herpes in Sheffield. *Int. J. STD and AIDS*, **1** (3), 184–186
16. Benedetti, J., Corey, L. and Ashley, R. (1994) Recurrence rates in genital herpes after symptomatic first-episode infection. *Ann. Intern. Med.*, **124**, 847–854.
17. Lafferty, W.E., Coombs, R.W., Benedetti, J. et al. (1987) Recurrences after oral and genital herpes simplex virus infection. *N. Engl. J. Med.*, **316** (23), 1444–1449.
18. Koelle, D.M., Benedetti, J., Langenberg, A. and Corey, L. (1992) Asymptomatic reactivation of herpes simplex virus in women after first episode of genital herpes. *Ann. Intern. Med.*, **116** (6), 433–437.
19. Holmes, K.K. (1990) *Sexually Transmitted Diseases*, 2nd edn, McGraw-Hill, New York.
20. Siegel, D., Golden, E., Washington, E. et al. (1992) Prevalence and correlates of herpes simplex infections. *J. Am. Med. Assoc.*, **268** (13), 1702–1708.
21. Kinghorn, G.R. (1994) Epidemiology of genital herpes. *J. Int. Med. Res.*, **22** (suppl 1), 14A–23A.
22. Johnson, R.E., Nahmias, A.J., Magder, L.S. et al. (1989) A seroepidemiologic survey of the prevalence of herpes simplex virus type 2 infection in the United States. *N. Engl. J. Med.*, **321** (1), 7–12.
23. Cowan, F.M., Johnson, A.M., Ashley, R. et al. (1994) Antibody to herpes simplex virus type 2 as serological marker of sexual lifestyle in populations. *Br. Med. J.*, **309**, 1325–1329.
24. Mertz, G.J., Benedetti, J., Ashley, R. et al. (1992) Risk factors for the sexual transmission of genital herpes. *Ann. Intern. Med.*, **116** (3), 197–202.
25. Schmidt, O.W., Fife, K.H. and Corey, L. (1984) Reinfection is an uncommon occurrence in patients with symptomatic genital herpes. *J. Inf. Dis.*, **149** (4), 645–646.
26. Thin, R.N. (1991) Does first episode genital herpes have an incubation period? A clinical study. *Int. J. STD and AIDS*, **2**, 285–286.
27. Nahmias, A.J. (1970) Disseminated herpes simplex virus infections. *N. Engl. J. Med.*, **282** (12), 684–685.
28. Sullender, W.M., Yasukawa, L.L., Shwartz, M. et al. (1998) Type specific antibodies to herpes simplex virus type 2 (HSV-2) glycoprotein G in pregnant women, infants exposed to maternal HSV-2 at delivery, and infants with neonatal herpes. *J. Inf. Dis.*, **157** (1), 164–171.
29. Wald, A., Benedetti, J., Davis, G. et al. (1994) A randomized double-blind comparative trial comparing high- and standard-dose oral acyclovir for first-episode genital herpes infections. *Antimicrob. Ag. and Chemother.*, **38** (2), 174–176.
30. Goodell, S.E., Quinn, T.C., Mertichian P.A.C. et al. (1983) Herpes simplex virus proctitis in homosexual men. *N. Engl. J. Med.*, **308** (15), 868–871.
31. Corey, L. and Spear, P.G. (1986) Infections with herpes simplex viruses. *N. Engl. J. Med.*, **324**, 686–691.

32. Sasadeusz, J.J. and Sacks, S.L. (1994) Herpes latency, meningitis, radiculomyelopathy and disseminated infection. *Genitourinary Med.*, **79** (6), 369–377.
33. Kost, R.G., Hill, E.L., Tigges, M. and Strauss, S.E. (1993) Brief report: recurrent acyclovir resistant genital herpes in an immunocompetent patient. *N. Engl. J. Med.*, **329**, 1777–1782.
34. Foley, F.D., Greenwald, K.A., Nash, G. *et al.* (1970) Herpesvirus infection in burned patients. *N. Engl. J. Med.*, **282**, 652.
35. Marks, G.L., Nolan, P.E., Erlich, K.S. and Ellis, M.N. (1989) Mucocutaneous dissemination of acyclovir resistant herpes simplex virus in a patient with AIDS. *Rev. Inf. Dis.*, **11**, 474–476.
36. Holmes, K.K. (1990) *Sexually Transmitted Diseases*, 2nd edn, McGraw-Hill, New York, p. 942.
37. Bryson, Y.J., Dillon, M., Lovett, M. *et al.* (1983) Treatment of first episodes of genital herpes simplex virus infections with oral acyclovir: a randomized double-blind controlled trial in normal subjects. *N. Engl. J. Med.*, **308**, 916–920.
38. Mertz, G.J., Critchlow, C.W., Benedetti, J. *et al.* (1984) Treatment of first episodes of genital herpes simplex virus infections with oral acyclovir: a randomized double-blind controlled trial in normal subjects. *N. Engl. J. Med.*, **252**, 1147–1151.
39. Reichman, R.C. *et al.* (1984) Treatment of recurrent genital herpes simplex infections with oral acyclovir: a controlled trial. *J. Am. Med. Assoc.*, **251**, 2103–7.
40. Wald, A., Benedetti, J., Davis, G. *et al.* (1994) A randomized double-blind comparative trial comparing high- and standard-dose acyclovir for first episode genital herpes infections. *Antimicrob. Agents and Chemother.*, **38** (2), 174–176.
41. Carney, O., Ross, E., Bunker, C. *et al.* (1994) A prospective study of the psychological impact on patients with a first episode of genital herpes. *Genitourinary Med.*, **79**, 40–45.
42. Brookes, J.L., Haywood, S. and Green, J. (1993) Adjustment to the psychological and social sequelae of recurrent genital herpes simplex infection. *Genitourinary Med.*, **69**, 384–387.
43. Carney, O., Ross, E., Ikos, G. and Mindel, A. (1993) The effect of suppressive oral acyclovir on the psychological morbidity associated with recurrent genital herpes. *Genitourinary Med.*, **69**, 457–459.
44. Bryson, Y., Dillon, M., Bernstein, D.I. *et al.* (1993) Risk of acquisition of genital herpes simplex virus type 2 in sex partners of persons with genital herpes: a prospective couple study. *J. Infect. Dis.*, **167**, 942–946.
45. Barton, S.E., Wright, L.K., Link, C.M. *et al.* (1986) Screening to detect asymptomatic shedding of herpes simplex virus (HSV) in women with recurrent genital HSV infection. *Genitourinary Med.*, **62**, 181–5.
46. Mertz, G.J., Coombs, R.W., Ashley, R. *et al.* (1988) Transmission of genital herpes in couples with one symptomtic and one asymptomatic partner: a prospective study. *J. Inf. Dis.*, **157** (6), 1169–1177.
47. Yamakawa, Y., Forslund, O., Chua, K.L. *et al.* (1994) Detection of the BC24 transforming fragment of the herpes simplex virus type 2 (HSV-2) in cervical carcinoma tissue by polymerase chain reaction. *APMIS*, **102** (6), 401–406.
48. Kulhanjian, J.A., Soroush, V., Au, D.S. *et al.* (1992) Identification of women at unsuspected risk of primary infection with herpes simplex virus type 2 during pregnancy. *N. Engl. J. Med.*, **326** (14), 916–920.
49. Arvin, A.M. (1991) Relationships between maternal immunity to herpes simplex virus and the risk of neonatal herpesvirus infection. *Rev. Inf. Dis.*, **13** (S11), S953–956.
50. Brown, Z.A., Benedetti, J., Ashley, R. *et al.* (1991) Neonatal herpes simplex virus infection in relation to asymptomatic maternal infection at the time of labour. *N. Engl. J. Med.*, **342** (18), 1247–1252.
51. Arvin, A.M., Hensleigh, P.A., Prober, C.G. *et al.* (1986) Failure of antepartum maternal culutures to predict the infant's risk of exposure to herpes simplex virus at delivery. *N. Engl. J. Med.*, **315**, 796–800.
52. Catalano, P.M., Merritt, R.N. and Mead, P.B. (1991) Incidence of genital herpes simplex virus at the time of delivery in women with known risk factors. *Am. J. Obstet. Gynecol.*, **164** (5)(i), 1303–1306.
53. Nahmias, A.J., Josey, W.E., Naiab, Z.M. *et al.* (1971) Perinatal risk associated with maternal genital herpes simplex virus infection. *Am. J. Obstet. Gynecol.*, **110**, 825–37.
54. Brown, Z.A., Vontver, L.A., Benedetti, J. *et al.* (1987) Effects on infants of a first episode of genital herpes during pregnancy. *N. Engl. J. Med.*, **317** (20), 1246–1251.
55. Corey, L. *et al.* (1988) Difference between herpes simplex virus type 1 and type 2 neonatal encephalitis in neurological outcome. *Lancet*, January 2/9, 1–4.
56. Whitley, R. *et al.* (1991) Predictors of morbidity and mortality in neonates with herpes simplex virus infections. The National Institute of Allergy

and Infectious Diseases Collaborative Antiviral Study Group. *N. Engl. J. Med.*, **324** (7), 450–454.
57. Gibbs, R.S. and Mead, P.B. (1992) Preventing neonatal herpes – current strategies. *N. Engl. J. Med.*, **326** (14), 956–947.
58. Hook III, E.W., Cannon, R.O., Nahmias, A.J. *et al.* (1991) Herpes simplex virus infection as a risk factor for human immunodeficiency virus infection in heterosexuals. *J. Inf. Dis.*, **165**, 251–255.
59. Skinner, G.R.B., Fink, C., Melling, J. *et al.* (1992) Report of twelve years experience in open study of Skinner herpes simplex vaccine towards prevention of herpes genitalis. *Med. Microbiol. Immunol.*, **180**, 305–320.
60. Skinner, G.R.B., Buchan, A., Davies, J. *et al.* (1991) A virus-particle vaccine prepared from bovine mammillitis virus against herpes genitalis. *Comp. Immunol. Microbiol. Infect. Dis.*, **14** (2), 163–168.
61. Mertz, G.J., Ashley, R., Burke, R.L. *et al.* (1989) Double-blind placebo-controlled trial of a herpes simplex virus type 2 glycoprotein vaccine in persons at high risk for genital herpes infection. *J. Infect. Dis.*, **161**, 653–660.
62. Strauss, S., Corey, L., Burke, R.L. *et al.* (1994) Placebo-controlled trial of vaccination with recombinant glycoprotein D of herpes simplex virus type 2 for immunotherapy of genital herpes. *Lancet*, **343**, 1460–1463.
63. Nelson, R.A. and Mayer, M.M. (1949) Immobilization of *Treponema pallidum in vitro* by antibody produced in syphilitic infection. *J. Exp. Med.*, **89**, 369.
64. Mindel, A., Tovey, S.J. and Williams, P. (1987) Primary and secondary syphilis, 20 years' experience. 1 Epidemiology. *Genitourinary Med.*, **63** (6), 361–364.
65. British Cooperative Clinical Group (1973) Homosexuality and venereal disease in the UK. *Br. J. Vener. Dis.*, **49**, 329–334.
66. PHLS/CDSC (1989) Sexually transmitted disease in Britain: 1985–6. *Genitourinary Med.*, **65** (2), 117–121.
67. CDC (1988) Syphilis and congenital syphilis. *Morbid. Mortal. Wkly Rep.*, **37** (32), 486–489.
68. Tang, A. and Barlow, D. (1989) Resurgence of heterosexually acquired early syphilis in London. *Lancet*, **ii**: 166–167.
69. Van den Hoek, J.A.R., van der Linden, M.M.D. and Coutinho, R.A. (1990) Increase in infectious syphilis amongst heterosexuals in Amsterdam: its relationship to prostitution and drug use. *Genitourinary Med.*, **66**, 31–32.
70. Magnuson, H.J. *et al.* (1948) The minimal infectious inoculum of *Spirochaeta pallida* (Nicholls strain) and a consideration of its rate of multiplication *in vivo*. *Am. J. Syph. Gon. Vener. Dis.*, **32**, 171.
71. Holmes, K.K. (1990) *Sexually Transmitted Diseases*, 2nd edn, Mcgraw-Hill, New York, p. 208.
72. Johns, D.R. *et al.* (1987) Alteration in the natural history of neurosyphilis by concurrent infection with the human immunodeficiency virus. *N. Engl. J. Med.*, **316**, 1569.
73. Holmes, K.K. (1990) *Sexually Transmitted Diseases*, 2nd edn, Mcgraw-Hill, New York, p. 222.
74. Fitzgerald, T.J. (1981) Accelerated lesion development in experimental syphilis. *Infect. Immun.*, **34**, 478.
75. Gjestland, T. (1955) The Oslo study of untreated syphilis: an epidemiologic investigation into the natural course of the syphilitic infections based upon a re-study of the Boeck-Bruusgard material. *Acta. Derm. Venereol.*, **35** (suppl 34), 1.
76. Mindel, A., Tovey, S.J., Timmins, D.J. and Williams, P. (1989) Primary and secondary syphilis, 20 years' experience. 2 Clinical features. *Genitourinary Med.*, **65**, 1–3.
77. Anderson, J., Mindel, A., Tovey, S.J. and Williams, P. (1989) Primary and secondary syphilis, 20 years' experience. 3 Diagnosis, treatment and follow-up. *Genitourinary Med.*, **65** (4), 239–243.
78. Dunlop, E.M.C., Al-Egaily, S.S. and Houang, E.T. (1981) Production of treponemicidal concentration of penicillin in CSF. *Br. Med. J.*, **283**, 646.
79. Van der Valk, P.G.M., Kraai, E.J., van Voorst Vader, P.C. *et al.* (1988) Penicillin concentrations in the CSF during repository treatment regimen for syphilis. *Genitourinary Med.*, **64**, 223–225.
80. Holmes, K.K. (1990) *Sexually Transmitted Diseases*, 2nd edn, Mcgraw-Hill, New York, p.243.
81. Dunlop, E.M.C. (1985) Survival of treponemes after treatment: comments, clinical conclusions and recommendations. *Genitourinary Med.*, **61**, 293.
82. Hicks, C.B., Benson, P.M., Lupton, G.F. and Tramont, E.C. (1987) Seronegative secondary syphilis in a patient infected with HIV with Kaposi's sarcoma. *Ann. Intern. Med.*, **107**, 492–495.
83. CDC (1988) Guidelines for the prevention and control of congenital syphilis. *Morbid. Mortal. Wkly Rep.*, **37** (S1), 1.
84. Stray-Pederson, B. (1983) Economic evaluation of maternal screening to prevent congenital syphilis. *Sexually Trans. Dis.*, **10**, 167.
85. Boot, J.M., Menke, H.E., Van Eijk, R.V.W. *et al.* (1988) Congenital syphilis in The Netherlands: cause and parental characteristics. *Genitourinary Med.*, **64**, 298–302.

86. Ingraham, N.R. (1951) The value of penicillin alone in the prevention and treatment of congenital syphilis. *Acta. Derm. Venereol.*, **31** (suppl 24), 60.
87. Mascola L., Pelosi, R. and Alexander, C.E. (1984) Inadequate treatment of syphilis in pregnancy. *Am. J. Obstet. Gynecol.*, **150**, 945–947.
88. Schofield, C.B.S. (1982) Sexually transmitted disease surveillance. *Br. Med. J.*, **284**, 825.
89. Holmes, K.K. (1990) *Sexually Transmitted Diseases*, 2nd edn, McGraw-Hill, New York, p. 782.
90. Talwar, S., Tutakne, M.A. and Tiwari, V.D. (1992) VDRL titres in early syphilis before and after treatment. *Genitourinary Med.*, **68**, 120–122.
91. Schulle, J.M., Martich, F.A. and Schmid, G.P. (1992) *Morbid Mortal. Wkly Rep.*, **41** (3), 57–61.
92. Schmid, G.P., Sanders, L.L., Blount, J.H. *et al.* (1987) Chancroid in the United States: reestablishment of an old disease. *J. Am. Med. Assoc.*, **258**, 3265–3268.
93. Kunimoto, D.Y. *et al.* (1988) Urethral infection with *Haemophilus ducreyi* in men. *Sexually Trans. Dis.*, **15**, 37.
94. Sottnek, F.O., Biddle, S.J., Kraus, S.J. *et al.* (1980) Isolation and identification of *H. ducreyi* in culture in a clinical study. *J. Clin. Microbiol.*, **12**, 170–174.
95. Anderson, B., Albritton, W.L., Biddle, J. and Johnson, S.R. (1984) Common B-lactamase specifying plasmid in *H. ducreyi* and *N. gonorrhoeae*. *Antimicrob. Agents and Chemother.*, **25**, 296–297.
96. Naamara, W., Kunimoto, D.Y., D'Costa, L.J. *et al.* (1988) Treating chancroid with enoxacin. *Genitourinary Med.*, **64** (3), 189–192.
97. Holmes, K.K. (1990) *Sexually Transmitted Diseases*, 2nd edn, McGraw-Hill, New York.
98. Schachter, J. and Osoba, A.O. (1983) Lymphogranuloma venereum, *Br. Med. Bull.*, **39**, 151–154.
99. Quinn, T.C. *et al.* (1986) Experimental proctitis due to rectal infection with *Chlamydia trachomatis* in nonhuman primates. *J. Infect. Dis.*, **154**, 833.
100. O'Farrell, N. (1992) Trends in reported cases of donovanosis in Durban, South Africa. *Genitourinary Med.*, **68**, 366–369.
101. Richens, J. (1991) The diagnosis and treatment of donovanosis (granuloma inguinale). *Genitourinary Med.*, **67**, 441–452.
102. Holmes, K.K. (1990) *Sexually Transmitted Diseases*, 2nd edn, Mcgraw-Hill, New York.
103. O'Farrell, N. (1993) Clinico-epidemiological study of donovanosis in Durban, South Africa. *Genitourinary Med.*, **69**, 108–111.
104. Sehgal, V.N., Shyamprasad, A.I. and Beohart, P.C. (1984) The histopathological diagnosis of donovanosis. *Br. J. Vener. Dis.*, **60**, 45–47.
105. O'Farrell, N. (1995) Global eradication of donovanosis: an opportunity for limiting the spread of HIV-1 infection. *Genitourinary Med.*, **71**: 27–31.
106. Viravan, C., Dance, D.A., Ariyarit, C. *et al.* (1996) A prospective clinical and bacteriologic study of inguinal buboes in Thai men. *Clin. Infect. Dis.*, **2** (2), 233–239.
107. Patel, R. and Barton, S.B. (1995) Antiviral chemotherapy in genital herpes simplex infections. *Int. J. STD and AIDS*, **6**, 320–328.
108. Barton, S.B., Munday, P.E. and Patel, R. on behalf of the Herpes Simplex Virus Advisory Panel (1996) *Int. J. STD and AIDS*, **7**, 229–232.
109. Sacks, S.L., Acki, F.Y., Diaz-Mitoma, E., Sellors, J., Shafran, S.D. for the Canadian Famiciclovir Study Group (1996). Patient initiated, twice-daily oral famiciclovir for early recurrent genital herpes. *J. Am. Med. Assoc.*, **276** (1), 44–49.
110. Spruance, S.L., Tyring, S.K., DeGregorio, B., Miller, C., Beutner, K. and the Valaciclovir HSV Study Group (1996) A large-scale placebo-controlled, dose-ranging trial of peroral valaciclovir for episodic treatment of recurrent herpes genitalis. *Arch. Intern. Med.*, **156**, 1729–1735.
111. Wald, A., Zeh, J., Barum, G., Davis, L.G. and Corey, L. (1996) Suppression of subclinical shedding of herpes simplex virus type 2 with acyclovir. *Ann. Intern. Med.*, **124** (1), 8–15.
112. Scott, L.L. Sanchez, P.J., Jackson, G.L., Zeray, F. and Wendel, G.D. (1996) Acylovir suppression to prevent caesarean delivery after first episode genital herpes. *Obstet. Gynecol.*, **87** (1), 69–73.
113. Dickerson, M.C., Johnston, J., Delea, T.E., White, A. and Andrews, E. (1996) The causal role for genital ulcer disease as a risk factor for transmission of human immunodeficiency virus: an application of the Bradford Hill criteria. *Sexually Trans. Dis.*,, **23** (5), 429–440.
114. Ndinya-Achola, J.O., Kihara, A.N., Fisher, L.D. *et al.* (1996) Presumptive specific clinical diagnosis of genital ulcer disease (GUD) in a primary health care setting in Nairobi. *Int. J. STD and AIDS*, **7**, 201–205.
115. Behets, E.M.T., Liomba, G., Lule, G. *et al.* (1995) Sexually transmitted diseases and human immunodeficiency virus control in Malawi: a field study of genital ulcer disease. *J. Infect. Dis.*, **171**, 451–455.

10 The diagnosis and management of urethral discharge in males

P. J. Horner and R. J. Coker

INTRODUCTION

In men a urethral discharge is generally believed to be characteristic of urethritis. However, physiological urethral discharge can occur in men, whilst urethritis can be present without an observable discharge. The diagnosis of urethritis is confirmed by demonstrating an excess of polymorphonuclear leucocytes (PMNLs) in the anterior urethra. This is usually undertaken using a urethral smear, but a first-pass urine specimen (FPU) can also be used. Urethritis is described as either gonococcal, when *Neisseria gonorrhoeae* is detected, or non-gonococcal (NGU) when it is not. Men with urethritis may have symptoms of dysuria and/or irritation at the tip of the penis. Some perceive no symptoms at all. There is some evidence that the aetiology of asymptomatic urethritis, without an observable discharge, differs from that of symptomatic urethritis.

Chlamydia trachomatis is the commonest cause of NGU, accounting for 30–50% of cases. *Ureaplasma urealyticum* probably causes 10–20% of cases, although the evidence is conflicting, and more recently, *Mycoplasma genitalium* has been found in men with NGU. Other aetiologies are identified less frequently, including *N. meningitidis*, herpes simplex virus, *Candida* spp., bacterial urinary tract infection, urethral stricture and foreign bodies. Between 20% and 30% of men with NGU have no organism detected.

EPIDEMIOLOGY

In the UK the incidence of gonococcal urethritis decreased markedly throughout the 1980s and in 1995, 6665 cases were diagnosed in England. The incidence of NGU, on the other hand, showed a sustained rise throughout the late 1970s and early 1980s, declined rapidly from 1986 to 1988 and more slowly until 1993, with a slight increase since then. In 1995 over 60 000 cases of NGU were diagnosed.[1]

The peak age group for both gonococcal and non-gonococcal urethritis is 20–24 years old, although patients with NGU tend to be older. No race is immune from urethritis. Studies from the USA indicate that NGU is more common among Caucasians than Afro-Caribbeans when compared to gonococcal urethritis. Similar results have been demonstrated in the United Kingdom.[2] Patients with NGU tend to be from a higher socioeconomic group, have fewer lifetime sexual partners and are older at the age of first sexual intercourse than patients with gonococcal urethritis. A previous history of gonorrhoea is more frequent among men with gonorrhoea and a past history of NGU is more common among men with NGU.

AETIOLOGY

Gonorrhoea

In the late 19th century *N. gonorrhoeae* was isolated and its typical Gram-stain appearance described. The association of *N. gonorrhoeae* with urethritis was demonstrated, allowing gonococcal urethritis to be distinguished from NGU. Bumm, in 1885, established the causal role of *N. gonorrhoeae* in urethritis by inoculating the organism into the human male urethra.

The majority of patients with *N. gonorrhoeae* infection of the urethra have either symptoms or signs of urethritis, but a proportion are asymptomatic and have no objective urethritis detectable.[2–6] Recently a population-based study from Tanzania found that the majority of men infected with *N. gonorrhoeae* had no symptoms or observable discharge on examination.[7] This suggests that outside a clinic setting, the proportion of gonorrhoea which is asymptomatic may be greater than that observed in an STD clinic.[7] Alternatively, these patients may be in a presymptomatic stage of gonorrhoea,[5] although this is likely to account for only a small proportion, given the short incubation period of *N. gonorrhoeae*[7]. Initial studies suggested that this asymptomatic carriage was associated with the arginine, hypoxanthine and uracil requiring auxotrophs, though this has recently been questioned.[4,6]

Non-gonococcal urethritis

The majority of studies have examined patients with symptomatic urethritis and compared them to asymptomatic controls without urethritis. It is often assumed that the aetiology of urethritis is independent of the presence of symptoms or signs and many studies have not distinguished between symptomatic and asymptomatic urethritis. Indeed, the terms symptomatic and asymptomatic are confusing in the context of urethritis as a significant minority of patients have an observable discharge in the absence of symptoms. We term these patients the 'genitally unaware'. A better term is 'clinically symptomatic', i.e. the patient has either symptoms of urethritis and/or an observable urethral discharge on examination. Swartz *et al.*[8] and Reitmeijer *et al.*[9] found that *C. trachomatis* was detected less frequently in patients with asymptomatic urethritis. More recently, Horner *et al.* had similar observations with *M. genitalium* and *U. urealyticum* in addition to *C. trachomatis*.[10,11] However, it may be that discharge is the primary distinguishing clinical feature as Janier *et al.*,[12] who studied consecutive symptomatic men attending an STD clinic, observed that all major pathogens were more common in men with a urethral discharge compared to other symptoms. Urethral discharge on examination was sought after gentle urethral stripping, i.e. urethral massage directed from the base of the penis to the meatus. Why the prevalence of the major pathogens in NGU should differ according to the presence of symptoms and/or signs is not known.

One interpretation is that the aetiology of the two conditions may differ, another is that men with lower grade infections have fewer symptoms and the organisms are more difficult to detect.

Chlamydia trachomatis

The aetiological role of *C. trachomatis* in NGU is now beyond doubt. In case control studies, it is detected significantly more often from men with NGU than controls and this cannot be accounted for by confounding variables. Treatment studies using antibiotics which are active against *C. trachomatis* but not *U. urealyticum* have shown that chlamydia-positive men respond significantly better than chlamydia-negative men.[13,14] In addition, significantly more chlamydia-positive men with NGU respond to tetracycline therapy than to treatment with placebo.[15,16] Although serological studies rarely demonstrate a fourfold rise in antibody titre during an episode of urethritis, positive serological tests for antichlamydial antibody are associated with NGU and in particular chlamydia-positive urethritis.[14,17] In non-humans, intraurethral inoculation does not cause a frank urethral exudate, but is associated with a polymorphonuclear leucocyte response and persistence of the microorganism for up to three months.[18]

Although *C. trachomatis* infection of the urethra causes objective urethritis, approximately 7% of men with chlamydial infection of the urethra examined in a GU clinic were asymptomatic and had no observable discharge.[9] The reason for this is unclear. It may be that the men were examined whilst incubating the infection and if they had been examined later, urethritis would have been detected. The authors found no evidence to support this hypothesis and concluded that these patients are probably asymptomatic. There is currently no evidence to suggest that this is related to phenotypic characteristics of *C. trachomatis*. It may be that host factors determine the degree of inflammation elicited by *C. trachomatis*. The incidence of asymptomatic carriage of *C. trachomatis* in the community has not been assessed in the United Kingdom, but undoubtedly a greater proportion is asymptomatic compared to *C. trachomatis* infection in a GU clinic.[19] In the USA asymptomatic chlamydial carriage was demonstrated in 11% of naval recruits.[20] In a population-based study in Tanzania, the majority of men infected with *C. trachomatis* were clinically asymptomatic.[7]

U. urealyticum

The evidence supporting *U. urealyticum* as a cause of NGU is conflicting. The most compelling evidence for its role in urethritis comes from a study in which two human volunteers inoculated themselves with an isolate that had been obtained from a patient with urethritis and subsequently passaged in vitro. Both men developed urethritis, which persisted for six months in one of them.[21] Treatment studies using antibiotics which are less active against ureaplasmas than chlamydia found that men who were ureaplasma positive failed to respond to treatment significantly more often than those who were.[13,14,22] In addition, Brown et al. demonstrated a serological response to U. urealyticum using an enzyme-linked immunosorbent assay in 12 (67%) of 18 men with NGU who were ureaplasma-positive.[23] Case control studies have demonstrated both significant and non-significant associations of U. urealyticum with NGU.[14,24,25] Why this should be so is unknown. It may reflect a failure to match the study and control groups for potential confounding variables. One confounding variable is the association of U. urealyticum carriage with an increasing number of lifetime sexual partners.[26] However, recent work from Japan suggests another explanation. Kawamura[27] inoculated himself four times with U. urealyticum and with each subsequent inoculation the inflammatory response was of a shorter and less severe degree. Thus, patients who are sexually active and who become infected with U. urealyticum may initially develop urethritis which subsequently resolves spontaneously. The failure to show an association of U. urealyticum with urethritis may be because of the high carriage rate in sexually active men attending departments of GU medicine who have become tolerant to this infection. In addition, there is some evidence that the number of organisms detected is important and therefore quantitative, rather than qualitative, studies of ureaplasma detection are required. Thus, Bowie et al. showed that $\geq$ 1000 organisms/ml of urine was associated with NGU.[14] There are more than 14 serovars of U. urealyticum and some may be more pathogenic than others. In one study serovar 4 was associated with NGU.[28] This finding has not been confirmed consistently in other studies.

In conclusion, U. urealyticum probably causes urethritis, but we do not fully understand why up to 60% of men carry the organism without having symptoms or signs of NGU.[14,24]

Mycoplasma genitalium

M. genitalium was isolated more than a decade ago in culture from the urethra of two of 13 men with NGU. Culture remains difficult and reliable detection has only become available with the advent of the polymerase chain reaction (PCR). Two case control studies have shown a significant association of M. genitalium with NGU. This was independent of the presence of C. trachomatis and could not be explained by differences in age, ethnicity and sexual behaviour between the study and control groups.[25,29] Significant associations of M. genitalium with NGU have now been demonstrated in five countries.[27] In addition, M. genitalium caused urethritis when inoculated intraurethrally into non-human primates.[30,31] A serological study, using a microimmunofluorescence technique, demonstrated antibodies to M. genitalium in 30% of men with NGU.[32] More recently, antibody to M. genitalium, measured by the ELISA technique, has been found significantly more often in men who are mycoplasma positive than those who are not.[27] Although there is significant serological crossreactivity between M. genitalium and M. pneumoniae, this is unlikely to account for the findings, as M. pneumoniae is not associated with NGU.

Some studies have not found evidence that M. genitalium is a cause of urethritis. One study using an ELISA failed to detect a serological response in men with NGU in whom M. genitalium was detected.[25] This may have been a reflection of the antigen preparation and needs to be confirmed. Hooten et al. did not detect M. genitalium significantly more often in men with urethritis than controls using a DNA probe.[33] This may have been because the DNA probe is less sensitive than PCR. In addition, as stated earlier, the aetiology of asymptomatic and symptomatic NGU probably differs. They did not differentiate between the two.[10]

Further studies are needed before it can be firmly established that M. genitalium is a cause of NGU.

Trichomonas vaginalis

This organism is seldom identified in men attending genitourinary medicine clinics as optimal diagnostic facilities are rarely available. It has been associated with NGU. A large study, using culture to isolate T. vaginalis, found it to be significantly associated with NGU and this association persisted after adjustment for potential confounding variables.[34] It is likely that the relative importance of T. vaginalis as a cause of NGU depends on the prevalence of the infection within the community. It is an uncommon cause in the United Kingdom but has been detected in up to 19% of men with urethral discharge in South Africa.[35]

Other causes of NGU

More recently, bacterial vaginosis in women has been associated with NGU in men.[27] This reported relationship requires further investigation. Generally, other identifiable causes of NGU each account for <1% of cases. These are detailed in Table 10.1. It is important to remember that a bacterial urinary tract infection can cause urethritis, as can prostatitis, urethral stricture, instrumentation of the urethra and Stevens–Johnson syndrome. There is no proof that urethritis can be induced by masturbation or dietary factors, such as caffeine. Frequent self-examination, often referred to

Table 10.1 Aetiology of non-gonococcal urethritis

Causes of NGU	Proportion of cases (%)
Major causes	
Chlamydia trachomatis	30–50
Ureaplasma urealyticum	?10–30
Mycoplasma genitalium	?20
None of the aforementioned	~30
Minor causes	
Viruses:	
Herpes simplex	<1
Adenovirus	?<1
Yeasts:	
Candida albicans	<1
Protozoa:	
Trichomonas vaginalis	<1
Bacteria:	
Neisseria meningitidis	<1
Haemophilus sp.	<1
Urinary tract infection	<1
Bacterial vaginosis	???
Other:	<1
Reiter's syndrome	
Instrumentation of the urethra	
Foreign bodies	
Urethral stricture	
Renal stones	
Chemical irritation	
Stevens–Johnson syndrome	
Schistosomiasis	

as 'squeezing', is sometimes invoked as a cause of persistent urethritis but again, there is no objective evidence for this assumption.

NGU in homosexual men

There are only limited data on the aetiology of NGU in homosexual men, as few studies have addressed this group specifically. C. trachomatis is detected in only 20–30% of homosexual men with NGU,[36,37,37a] which is less than in heterosexual men. Insertive genital-oral intercourse is associated with chlamydia-negative, ureaplasma-negative urethritis in homosexual men.[37a] Hooton et al. detected M. genitalium more frequently in homosexual men with NGU than heterosexual men.[33] M. genitalium has been isolated from the respiratory tract and a possible preference for the intestinal tract needs to be considered.[37b] The aetiology of NGU in homosexual men requires further investigation, including the role of M. genitalium with particular reference to insertive genital-oral intercourse.

CLINICAL FEATURES

Although the symptoms and signs of gonococcal urethritis are usually more pronounced than those of NGU, the severity of urethritis cannot differentiate reliably between the two conditions.[38] A recent study by Sherrard et al. found that the classic symptoms of discharge in association with dysuria for gonorrhoea may be occurring less frequently than previously.[5] This was only present in 48% of men with gonorrhoea studied. The incubation period of gonorrhoea is usually 2–6 days but up to 60 days has been reported.[5,39] Men with a previous history of gonorrhoea are more likely to present earlier. There is evidence that the incubation period is increasing.[5,39] Whether this observation reflects an evolutionary adaptation of the organism or is just a reflection of differences in previous exposure to N. gonorrhoeae is unknown.[40] The incubation period of NGU is between one and five weeks and can be longer. The onset of symptoms with gonorrhoea is usually more abrupt than that of NGU. Both can cause urethral discharge, dysuria, penile tip irritation and occasionally frequency. As stated previously, both can cause no symptoms or signs. The discharge in patients with gonorrhoea tends to be profuse and purulent or mucopurulent, whereas the discharge in men with NGU tends to be mucopurulent or mucoid and may only be noticeable on waking in the morning.

As there is mounting evidence that the detection of pathogens is less likely in the absence of symptoms or signs in men (clinically asymptomatic) with urethritis, there is a strong argument to examine for urethral discharge carefully. If none is visible the urethra should be gently massaged to express any discharge and the meatal orifice inspected. Whilst this is usually a relatively straightforward procedure it may be problematic when gender or sexuality is an issue. The presence of meatal inflammation, shown by reddening, is a relatively non-specific finding. Balanoposthitis can be a cause of a positive urethral smear. It is often very difficult to tell whether the balanoposthitis is causing the urethritis or vice versa.

COMPLICATIONS

Complications such as epididymoorchitis, sexually acquired reactive arthritis, Reiter's disease and disseminated gonococcal infection are infrequent and occur in fewer than 3% of cases.

DIAGNOSIS OF URETHRITIS

Urethritis

Urethritis must be confirmed by demonstrating PMNLs in the anterior urethra. The use of either a urethral smear or a first-pass urine specimen is a valid method of sampling. The former appears to be more sensitive although both tests will identify cases missed by the other test.[11] The quality of the smear is heavily dependent on how the smear is taken. Either a plastic loop or cotton-tipped swab can be used. The plastic loop has, in our experience, the advantage of being less traumatic to the patient. Urethritis is present if there are >=5 PMNL per high-power (×1000) microscopic fields (averaged over five fields with greatest concentration of PMNLs).[8] Gonorrhoea is suggested by the presence of Gram-negative diplococci within the PMNLs. The reported sensitivity of microscopy for diagnosing gonorrhoea is between 90–99% compared to culture and is operator dependent. In any case, all patients should have a urethral culture for *N. gonorrhoeae* to confirm or exclude the diagnosis of gonorrhoea and to determine the antibiotic sensitivity of the isolate. Ideally, *C. trachomatis* should also be sought either from a swab or from a first-pass urine sample. A FPU sample is as sensitive as a urethral swab (see below) and avoids an uncomfortable second urethral swab.

The patient is asked to collect two specimens of urine, a first pass and a midstream. Urethritis is characterized by finding threads in the FPU-specimen and none in the midstream specimen.[41] If threads are found in both, the most likely reason is that the patient has divided the FPU specimen in two, although a urinary tract infection must be considered. In the presence of a negative smear and threads in the FPU specimen, one should be removed and Gram stained. The presence of PMNLs in the threads is indicative of urethritis. However, few studies have investigated this and the exact number of PMNLs has not been defined. We use >=10 PMNLs/high power field.[42] Previous studies using a urinary sediment defined urethritis as >=15 PMNLs per × = 400 objective in two or more of five random fields which is equivalent to >=5 PMNLs/HPF of a Gram-stained urethral smear. These are not strictly comparable as threads can float and are not always deposited in the urinary sediment during centrifugation and not all patients with pyuria will have urinary threads. Haze or cloudiness in both specimens is usually indicative of phosphates in the urine and can be cleared using acetic acid. However, it can be indicative of a urinary tract infection. The midstream urine specimen should be assessed by urinalysis. We believe a dipstick which contains leucocyte esterase and nitrites in addition to blood, protein and glucose should be used. The sensitivity of these dipsticks for detecting a bacterial urinary tract infection is good, but they have not been assessed specifically in a genitourinary medicine clinic.[43] If the dipstick is positive, the MSU specimen should be sent for microscopy and culture. The presence of blood in the MSU is usually a result of trauma whilst taking a urethral smear, but the test should be repeated without urethral swabbing if positive.

Those patients who report symptoms but have no objective evidence of urethritis should be asked to reattend for an early morning urethral smear, having held their urine overnight. If this test and the microbiology are negative, the patient should be reassured. Occasionally the patient will remain concerned about his symptoms. The patient may have abacterial prostatitis, prostadynia[44,45] or alternatively there may be a psychosexual cause underlying the patient's anxiety.

Isolation of *N. gonorrhoeae*

N. gonorrhoeae is usually cultured on standard selective media. It is a fastidious organism, highly susceptible to environmental influences, and requires CO_2 for its initial growth. The most sensitive means of culturing *N. gonorrhoeae* is to inoculate the specimen directly from the patient onto the culture plate. Alternatively, transport medium can be used, such as Amie medium or modified Stuart medium. If the specimen is inoculated onto the culture plate within 24 h, there is only a 5–10% loss in sensitivity. However, thereafter its sensitivity decreases markedly.

Detection of *C. trachomatis*

C. trachomatis infection was identified initially by culture. This is slow and labour intensive and has now been superseded by non-culture techniques. Direct fluorescent antibody (DFA) staining is rapid although still labour intensive and requires a skilled observer to achieve sensitivities at the top end of its reported range, 70–100%.[46] Centrifugation of specimens before staining is also associated with an increased sensitivity. Enzyme immunoassays (EIA) are the most commonly used method to detect *C. trachomatis*. A wide range of sensitivities and specificities has been reported in the literature and caution is urged in interpreting these data as the sensitivities and specificities may be artificially high. Often the test is compared against a poor reference test and many studies have failed to confirm that all negative tests were 'true negatives' and only confirmed the 'positives'.[46] As EIAs vary significantly in both sensitivity and specificity it is important to use assays with the best performance[46,47] and to ensure that all positives are confirmed. The importance of confirming samples testing positive by EIA was stressed by a report from Norway, of three women who filed complaints because of allegedly false-positive results.[48]

More recently, PCR and the ligase chain reaction (LCR) have been used to develop specific automated tests for routine laboratory use. However, these are currently considerably more expensive than EIAs. The LCR and PCR have been demonstrated to have a high sensitivity and specificity in FPU specimens.[49,50] The identification of additional positives by LCR which cannot be confirmed by either culture or DFA are assumed to be true positives, although this is contentious.[51] This may cause problems in clinical practice.

C. trachomatis can be detected in both a urethral swab and a FPU specimen in men with chlamydial urethritis.[46,52,53] Detection of C. trachomatis in the FPU is no less sensitive than that of a urethral swab[46,52] and, indeed, may be more sensitive.[54] The LCR and PCR tests utilize FPU specimens in men. The IDEIA immunoassay performs reliably on urine.[54,55] However, not all studies have found such reliable results with EIA on urine.[56] This may be a reflection of the relative overall performance of the respective immunoassays. Recently Deguchi et al. demonstrated that the sensitivity of the LCR was similar to IDEIA (EIA) and greater than chlamydiazyme (EIA) in FPU specimens from men with NGU.[57]

Detection of U. urealyticum

Because this organism may be detected in many men without urethritis, isolation is rarely undertaken routinely in men. The diagnostic yields of cultures of U. urealyticum from urethral swabs or FPU specimens are similar.

Detection of M. genitalium

M. genitalium grows poorly in culture and this is not a reliable means of detecting it. Both the PCR test and DNA probe have been used to detect it in research studies. PCR is the more sensitive technique.

THERAPY

The gonococcus is very sensitive to a wide range of antibiotics. However, it has the ability to acquire antibiotic resistance. In the early 1970s the first penicillinase-producing N. gonorrhoeae (PPNG) was isolated and was soon found worldwide. This was totally resistant to penicillin and was mediated by a plasmid containing the β-lactamase gene. High-level chromosomally mediated resistance to penicillin can also result in treatment failure with some isolates. Tetracycline-resistant gonococcal strains, which were plasmid mediated, were first detected in 1985 in the USA and have now spread worldwide. Initially in developing countries and subsequently in the United Kingdom, gonococcal strains with decreased susceptibility to the quinolones have been detected. Recently a clinical isolate resistant to penicillin, the quinolones and tetracycline has been identified in the United Kingdom.[58]

Thus, all isolates need to be assessed for antibiotic resistance and the initial treatment regimen depends on local sensitivity patterns. In addition to the antimicrobial activity of drugs, other factors are important, including ease of administration, compliance, cost, toxicity and the probability that patients with acute gonococcal infection are coinfected or have recently been exposed to other sexually transmitted agents. Single-dose therapy with medications effective for eradication of N. gonorrhoeae followed by therapy expected to eradicate C. trachomatis infections are recommended.[59] In areas where antimicrobial resistance of N. gonorrhoeae is not a problem, single-dose therapy with ampicillin (2.0–3.5 g) or amoxycillin (2.0–3.5 g) orally plus probenecid (1.0 g) is suitable. Where resistance is a problem there is a number of choices. The fluoroquinolones, in particular ciprofloxacin, are widely used in the UK. They are effective as single-dose treatments, but vigilance is needed because isolates with reduced susceptibility are emerging, particularly in South East Asia. Some clinics have adopted a higher dose of ciprofloxacin – 500 mg orally, rather than 250 mg – because of rare reports of treatment failures with the lower dose. Spectinomycin (2.0 g) IM or a third-generation cephalosporin (e.g. ceftriaxone 250 mg IM) can be used as alternatives. Chromosomally mediated resistance to spectinomycin and cephalosporins has been described.

In the USA concomitant therapy (see below) for C. trachomatis is recommended for all patients with gonorrhoea. This also reduces the incidence of postgonococcal urethritis. Many clinics in the UK, including our own, follow this policy. Azithromycin as a single dose is effective against both N. gonorrhoeae and C. trachomatis.[59] It is, however, relatively expensive and it does not currently have a product licence for the treatment of gonorrhoea in the UK.

The optimal drug, dosage and duration of treatment for NGU are still uncertain, largely because the aetiology of all cases of NGU is still not clear. Early studies did not use cultures for both C. trachomatis and U. urealyticum. Compliance is often uncertain and relapse cannot be distinguished from reinfection. Because tetracyclines were originally shown to be the most effective treatment for NGU and later demonstrated to be capable of eradicating C. trachomatis and are also active against most strains of U. urealyticum, they have become the mainstay of treatment of NGU.[60]

Current recommendation for initial treatment of NGU is one week of tetracycline (500 mg qds orally) or doxycycline (100 mg bd orally). An alternative is seven days of erythromycin stearate (500 mg qds orally[61]). Many other 'tetracycline' regimes have been evaluated, including minocycline and oxytetracycline, and these appear to have similar success rates.[36,42,62–64] More recently, the quinolone ofloxacin has also been shown to be effective in the treatment of NGU if given for one week.[65,66] Ciprofloxacin and norfloxacin are not sufficiently effective.[66] Recently, interest has focused on azithromycin, an azalide antibiotic that achieves high intracellular concentrations. This may be beneficial for eradicating *C. trachomatis*, an obligate intracellular pathogen. It has a high bioavailability and a tissue half-life of between two and four days which allows sustained antimicrobial activity at sites of infection after single-dose therapy. A single 1 g oral dose has been shown to be as effective as a standard one-week regimen of twice-daily doxycycline.[67,68] Single-dose therapy is attractive as it overcomes problems of poor compliance and is now the preferred mode of treatment for gonorrhoea. However, azithromycin is more expensive than doxycycline. A recent study by Carlin *et al.*[69] suggested that this may be offset by savings elsewhere, although this needs further study. In addition they found that patients preferred single-agent therapy using azithromycin.

TREATMENT OF SEXUAL PARTNERS

Patients with urethritis should be encouraged to persuade all at-risk sexual contacts to attend for investigation and treatment and to abstain from sexual intercourse until given the 'all-clear' at follow-up. Those patients with gonorrhoea or chlamydial urethritis should ideally be interviewed by the health adviser to establish details of contacts. Notification of the contacts can then be undertaken by the health adviser if the patient chooses not to inform them directly. Contacts should be interviewed, examined and appropriate tests taken. Epidemiological treatment is treatment given to named contacts of patients after a history of exposure to disease without or in advance of confirmatory pathological findings. Epidemiological treatment may be given when the clinician considers that the risk to the patient of unnecessary treatment is outweighed by the risk of complications of the infection or the probability of transmission of the infection to other contacts.[70] The diagnostic tests for *N. gonorrhoeae* and *C. trachomatis* are not perfect and there is no test for NGU in women. There is therefore a strong argument for treating all contacts of *C. trachomatis* and *N. gonorrhoeae* epidemiologically.[70] It is accepted practice to treat all contacts of NGU epidemiologically. This needs to be handled sensitively and the confidentiality of the index patient maintained.

FOLLOW-UP

Tests for *N. gonorrhoeae* should be repeated one and two weeks after treatment of gonorrhoea. Men with NGU should be reviewed after completion of initial therapy when results of laboratory investigations should be known.

Postgonococcal urethritis (PGU)

Postgonococcal urethritis occurs after single-dose treatment of urethral gonorrhoea in 30–70% of patients. A course of tetracycline in addition to single therapy for gonorrhoea reduces the incidence of PGU.[63]

It is assumed that the aetiology of PGU is similar to that of NGU and that those patients who developed PGU had a dual infection. This is supported by the strong association between coinfection with *C. trachomatis* and the subsequent development of PGU.[63] However, the cause of chlamydia-negative PGU remains unknown. *U. urealyticum* has not been associated with PGU. Recently *M. genitalium* was detected in 4.4% of men with gonococcal urethritis, suggesting that it may play a role in PGU.[71] There are few data on the natural history of this condition and it is not possible to comment on the risk of persistent or recurrent urethritis. It is unusual for patients with PGU to be symptomatic. It is possible, therefore, that the aetiology of chlamydia-negative PGU differs from that of chlamydia-negative NGU in which symptoms or signs are present.

Persistent and recurrent NGU

This is defined as the persistence or recurrence of urethritis despite appropriate treatment of acute NGU. It is one of the most difficult clinical problems in genitourinary medicine and occurs in 30–60% of men treated for acute NGU.[29,36,42,64,72,73] Chronic NGU is a complex condition, the aetiology of which is probably multifactorial.[74] It does not appear to cause any long-term chronic inflammatory damage in men and its aetiology(ies) may therefore differ from that of the chronic sequelae of pelvic inflammatory disease. There is no precise definition of chronic NGU, some workers relying solely on examination of the urethral smear, while others have used the FPU. Horner *et al.*[42] reported a prevalence of 69% using both methods to detect chronic NGU. This only partially explains the increased prevalence compared to other studies. Lomas et al. recently demonstrated, using an *in vitro* assay, that the PMNL stimulatory response persists many weeks after treatment of acute NGU.[75] This occurs in both chlamydia-positive and chlamydia-negative NGU, the stimulatory response in the former being less marked. Their findings were consistent with this and they speculated that the discrepancy between their findings and those of other studies was probably a reflection of the sensitivities of the tests used for assessing urethral inflammation during follow-up.[42]

Understanding of this condition has been hampered by a number of factors, not least that we still do not know the aetiology of a significant proportion of men with acute NGU. Prospective studies of men with acute NGU are difficult as many default from follow-up, preventing reliable analysis of the findings. In general, no distinction has been made between symptomatic or asymptomatic and persistent or recurrent urethritis in prospective studies.

EPIDEMIOLOGY AND DEMOGRAPHY

For a more detailed discussion, readers should refer to the excellent review by Munday,[74] which I have summarized briefly below. There appears to be no association between age, ethnic origin and the development of persistent or recurrent NGU. Men with a previous history of NGU do not appear to be at increased risk of treatment failure unless they have previously had persistent/recurrent urethritis or >5 episodes of acute NGU. The duration of symptoms does not appear to influence the response to treatment.

Microbiological causes and antibiotic treatment of NGU

Some workers have found chlamydia-negative men with acute NGU to have a worse outcome than chlamydia-positive men, but this has not been shown by others and, indeed, some studies have found the opposite.[15,73,74] The reason for the variance in these observations is unclear. Horner et al.[42] observed that chronic NGU at 10–29 days after treatment was associated with chlamydial urethritis, but not at 30–92 days or 10–92 days. They speculated that it may either be due to selection bias or a reflection of the sensitivity of the tests used to detect persistent urethritis. The persistent inflammatory response following chlamydial urethritis is less marked than that following non-chlamydial urethritis.[75]

The choice of antibiotic, whether a tetracycline or erythromycin, also does not appear to influence the response to treatment and neither does the duration of therapy.[74]

Treatment of sexual partner

No randomized prospective studies have evaluated the effect of treating the sexual partners of men with NGU. This would be considered unethical now that the pathogenic role of *C. trachomatis* in women has been established. Evans, in an uncontrolled study, found no apparent benefit from empirical treatment of contacts.[76] Bowie et al. found that the response to treatment was not influenced by whether the patient was sexually active or not during treatment. They did report that NGU recurred more often in men who did not use condoms with either a new or untreated sex partner than in those who had no reexposure, were reexposed only to a treated partner or used a condom.[64] There are reports of patients with persistent or recurrent urethritis being cured only after their sexual partner received appropriate treatment.[77] It seems likely, therefore, that concurrent treatment of the sexual partner may result in an improved response to treatment in some, but not all, patients with acute NGU.

PATHOGENESIS

A number of theories have been put forward to account for persistent and recurrent urethritis. No single theory can explain the pathogenesis(es) of this condition and the aetiology is probably multifactorial.

Inadequate treatment

Persistence of C. trachomatis

Tests to detect *C. trachomatis* have been available widely for well over a decade. These include culture, direct fluorescent antibody (DFA) staining and enzyme immunoassays. It is uncommon for *C. trachomatis* to be found in the urethra of treated patients unless reinfection has occurred. It has been proposed that *C. trachomatis* may remain in protected sites such as the prostate, but many studies have failed to demonstrate it in prostatic tissue.[45] However, the sensitivity of these techniques to detect infection after treatment has been questioned.[78]

In vitro studies have shown that *C. trachomatis* can exist in a latent state as abnormal reticulate bodies, with no production of the infectious elementary bodies. This state can be induced by the addition of penicillin γ-interferon to the culture media. As there are no elementary bodies, culture would be insensitive. The antigen detection tests detect either the major outer membrane protein or chlamydial lipopolysaccharide. The production of these two antigens appears to be down-regulated in the abnormal reticulate bodies.[78] Thus, the sensitivities of the conventional *C. trachomatis* tests (culture, DFA and EIA) to detect such persistent infection are likely to be low. The possibility remains, therefore, that chlamydiae may be persisting, as abnormal reticulate bodies, in patients with persistent and recurrent NGU. However, Hay *et al.*, Vogels *et al.*, and, more recently, Horner *et al.* using the polymerase chain reaction (PCR) only rarely detected *C. trachomatis* in chlamydia-positive patients treated with doxycycline.[42,73,79] These studies indicate that persistent *C. trachomatis* infection is an uncommon cause of chronic NGU.

Ureaplasma urealyticum
In vitro, about 10% of ureaplasma strains exhibit resistance to tetracyclines. The persistence of *U. urealyticum* has been associated with chronic NGU.[64,80] Although Bowie *et al.* detected it in only a minority of cases.[64] It may therefore account for some cases of persistent or recurrent urethritis.

Mycoplasma genitalium
M. genitalium has only recently been associated with NGU in humans and probably also causes chronic NGU. Hooton *et al.* found that *M. genitalium* was associated with persistent or recurrent urethritis.[33] Horner *et al.* detected persistent *M. genitalium* DNA by PCR in four out of 14 mycoplasma-positive patients after two weeks of treatment with doxycycline.[29] They extended this work further and between 10–92 days follow-up, eight patients were mycoplasma positive, all of whom had chronic NGU.[80] Little is known about antibiotic resistance as it is a fastidious organism. Whether the DNA came from viable organisms is uncertain, but there is a possibility that tetracycline resistance might also be present in *M. genitalium*.

Other organism(s)
This question can only be properly addressed when the infective aetiologies of acute NGU have been elucidated fully. Other organisms probably only have a minor role in chronic NGU.

Reinfection

Although reinfection with *C. trachomatis* and *U. urealyticum* does occur, studies indicate that this accounts for only a minority of cases of recurrent NGU.[42,73,77,81]

Immune mediated

The complexities of the immune response to infection and injury are only now being understood. Inflammation, which is immune mediated, may persist after the initial infective insult has disappeared. For example, in rheumatic fever, streptococcal and myocardial antigens contain homologous epitopes. Infection with group A streptococci in a susceptible host is believed to lead to an autoimmune response to epitopes in the organisms which are immunologically crossreactive with similar epitopes in human tissue.[82] Whether a similar mechanism accounts for the persistence or recurrence of urethritis in some cases is unknown. Horner *et al.* have recently provided some evidence to support this hypothesis.[42]

The chronic inflammation induced by *C. trachomatis* is believed to be primarily immunologically mediated and not a direct consequence of tissue destruction by the organism.[83] Morrison *et al.*[84] showed, in an animal model, that the chlamydial 60 kD heat-shock protein (hsp 60) was an important target for this type of delayed hypersensitivity reaction.

Heat-shock proteins, popularly referred to as molecular chaperones, have essential roles in the synthesis, transport and folding of proteins. They are among the most conserved proteins in phylogeny with respect to function and structure.[85] Many heat-shock proteins are essential for life. Their synthesis increases in response to a variety of insults and enhances survival under stressful conditions. The chlamydial hsp 60 can be isolated from the outer membrane of elementary bodies and it has been proposed that it participates in the assembly of the chlamydial cell wall.[86] This protein has been sequenced and has been shown to be a member of the hsp 60 family.[84]

Wagar *et al.*[87] and Toye *et al.*[88] showed that a serological response to the chlamydial hsp 60 was associated with chronic sequelae of pelvic inflammatory disease in patients with positive chlamydial serology. This suggests that the immune response to chlamydial hsp 60 may be important in the immunopathogenesis of persistent genital tract inflammation in humans. Whether this is the result of a breakdown in self-tolerance or a consequence of the persistence of chlamydial hsp 60 in abnormal reticulate bodies is unknown.[78] Horner *et al.*[42] recently demonstrated that chlamydial hsp 60 antibody was associated with the development of chronic urethritis between 10 and 92 days (both at 10–29 days and at 30–92 days). These results are consistent with the immune response to hsp 60 being important in the development of this chronic disease. This was not as a consequence of continued production of hsp 60 by *C. trachomatis*. The role of the immune response to chlamydial hsp 60 in persistent or recurrent NGU needs further investigation.

Other theories

Numerous other theories have been put forward to explain persistent and recurrent NGU. These include excessive alcohol intake, frequent self-examination and trauma. There is no evidence to support these theories although, in the authors' experience, some patients do indeed associate recurrence of NGU with the consumption of alcohol. However, these are a minority of cases.

CLINICAL FEATURES

The vast majority of patients with NGU respond to treatment with amelioration or disappearance of their symptoms and a reduction in urethral inflammation. At follow-up, after treatment for acute NGU, patients can be divided into four broad clinical categories:

- clinically asymptomatic (no symptoms or discharge) and no objective evidence of urethral inflammation;
- clinically symptomatic (symptoms and/or discharge) but no objective evidence of urethral inflammation;
- clinically asymptomatic but objective evidence of urethral inflammation;
- clinically symptomatic and objective evidence of urethritis.

In patients with clinically symptomatic recurrent urethritis, symptoms are usually milder or may have disappeared after initial treatment, only to recur after a short period.

COMPLICATIONS

These are both physical and psychological. All the research so far has only dealt with the former. The evidence to date shows that despite the persistence or recurrence of urethritis, this is not associated with any long-term inflammatory damage to the genital tract. A proportion of patients have evidence of chronic abacterial prostatitis.[81,89] Although structural abnormalities of the lower genital tract are often sought, this is rarely fruitful. Some patients may have evidence of bladder outflow obstruction. Whether this is a result of the persistent or recurrent urethritis or a premorbid condition is unknown. Krieger et al. showed that those who required endoscopic evaluation could be identified from physical examination and urinary flow studies.[89] There does not appear to be any adverse effect on fertility although this requires further investigation. The potentially devastating effect of chronic NGU on the sexual relationship needs to be evaluated.

TREATMENT AND MANAGEMENT OF FOLLOW-UP

Patients with postgonococcal urethritis should be managed in a similar way to patients with NGU (see below). If the patient is asymptomatic and the urethral smear is negative, he can probably be considered cured. Patients who continue to be clinically symptomatic with a positive urethral smear should be treated with a two-week course of erythromycin 500 mg qds (enteric coated is better tolerated) and consideration given to a simultaneous five-day course of metronidazole 400 mg bd. This will cover tetracycline-resistant *U. urealyticum* strains and patients with trichomonal urethritis.[34,64,81] Use of erythromycin in patients with persistent or recurrent urethritis has been shown to be effective in 50% of cases.[90] If at the second follow-up visit clinically symptomatic urethritis still persists, the patient should be asked to return 1–2 weeks later. Tests for *C. trachomatis*, *N. gonorrhoeae*, *T. vaginalis* and a bacterial urinary tract infection should then be undertaken. The presence of prostatitis should be evaluated using the Stamey technique and urinary flow rate assessed to exclude a urethral stricture.[44,45,81,89] More recently, prostatic ultrasound has been used to investigate for prostatitis.[44,45] Its use in the management of patients with persistent or recurrent urethritis needs to be evaluated. A decision then needs to be taken on whether further antibiotic therapy is indicated. Patients whose primary complaint is a urethral discharge may often respond to a further course of erythromycin 500 mg qds for four weeks. This condition may persist for many months or even years and often tends to remit and then relapse. Those patients with abacterial prostatitis or prostadynia have a variable response to treatment. Luzzi describes in detail the management of these conditions.[45] The nature of chronic NGU needs to be explained to the patient and this often requires in-depth counselling. It should be stressed that chronic NGU has not been associated with any significant long-term chronic inflammatory consequences and that its aetiology is not believed to be infective.

The management of these patients is complicated by a number of factors. Often they continue to be sexually active and their partner may or may not have been treated. This needs to be established for each patient at follow-up. If there is a history of sexual intercourse with an untreated partner, this needs to be addressed. Bowie et al. found some evidence to show that condoms do offer a degree of protection.[64] However, although condoms reduce infection by STDs, they do not completely prevent them and thus one cannot exclude potential reinfection. In addition, many patients at follow-up still have either persistent symptoms, albeit much improved, with a negative urethral smear or are clinically asymptomatic with a positive urethral smear. Those patients with persistent symptoms and a negative urethral smear should be reassured that the symptoms will settle. If they have not settled within one week, an early morning smear is indicated.

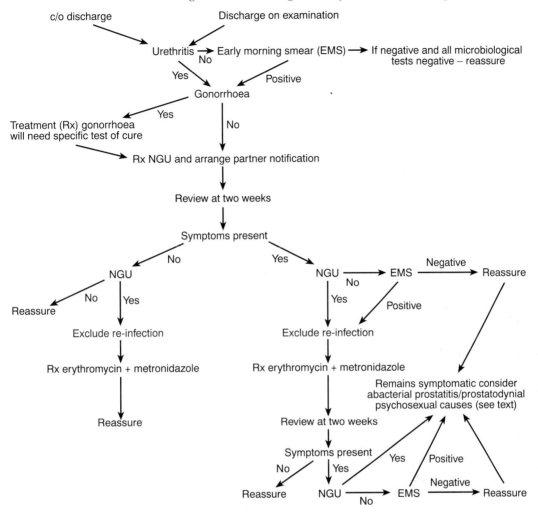

Fig. 10.1 Treatment (Rx) and management algorithm of urethral discharge in males.

The management of persistent symptoms in the presence of a negative urethral smear is complex and requires further evaluation. A diagnosis of prostadynia or abacterial prostatitis should be considered, as should a psychosexual aetiology. The management of clinically asymptomatic patients with a positive urethral smear is controversial. Some authors believe it does not require further treatment whilst others believe it should be managed the same as clinically symptomatic persistent urethritis. Until the aetiology of persistent or recurrent urethritis in all its forms is established, we believe that patients with clinically asymptomatic urethritis at the first follow-up visit should be treated with both erythromycin and metronidazole. We use erythromycin (EC) 500 mg bd for two weeks and metronidazole 400 mg bd for five days. If urethritis, which is clinically asymptomatic, still persists after a second course of treatment, those patients should be reassured.

Follow-up of the sexual partner

There are few data on the management of partners of patients with recurrent or persistent urethritis. The recommendation is that the partner should be reviewed and treatment given at presentation only. However, occasionally in our experience, retreating the partner with erythromycin has proved beneficial in some patients with recurrent urethritis. This requires further evaluation in well-designed prospective studies.

In our experience it is often more satisfactory to deal with the patient and the partner together so that both can understand the nature of the condition and its expected clinical course. An empathic approach early in the course of this disease may address and defuse potentially disruptive underlying anxieties of a sexual nature.

URETHRITIS AND HIV DISEASE

Worldwide, and particularly in Africa, human immunodeficiency virus type 1 (HIV-1) infection is largely a heterosexually transmitted disease and sexually transmitted diseases are important cofactors for enhancing HIV transmission.[91] Several studies have shown that non-ulcerative sexually transmitted diseases are risk factors for the acquisition of HIV.[91-93] Men who frequent prostitutes and those who have multiple sex partners are major groups at risk of acquisition of both HIV and other sexually transmitted diseases, including NGU.

HIV undergoes abortive infection in T-helper lymphocytes unless they are activated. STDs induce an inflammatory response, resulting in an increase in the number of activated lymphocytes in both men and women. Thus there is a potential for an increased infectious load in the genital tract of HIV-positive patients with an STD compared to those without an STD. This was elegantly shown by Atkins et al.[94] in men with urethritis. They demonstrated a significant fall in HIV load in the semen but not blood following successful treatment. In addition, STDs compromise the integrity of the epithelial or mucosal barrier. This classically occurs with genital ulcer disease but also with chlamydial infection where microulceration of the cervix has been demonstrated. There is, therefore, also an increased susceptibility to HIV infection for those who have an STD.

The first study to examine prospectively the possibility of reducing HIV-transmission through enhanced control of STDs was reported from rural Tanzania.[95] Twelve community clinics were randomized to be control centres or to have an STD intervention programme. A syndromic approach was used for initial management of patients. There was a statistically significant 42% reduction in the incidence of new HIV infections in the intervention areas compared to that in the control areas, yet there was no statistically significant reduction in symptomatic urethritis achieved by the intervention. This study was the subject of considerable correspondence.[96-98] A high prevalence of urethral infection was reported separately: gonorrhoea prevalence 2.2%; chlamydia 0.7%. Only 15% of men with those urethral pathogens were symptomatic and a further 19% had an observable discharge.[7] This was thought to be an important reservoir for infection in the community.

These studies provide support for an aggressive approach to treating STDs in HIV-infected patients, as a means of reducing transmission of HIV and for reinforcing the benefits of using condoms.

There are no data to suggest that the course of urethritis in men is altered by coexistent infection with HIV.

PSYCHOLOGICAL ASPECTS OF URETHRITIS

Appreciable levels of psychological disturbance are found in patients attending clinics for sexually transmitted diseases. Few studies have attempted to correlate psychiatric morbidity with subsequent diagnosis of urethritis or adaptive psychological behaviour. Men who experience chronic/recurrent symptoms at the lower genital tract which necessitate repeated examinations, investigations and treatment may be expected to develop various degrees of sexual dysfunction. A psychosexual assessment should be considered in all cases of chronic NGU. We recently looked at psychiatric morbidity in men with persistent NGU and found a similar prevalence of psychiatric morbidity (27%) to that found in other studies assessing rates of psychiatric morbidity in male GUM clinic attenders. Further studies are needed to determine whether the psychological morbidity is a reflection only of having urethritis or is a premorbid condition in some clinic attenders.

CONCLUSION

Urethritis is one of the most common problems encountered in genitourinary medicine. Despite considerable research efforts, much still remains unclear. There is evidence that symptoms and signs are important in determining the aetiology of urethritis. Clearly this is an important issue, especially from the patient's perspective, which requires further investigation. *C. trachomatis* is still the only pathogen conclusively shown to be causal. However, the evidence supporting *M. genitalium* as a cause is mounting and although the evidence is conflicting, the findings of Kawamura may explain why *U. urealyticum* is not always associated with urethritis (see [27]). The aetiology of persistent or chronic urethritis is still unclear but new insights have recently occurred. It is a complex condition, the aetiology of which is undoubtedly multifactorial. Investigation is hampered by variation in the criteria and methodology used to define it. This needs to be standardized. Horner *et al.* have provided evidence that some cases may be immunologically mediated and have implicated the immune response to hsp 60. They have also presented data implicating both *M. genitalium* and *U. urealyticum* as important causes, the full details of which are in preparation. These findings need to be confirmed and further investigation undertaken.

We have attempted to highlight aspects of urethritis which need further investigation and research. Research in medicine now has at its disposal a vast array of investigative techniques. What is required are well-designed studies targeting organic and psychological aspects which incorporate these new techniques to address specific aspects of the aetiology, pathogenesis and management of urethritis. Only then will our clinical practice marry the 'science' of medicine with the 'art'. We must avoid poorly designed studies which use insensitive microbiological techniques and which repeat previous work. These add nothing to our knowledge and often only serve to confuse the issues involved.

REFERENCES

1. Statistics and Research Division (1995/6) *New cases seen at NHS genitourinary Medicine Clinics in England. Summary Information from Form KC60.* Department of Health, London.
2. Daker-White, G. and Barlow, D. (1997) Heterosexual gonorrhoea at St Thomas'. I: patient characteristics and implications for targeted STD and HIV prevention strategies. *Int. J. STD and AIDS*, **8**, 32–35.
3. Handsfield, H.H., Lipman, T.O., Harnisch, J.P. *et al.* (1974) Asymptomatic gonorrhoea in men. Diagnosis, natural course, prevalence, and significance. *N. Engl. J. Med.*, **290**, 117–123.
4. Crawford, G., Knapp, J.S., Hale, J. and Holmes, K.K. (1977) Asymptomatic gonorrhoea in men: caused by gonococci, with unique nutritional requirements. *Science*, **196**, 1352–1353.
5. Sherrard, J. and Barlow, D. (1996) Gonorrhoea in men: clinical and diagnostic aspects. *Genitourinary Med.*, **72**, 422–426.
6. Horner, P.J., Coker, R.J., Turner, A. *et al.* (1992) Gonorrhoea: signs, symptoms and serovars. *Int. J. STD and AIDS*, **3**, 430–433.
7. Grosskurth, H., Mayaud, P., Mosha, F. *et al.* (1996) Asymptomatic gonorrhoea and chlamydial infection in rural Tanzanian men. *Br. Med. J.*, **312**, 277–280.
8. Swartz, S.L., Kraus, S.J., Herrmann, K.L. *et al.* (1978) Diagnosis and etiology of nongonococcal urethritis. *J. Infect. Dis.*, **138**, 445–454.
9. Rietmeijer, C.A.M., Judson, F.N., van Hensbroek, M.B. *et al.* (1991) Unsuspected *Chlamydia trachomatis* infection in heterosexual men attending a sexually transmitted disease clinic: evaluation of risk factors and screening methods. *J. Infect. Dis.*, **18**, 28–35.
10. Horner, P.J. and Taylor-Robinson, D. (1994) *Mycoplasma genitalium* and non-gonococcal urethritis. *Lancet*, **343**, 790–791.
11. Horner, P.J., Thomas, B.J., Gilroy, C. *et al.* (1996) Objective urethritis as a screening test for infection in men. Abstract, MSSVD Spring Meeting Edinburgh.
12. Janier, M., Lassau, F., Casin, I. *et al.* (1995) Male urethritis with and without discharge: a clinical and microbiological study. *Sexually Trans. Dis.*, **22**, 244–252.
13. Bowie, W.R., Alexander, E.R., Floyd, J.F. *et al.* (1976) Differential response to chlamydial and ureaplasma-associated urethritis to sulphafurazole (sulfisoxazole) and aminocyclitols. *Lancet*, **ii**, 1276–1278.
14. Bowie, W.R., Wang, S-P., Alexander, E.R. *et al.* (1977) Aetiology of non-gonococcal urethritis. Evidence for *Chlamydia trachomatis* and *Ureaplasma urealyticum*. *J. Clin. Invest.*, **59**, 735–742.
15. Handsfield, H.H., Alexander, E.R., Wang, S.P. *et al.* (1976) Differences in the therapeutic response of chlamydia-positive and chlamydia-negative forms of nongonococcal urethritis. *J. Am. Vener. Dis. Assoc.*, **2**, 5–9.
16. Prentice, M.J., Taylor-Robinson, D. and Csonka, G.W. (1976) Non-specific urethritis: a placebo controlled trial of minocycline in conjunction with laboratory investigations. *Br. J. Vener. Dis.*, **52**, 269–275.
17. Taylor-Robinson, D. and Thomas, B.J. (1980) The role of *Chlamydia trachomatis* in genital-tract and associated diseases. *J. Clin. Pathol.*, **33**, 205–233.
18. DiGiacomo, R.F., Gale, J.L., Wang, S-P. and Kiviat, M.D. (1975) Chlamydial infections of the male baboon urethra. *Br. J. Vener. Dis.*, **51**, 310–313.
19. Singh, G. and Blackwell, A. (1994) Morbidity in male partners of women who have chlamydial infection before termination of pregnancy. *Lancet*, **344**, 1438.
20. Podgore, J.K., Holmes, K.K. and Alexander, E.R. (1982) Asymptomatic urethral infections due to *Chlamydia trachomatis* in male U.S. military personnel. *J. Infect. Dis.*, **146**, 828.
21. Taylor-Robinson, D., Csonka, G.W. and Prentice,

M.J. (1977) Human intraurethral inoculation of ureaplasmas. *Q. J. Med.*, **46**, 309–326.
22. Coufalik, E.D., Taylor-Robinson, D. and Csonka, G.W. (1979) Treatment of nongonococcal urethritis with rifampicin as a means of defining the role of *Ureaplasma urealyticum.*, *Br. J. Vener. Dis.*, **55**, 36–43.
23. Brown, M.B., Cassell, G.H., Taylor-Robinson, D. and Shepard, M.C. (1983) Measurement of antibody to *Ureaplasma urealyticum* by an enzyme-linked immunosorbent assay and detection of antibody responses in patients with nongonococcal urethritis. *J. Clin. Microbiol.*, **17**, 288–295.
24. Holmes, K.K., Handsfield, H., Wang, S.P. *et al.* (1975) Etiology of non-gonococcal urethritis. *N. Engl. J. Med.*, **292**, 1199–1205.
25. Jensen, J.S., Orsum, R., Dohn, B. *et al.* (1993) *Mycoplasma genitalium*: a cause of male urethritis? *Genitourinary Med.*, **69**, 265–269.
26. McCormack, W.M., Lee, Y-H. and Zinner, S.H. (1973) Sexual experience and urethral colonisation with genital mycoplasmas. A study of normal men. *Ann. Intern. Med.*, **78**, 696–698.
27. Taylor-Robinson, D. (1996) The history of non-gonococcal urethritis. *Sexually Trans. Dis.*, **23**, 86–91.
28. Shepard, M.C. and Lunceford, C.D. (1978) Serological typing of *Ureaplasma urealyticum* isolates from urethritis patients by an agar growth inhibition method. *J. Clin. Microbiol.*, **8**, 566–574.
29. Horner, P.J., Gilroy, C.B., Thomas, B.J. *et al.* (1993) Association of *Mycoplasma genitalium* with acute non-gonococcal urethritis. *Lancet*, **342**, 582–585.
30. Tully, J.G., Taylor-Robinson, D., Rose, D.L. *et al.* (1986) Urogenital challenge of primate species with *Mycoplasma genitalium* and characteristics of infection induced in chimpanzees. *J. Infect. Dis.*, **153**, 1046–1054.
31. Taylor-Robinson, D., Furr, P.M., Tully, J.G. *et al.* (1987) Animal models of *Mycoplasma genitalium* urogenital infection. *Isr. J. Med. Sci.*, **23**, 561–564.
32. Taylor-Robinson, D., Furr, P.M. and Hanna, N.F. (1985) Microbiological and serological study of nongonococcal urethritis with special reference to *Mycoplasma genitalium*. *Genitourinary Med.*, **61**, 319–324.
33. Hooton, T.M., Roberts, M.C., Roberts, P.L. *et al.* (1988) Prevalence of *Mycoplasma genitalium* determined by DNA probe in men with urethritis. *Lancet*, **i**, 266–268.
34. Krieger, J.N., Jenny, C., Verdon, M. *et al.* (1993) Clinical manifestations of trichomoniasis in men. *Ann. Intern. Med.*, **118**, 844–849.
35. Pillay, D.G., Hoosen, A.A., Vezy, B. and Moodley, C. (1994) Diagnosis of *Trichomonas vaginalis* in male urethritis. *Trop. and Geogr. Med.*, **46**, 44–45.
36. Bowie, W.R., Yu, J.S., Fawcett, A. and Jones, H.D. (1980) Tetracycline in nongonococcal urethritis: comparison of 2 g and 1 g daily for seven days. *Br. J. Vener. Dis.*, **56**, 332–336.
37. Oriel, J.D., Reeve, P., Wright, J.T. and Owen, J. (1976) Chlamydial infection of the male urethra. *Br. J. Vener. Dis.*, **52**, 46–51.
37a. Hernandez-Aguado, I., Alvarez-Dardet, C., Gili, M. *et al.* (1988). Oral sex as a risk factor for chlamydia-negative ureaplasma-negative nongonococcal urethritis. *Sexually Trans. Dis.*, **15**, 100–102.
37b. Taylor-Robinson, D. (1995) The history and role of *Mycoplasma genitalium* in sexually transmitted diseases. *Genitourinary Med.*, **71**, 1–8.
38. Jacobs, N.F. and Kraus, S.J. (1975) Gonococcal and nongonococcal urethritis in men: clinical and laboratory differentiation. *Ann. Intern. Med.*, **82**, 7–12.
39. Sherrard, J. and Barlow, D. (1993) Gonorrhoea in men. *Lancet*, **341**, 245.
40. Horner, P.J., Coker, R.J., McBride, M.M. and Murphy, S.M. (1993) Changing gonococcal urethritis in men. *Lancet*, **341**, 761.
41. Munday, P.E., Altman, D.G. and Taylor-Robinson, D. (1981) Urinary abnormalities in non-gonococcal urethritis. *Br. J. Vener. Dis.*, **57**, 387–390.
42. Horner, P.J., Cain, D., McClure, M. *et al.* (1997) Association of antibodies to *Chlamydia trachomatis* heat-shock protein 60 kDa with chronic non-gonococcal urethritis. *Clin. Infect. Dis.*, **24**, 653–660.
43. Flanagan, P.G., Rooney, P.G., Davies, E.A. and Stout, R.W. (1989) Evaluation of four screening tests for bacteriuria in elderly people. *Lancet*, **i**, 1117–1119.
44. Thin, R.N. (1991) The diagnosis of prostatitis: a review. *Genitourinary Med.*, **67**, 279–283.
45. Luzzi, G (1996). The prostatitis syndromes. *Int. J. STD and AIDS*, **7**, 471–478.
46. Taylor-Robinson, D. and Thomas, B.J. (1991) Laboratory techniques for the diagnosis of chlamydial infections. *Genitourinary Med.*, **67**, 256–266.
47. Paul, I.D. and Caul, E.O. (1990) Evaluation of three *Chlamydia trachomatis* immunoassays with an unbiased, noninvasive clinical sample. *J. Clin. Microbiol.*, **28**, 220–222.
48. Ezazi, S. and Aavitsland, P. (1995) False positive tests of sexually transmitted *Chlamydia trachomatis* infections [Norwegian]. *Tidsskrift for Den Norske Laegeforening*, **115**, 3145–3147.

49. Chernesky, M.A., Lee, H., Schachter, J. et al. (1994) Diagnosis of Chlamydia trachomatis urethral infection in symptomatic and asymptomatic men by testing first-void urine in a ligase chain reaction assay. *J. Infect. Dis.*, **170**, 1308–1311.
50. De Barbeyrac, E., Rodriguez, P., Dutilh, B., Le Roux P. and Bebear, C. (1995) Detection of *C. trachomatis* by ligase chain reaction compared with polymerase chain reaction and cell culture in urogenital specimens. *Genitourinary Med.*, **71**, 382–384.
51. Hadgu, A. (1996) The discrepancy in discrepant analysis. *Lancet*, **348**, 592–593.
52. Hay, P.E., Thomas, B.J., Gilchrist, C. et al. (1991) The value of urine samples from men with non-gonococcal urethritis for the detection of *Chlamydia trachomatis*. *Genitourinary Med.*, **67**, 124–128.
53. Caul, E.O., Paul, I.D., Milne, J.D. and Crowley, T. (1988) Non-invasive sampling methods for detecting *Chlamydia trachomatis*. *Lancet*, **ii**, 246–247.
54. Crowley, T., Milne, D., Arumainayagam, J.T., Paul, I.D. and Caul, E.O. (1992) The laboratory diagnosis of male *Chlamydia trachomatis* infections – a time for change? *J. Infect.*, **25**, 69–75.
55. Hay, P.E., Thomas, B.J., McKenzie, P. and Taylor-Robinson, D. (1993) Detection of *Chlamydia trachomatis* in men. *Sexually Trans. Dis.* **20**, 1–4.
56. Matthews, R.S., Pandit, P.G., Bonigal, S.D., Wise, R. and Radcliffe, K.W. (1993) Evaluation of an enzyme-linked immunoassay and confirmatory test for the detection of *Chlamydia trachomatis* in male urine samples. *Genitourinary Med.*, **69**, 47–50.
57. Deguchi, T., Yasada, M., Uno, M. et al. (1996) Comparison among performances of a ligase chain reaction-based assay and two enzyme immunoassays in detecting *Chlamydia trachomatis* in urine specimens from men with nongonococcal urethritis. *J. Clin. Microb.*, **3**, 1708–1710.
58. Turner, A., Gough, K.R., Jephcott, A.E. and McLean, A.N. (1995) Importation into the UK of a strain of *Neisseria gonorrhoeae* resistant to penicillin, ciprofloxacin and tetracycline. *Genitourinary Med.*, **71**, 265–266.
59. Bignell, C. (1996) Antibiotic treatment of gonorrhoea – clinical evidence for choice. *Genitourinary Med.*, **72**, 315–320.
60. Oriel, J.D. (1996). The history of non-gonococcal urethritis. *Genitourinary Med.*, **72**, 374–379.
61. Centers for Disease Control and Prevention (1993) Sexually transmitted diseases treatment guidelines. *Morbid. Mortal. Wkly Rep.*, **4**, 47–49.
62. Grimble, A.S. and Amarasuriya, K.L. (1975) Non-specific urethritis and the teracyclines. *Br. J. Vener. Dis.*, **51**, 198–205.
63. McClean, K.A., Evans, B.A., Lim, J.M.H. and Azadin, B.S. (1990) Postgonococcal urethritis: a double-blind study of doxycycline vs placebo. *Genitourinary Med.*, **66**, 20–23.
64. Bowie, W.R., Alexander, E.R., Stimson, J.B. et al. (1981) Therapy for nongonococcal urethritis: double-blind randomized comparison of two doses and two durations of minocycline therapy for non-gonococcal urethritis. *Ann. Intern. Med.*, **95**, 306–311.
65. Ibsen, H.W., Møller, B.R., Halkier-Sørensen, L. and From, E. (1989) Treatment of nongonococcal urethritis: comparison of ofloxacin and erythromycin. *Sexually Trans. Dis.* **16**, 32–35.
66. Segretti, J. (1991) Fluoroquinolones for the treatment of nongonococcal urethritis/cervicitis. *Am. J. Med.*, **91**, (suppl 6a): 150S–152S.
67. Nilsen, A., Halsos, A., Johansen, A. et al. (1992) A double blind study of single dose azithromycin and doxycycline in the treatment of chlamydial urethritis in males. *Genitourinary Med.*, **68**, 325–327.
68. Stamm, W.E. (1991) Azithromycin in the treatment of uncomplicated genital chlamydial infections. *Am. J. Med.*, **91** (suppl 3a): 19s–22s.
69. Carlin, E.M. and Barton, S.E. (1996) Azithromycin as the first-line treatment of non-gonococcal urethritis (NGU): a study of follow-up rates, contact attendance and patients' treatment preference. *Int. J. STD and AIDS*, **7**, 185–189.
70. Carne, C.A. (1997) Epidemiological treatment and tests of cure in gonococcal infection: evidence for value. *Genitourinary Med.*, **73**, 12–15.
71. Uno, M., Deguchi, T., Komeda, H. et al. (1996) Prevalence of *Mycoplasma genitalium* in men with gonococcal urethritis. *Int. J. STD and AIDS*, **7**, 443–444.
72. Munday, P.E., Thomas, B.J., Johnson, A.P. et al. (1981) Clinical and microbiological study of non-gonococcal urethritis with particular reference to non-chlamydial disease. *Br. J. Vener. Dis.*, **57**, 327–333.
73. Hay, P.E., Thomas, B.J., Gilchrist, C. et al. (1992) A reappraisal of chlamydial and non-chlamydial acute non-gonococcal urethritis. *Int. J. STD and AIDS*, **3**, 191–195.
74. Munday, P.E. (1985) Persistent and recurrent non-gonococcal urethritis, in *Clinical Problems in Sexually Transmitted Diseases*, (ed. D. Taylor-Robinson), Martinus Nijhoff, Dordrecht, pp. 15–34.
75. Lomas, D.A., Natin, D., Stackley, R.A. and Shahmanesh, M. (1993) Chemotactic activity of

urethral secretions in men with urethritis and the effect of treatment. *J. Infect. Dis.*, **167**, 233–236.
76. Evans, B.A. (1978) Treatment and prognosis of non-specific genital infection. *Br. J. Vener. Dis.*, **54**, 107–111.
77. Ford, D.K. and Henderson, E. (1976) Nongonococcal urethritis due to T-mycoplasma (*Ureaplasma urealyticum*) serotype 2 in a conjugal sexual partnership. *Br. J. Vener. Dis.*, **52**, 341–342.
78. Beaty, W.L., Byrne, G.I. and Morrison, R.P. (1994) Repeated and persistent infection with *Chlamydia* and the development of chronic inflammation and disease. *Trends Microbiol.*, **2**, 94–98.
79. Vogels, W.H.M., van Voorst Vader, P.C. and Schroder, F.P. (1993) *Chlamydia trachomatis* infection in a high-risk population: comparison of polymerase chain reaction and cell culture for diagnosis and follow-up. *J. Clin. Microbiol.*, **31**, 1103–1107.
80. Horner, P., Thomas, B.J., Gilroy, G. et al. (1996) The role of *Ureaplasma urealyticum* and *Mycoplasma genitalium* in acute and chronic nongonococcal urethritis. Abstract, MSSVD Spring Meeting, Edinburgh.
81. Wong, E.S., Hooton, T.M., Hill, C.C. et al. (1988) Clinical and microbiological features of persistent or recurrent nongonococcal urethritis in men. *J. Infect. Dis.*, **158**, 1098–1101.
82. Stollerman, G.H. (1997) Rheumatic fever. *Lancet*, **349**, 935–942.
83. Schachter, J. (1989) Pathogenesis of chlamydial infections. *Pathol. Immunopathol. Res.*, **8**, 206–220.
84. Morrison, R.P., Belland, R.J., Lyng, K. and Caldwell, H.D. (1989) Chlamydial disease pathogenesis: the 57-kD chlamydial hypersensitivity antigen is a stress response protein. *J. Exp. Med.*, **170**, 1271–1283.
85. Craig, E.A. (1993) Chaperones: helpers along the pathways to protein folding. *Science*, **260**, 1902–1904.
86. Bavoil, P., Stephens, R.S. and Falkow, S. (1990) A soluble 60 kiloDalton antigen of *Chlamydia* spp. is a homologue of *Escherichia coli*. GroEL. *Mol. Microbiol.*, **4**, 461–469.
87. Wagar, E.A., Schachter, J., Bavoil, P. and Stephens, R.S. (1990) Differential human serological response to two 60,000 molecular weight *Chlamydia trachomatis* antigens. *J. Infect. Dis.*, **162**, 922–927.
88. Toye, B., Laferriere, C. and Claman, P. (1993) Association between antibody to the chlamydial heat-shock protein and tubal infertility. *J. Infect. Dis.*, **168**, 1236–1240.
89. Krieger, J.N., Hooton, T.M. and Brust, P.J. (1988) Evaluation of chronic urethritis: defining the role for endoscopic procedures. *Arch. Intern. Med.*, **148**, 703–707.
90. Hooton, T.M., Wong, E.S., Barnes, R.C. et al. (1990) Erythromycin for persistent or recurrent nongonococcal urethritis: a randomized, placebo controlled trial. *Ann. Intern. Med.*, **113**, 21–26.
91. Laga, M., Diallo, M.O. and Buve, A. (1994) Interrelationship of sexually transmitted diseases and HIV: where are we now? *AIDS*, **8** (suppl 1): S119–124.
92. Craib, K.J.P., Meddings, D.R., Strathdee, S.A. et al. (1995) Rectal gonorrhoea as an independent risk factor for HIV in a cohort of homosexual men. *Genitourinary Med.*, **71**, 150–154.
93. Burn, S. and Horner, P.J. (1995) Rectal gonorrhoea as an independent risk factor for HIV infection in homosexual males. *Genitourinary Med.*, **71**, 335–336.
94. Atkins, M.C., Carlin, E.M., Emery, V.C., Griffiths, P.D. and Boag, F. (1996) Fluctuations of HIV load in semen of HIV positive patients with newly acquired sexually transmitted diseases. *Br. Med. J.*, **313**, 341–342.
95. Grosskurth, H., Mosha, F., Todd, J. et al. (1995) Impact of improved treatment of sexually transmitted diseases on HIV infection in rural Tanzania: randomised controlled trial. *Lancet*, **346**, 530–536.
96. Whitaker, L. and Renton, A. (1995) Impact of improved treatment of sexually transmitted disease on HIV infection. *Lancet*, **34**, 1158–1159.
97. Dik, J., Habbema, F. and de Vlas, S.J. (1995) Impact of improved treatment of sexually transmitted disease on HIV infection. *Lancet*, **34**, 1157–1158.
98. Rygnestad, T., Smabrekke, L., Nesje, L. et al. (1995) Impact of improved treatment of sexually transmitted disease on HIV infection. *Lancet*, **34**.

11 Lumps and bumps of the external genitalia: diagnosis and management

M. Murphy and C.J.N. Lacey

INTRODUCTION

Papules and tumours of the external genitalia are a frequent reason for presentation to genitourinary medicine clinics. Although there is broad diversity of both infectious and non-infectious causes of such lesions, for many patients they signify the acquisition of a sexually transmitted disease. In this chapter we discuss the diagnosis and management of infectious lumps of the external genitalia in men and women (Table 11.1), with particular emphasis on genital warts and related conditions and molluscum contagiosum. Non-infectious lesions, genital dermatoses and other causes of infectious genital tumours arising from complications of lower genital tract infections (e.g. epididymo-orchitis) will be dealt with elsewhere in this book (see Chapters 10 and 12).

CONDYLOMATA ACUMINATA – GENITAL WARTS AND RELATED CONDITIONS

Genital warts or condylomata acuminata were a well-described and probably common clinical condition in ancient times.[1,2] The word *condyloma* is of Greek origin, meaning something like 'round tumour'. Genital warts were widely believed to be sexually transmitted – hence the name 'venereal' warts. In recent years the incidence of reported cases of genital warts has risen substantially and it is now one of the commonest sexually transmitted diseases worldwide.[3,4] The management of this problematic condition poses an increasing burden on health services, particularly GUM clinics. Moreover, human papillomavirus (HPV), the causative agent of genital warts, has been impli-

Table 11.1 Infectious causes of lumps of the external genitalia

Viral
Condylomata acuminata ⎫
Bowenoid papulosis ⎬ HPV associated
Buschke–Lowenstein tumour ⎪
  (giant condyloma) ⎭
Molluscum contagiosum

Bacterial
Furuncle/carbuncle
Bartholin's abscess
Periurethral abscess
Hidradenitis suppurativa

Miscellaneous
Condylomata lata
Scabies
Epididymoorchitis

cated in the genesis of genital tract neoplasia in both men and women.

Aetiology

Although long suspected of being of infectious aetiology, viral particles were first identified in genital warts in 1968.[5] Genital warts are caused by various genotypes of the human papillomaviruses, site-specific DNA viruses which produce characteristic proliferations on epidermal and mucosal surfaces. HPV virion particles are small (55 nm), icosahedral in shape and have no envelope. They contain a double-stranded circular DNA genome of around 8000 base pairs. The genome is arranged into two regions: an 'early' region that encodes the viral proteins involved in DNA replication, transcription and cellular transformation; and a 'late' region that encodes the viral capsid proteins.[6] More than 70 different types of HPV have been identified on the basis of molecular typing[7] and around 20 of these are associated with infection of the lower genital tract. HPV 6 and 11 account for over 90% of typical condylomatous lesions of the genital tract.[8,9]

Epidemiology

The incidence of reported (symptomatic) cases of genital warts has increased dramatically in the past two decades. The number of patients attending GUM clinics in the UK with genital warts rose by 130% for males and 190% for females between 1981 and 1990.[3] Genital warts occur more commonly in early sexual life, with a peak age incidence of 16–25.[4,10] Independent risk factors for the acquisition of genital warts include the number of sexual partners, cigarette smoking and long-term use of oral contraceptives.[11]

However, the number of individuals who develop genital warts substantially underestimates the prevalence of asymptomatic HPV infection. Subclinical and latent HPV infection of the anogenital tract has been demonstrated in a considerable proportion of young sexually active men and women. Using DNA hybridization techniques, HPV sequences have been found in 10–30% of exfoliated cervical cells from healthy women.[12,13] With the more sensitive technique of polymerase chain reaction DNA amplification in vulval and cervical specimens, up to 46% of sexually active young women have been found to be infected with HPV.[14] However, HPV carriage appears to be age dependent, with peak rates occurring in young sexually active women (age 20–24) with a gradual decline thereafter.[15] Prevalence studies in men are less well evaluated, but rates appear to be comparable.

Transmission

Although long suspected of being a sexually transmitted disease, it was not until the 1950s that sexual transmission of genital warts was demonstrated in a study of US servicemen returning from the Korean war.[16] High rates of genital warts were observed in partners of servicemen who had acquired genital warts whilst in Korea. In his study of the natural history of genital warts, Oriel confirmed the infective nature of genital warts and found a transmission rate to sexual partners of 64%.[10] The incubation period for the development of genital warts varies from three weeks to eight months[10,16] and perhaps longer. Moreover, subclinical HPV infection can be demonstrated in a high proportion of sexual partners of individuals with HPV-associated disease.[17,18] Although overt genital warts are clearly infectious, the degree of infectivity of non-condylomatous HPV infection is uncertain.

Pathogenesis and pathology

Although recent advances in molecular biology have led to increased understanding of the organization and function of the papillomavirus genome, the pathogenesis of HPV infection and the host response to disease remain poorly understood. Difficulties in propagating the virus in cell culture have hampered efforts to study the pathogenesis of HPV. The precise mechanisms by which HPV induces the formation of warts are ill defined. The virus probably gains access to the basal epidermal layers through microabrasions or microlacerations. There it usually persists in an extrachromosomal form.[19,20] Keratinocyte differentiation is a major factor in the lifecycle of HPV.[20] The amount of viral DNA increases towards the epithelial surface with virion synthesis occurring in the most differentiated superficial cells. However, the control of viral gene expression and replication within the keratinocytes is not fully understood.

The duration of HPV-induced disease is variable and genital warts may last months to years. No systematic evaluation of the spontaneous regression rate of condylomata has been carried out. In placebo-controlled studies, this rate varies between 0% and 69%.[21,24] Even after regression of disease, latent infection persists in a large proportion of individuals.[25] Competent cell-mediated immunity is central to the control of HPV infection and its manifestations.[19] The immune response in HPV-infected tissue is characterized by depletion of CD4+ lymphocytes and Langerhan's cells and impaired immunological function of natural killer (NK) cells.[26] Epidemiological studies in immunosuppressed individuals demonstrate an increased incidence of genital tract HPV infection and associated disease.

HPV infection of the genital tract manifests in three ways: clinically overt disease, subclinical disease and latent infection. Histologically, the clinical exophytic wart is characterized by papillomatosis, acanthosis, elongation and thickening of the rete ridges, parakeratosis and cytoplasmic vacuolation or koilocytosis.[27] The koilocyte is the diagnostic hallmark of HPV infection on light microscopy and is a result of the interaction of viral proteins with cytoplasmic filaments. Subclinical infection of the external genitalia or the cervix has similar histological features apart from papillomatosis. In latent infection, HPV can be detected by various hybridization techniques, but morphological abnormalities are absent or minimal.

Clinical features and diagnosis

Genital warts show considerable morphological variation. To a large extent their appearance is determined by the anatomical site affected and the subtype of HPV involved. Three principal clinical patterns of warts exist: acuminate, papular and flat subclinical.[28] Acuminate warts predominate in moist areas and are papilliferous in form. Colour ranges from pink to grey-white depending on the degree of keratinization present. Acuminate warts may coalesce to form large cauliflower-type lesions. Low-risk HPV types, e.g. 6 and 11, tend to be associated with acuminate lesions. Papular lesions occur most commonly on fully keratinized epithelium, e.g. shaft of penis, perineum and the perianal area. They appear smooth and tend to be multiple. Flat condylomatous (subclinical) lesions were first described by Meisels[29] and can occur anywhere on the external genitalia. Flat lesions are usually demonstrated only after the application of 5% acetic acid (HPV-infected epithelium stains white) and are far commoner than clinical exophytic lesions.

In uncircumcised men, the preputial cavity is most commonly affected, followed by the shaft and urinary meatus, whereas in circumcised men, the shaft is the commonest site. In women, the vestibule and posterior fourchette are the commonest sites affected. Warts may develop in the vagina and exophytic warts of the cervix occur in 6–10% of women with external genital warts.[10,30] Perianal and anal warts occur in both sexes. Infection at this site results from direct transmission or from indirect spread from genital lesions.

The diagnosis of genital warts is readily made on clinical grounds in most cases. In women, prominent physiological mucosal folds in the vestibule and around the urinary meatus can give rise to confusion. In males, prominent coronal papillae (pearly penile papules) occur in approximately 20% of young men and are often mistaken for genital warts by the untrained observer. Histologically, they consist of fibropapillomata and lack the characteristic morphological features of HPV infection.[31] Other differential diagnoses include molluscum contagiosum, sebaceous cysts and condylomata lata of secondary syphilis. If there is any doubt or the lesion is atypical, an excisional or punch biopsy should be performed to confirm the diagnosis. Examination of the distal urinary meatus can be performed satisfactorily with an auroscope.[32] Proctoscopy is essential when perianal warts are present and/or a history of anal sex obtained. Flat cervical condyloma can be identified on colposcopic examination following the application of 5% acetic acid but biopsy is necessary to distinguish them from cervical intraepithelial neoplasia (CIN).

HPV and anogenital neoplasia

In recent years, considerable attention has been focused on the carcinogenic potential of HPV. There is now a substantial body of epidemiological and laboratory evidence implicating HPV infection in the development of anogenital neoplasia.[33,34] The frequent identification of HPV DNA in specimens from squamous carcinoma of the cervix, penis, vulva and anus supports the theory that HPV may have an aetiological role in these malignancies. The role of HPV in cervical neoplasia has been most extensively studied. Infection with 'high-risk' HPV type 16 is associated with higher rates of progression of CIN[35] and in the majority of cases of cervical carcinoma, high-risk types HPV 16 or 18 are found integrated into the host genome.[36] Because women with genital warts and HPV infection may be at increased risk of developing cervical neoplasia, it is recommended to perform cervical cytology at initial presentation. If normal, patients continue with routine cytological surveillance.

Treatment and management

The management of condylomata acuminata remains a formidable challenge. The multiplicity of available treatments is testament to the fact that no single method is universally effective (Table 11.2). Management is predominantly clinic based, treatment courses are often protracted and recurrence rates following resolution are high. Condylomata acuminata in individual patients vary in responsiveness to different treatments. Particular problems arise depending on the number of lesions and the anatomical site(s) involved.[37] Newer lesions tend to be associated with a more vigorous immune response[20] and are more-likely to be responsive to treatment than persistent warts. Initial management

Table 11.2 Therapeutic options for condylomata acuminata

Chemotherapeutic/chemical agents
Podophyllin (10–30%) solutions
Podophyllotoxin
Trichloroacetic acid 90%
5-Fluorouracil

Cryotherapy
Nitrous oxide
Liquid nitrogen

Surgical techniques
Curettage
Excision
Electrocautery
Laser surgery

Immunomodulators
Interferons
Imiquimod

should be tailored to the individual clinical situation and, where treatment failure occurs, second-line therapies chosen in an orderly progression. Coexistent genital tract infections are common and both patients and their partners require screening and treatment where appropriate.[38]

Podophyllin, a crude resinous plant extract, was first used for the treatment of genital warts in the 1940s.[39] Since then, variable-strength (10–30%) solutions of podophyllin have been widely employed as first-line therapy and it remains the treatment of first choice in genitourinary medicine clinics in the UK.[40] In the clinic setting, a 25% solution in tincture of benzoin compound is most frequently employed. Its cytodestructive mechanism of action involves disruption of cellular mitoses and inhibition of nucleoside transport.[41] However, success rates using podophyllin are variable and may be as low as 22% following a course of treatment.[42] Acuminate warts respond best, with flatter hyperkeratotic warts responding poorly.

In addition to podophyllin's limited efficacy, solutions are poorly standardized, have significant local and systemic toxicity and are contraindicated in pregnancy.[43,44]

Podophyllotoxin, the active component of podophyllin, appears to be more effective and safer than standard podophyllin solutions and is licensed for self-application by the patient.[45,46]

Trichloroacetic acid is a caustic agent which causes necrosis of the superficial skin layers. It is particularly useful for small keratotic warts. Injudicious application may result in deep ulceration and scarring. 5-Fluorouracil cream is an antimitotic which inhibits cellular proliferation by interfering with DNA and RNA synthesis. It has been used particularly for the management of intraurethral warts and vaginal disease with high resolution rates.[47,48] Application to other sites is poorly tolerated due to severe irritation.

In general, simple surgical techniques are indicated when few or small lesions are present and where topical treatment of warts has failed. These techniques include curettage, cryotherapy, electrocautery and excision. Curettage rapidly clears small keratinized lesions and both cryotherapy and electrocautery have greater reported clearance and lower relapse rates than topical treatments.[49] More extensive lesions may require surgical excision under anaesthesia or, in specialist centres, laser surgery.

Interferons (α and β) – a group of biologically active glycoproteins with antiviral, antiproliferative and immunomodulatory properties – have been used topically, intralesionally and systemically for the treatment of genital warts.[50,51,52] Systemic administration has the theoretical advantage of treating all infected epithelia. However, success rates of intralesional and systemic treatment, alone or in combination with other treatment modalities, have been variable and no better than traditional therapies.[53] Moreover, systemic treatment is expensive and produces frequent side effects.

Apart from interferons, current modes of treatment for genital warts are directed towards symptomatic or manifest wart disease. Whatever the mode of treatment, latent infection remains in many patients and recurrence rates are high. For persistent or recurrent disease, counselling and psychosexual support are often required.

Bowenoid papulosis

Bowenoid papulosis (BP) is a relatively recently described condition occurring most commonly in young adults. Patients typically have multiple, often pigmented papules or nodules of the external genitalia which may coalesce to form plaques.[54] Lesions may be associated with pruritus. Based on histological features, BP was initially interpreted as multicentric Bowen's disease.[55] However, the two conditions are readily distinguished on the basis of clinical features and age distribution. Bowen's disease tends to develop from middle age onwards and involves any cutaneous site. Histologically, BP represents a form of high-grade intraepithelial neoplasia and displays varying degrees of hyperkeratosis, parakeratosis, vacuolated keratinocytes, irregular acanthosis, papillomatosis and inflammation. There tends to be a lesser degree of cellular atypia in BP compared to Bowen's disease.[55]

The clinical, histological and ultrastructural features of BP indicate an aetiological role for HPV and HPV

16 sequences can be detected in the majority of lesions.[56,57] Women affected by BP may have coexistent cervical HPV infection and therefore may be at increased risk of CIN.

Clinically, BP follows a largely benign course in young patients and spontaneous regression can occur. Treatment is usually conservative and local excision, electrodessication, cryotherapy and topical 5-fluorouracil have all been successfully employed.[54] However, lesions may recur following treatment.

Buschke–Lowenstein tumour

The Buschke–Lowenstein tumour or giant condyloma is a rare manifestation of HPV infection affecting the anogenital region. A number of cases affecting the penis, vulva and anorectal areas have been described.[58,59] Typical lesions usually develop slowly over a number of years into extensive cauliflower-like growths up to several centimetres in diameter. An inherent feature of this condition is a simultaneous exophytic and endophytic growth pattern. This may result in penetration of the underlying tissues by compression and may mimic microinvasion. In some instances, carcinoma *in situ* or invasive carcinoma may develop. Treatment requires surgical excision.

MOLLUSCUM CONTAGIOSUM

Molluscum contagiosum is a benign viral skin infection affecting predominantly children and sexually active young adults.[60] It causes characteristic lesions consisting of multiple, small (2–5 mm) flesh-coloured or translucent umbilicated papules.

Aetiology

Molluscum contagiosum is caused by a double-stranded DNA virus – molluscum contagiosum virus (MCV). On the basis of its size, ultrastructure, physical characteristics and cytoplasmic site of replication, MCV is classed as a member of the poxvirus family. Although inducing a characteristic cytopathic effect in a variety of cell lines, successful replication of MCV has not been achieved *in vitro*. This has limited characterization of viral strains. However, molecular epidemiology of molluscum contagiosum using genetic analysis of viral DNA has identified at least two major subtypes – MCV 1 and MCV 2.[61,62]

Epidemiology

The epidemiology of molluscum contagiosum infection displays a biphasic pattern with peak infection rates occurring among children and young adults. In children, infection is spread by direct contact or through fomites, with a generalized distribution of lesions affecting predominantly the trunk, extremities and face.[60,63] In young adults, molluscum contagiosum lesions are typically on or near the genitalia and appear to be spread primarily by sexual contact. The evidence that molluscum contagiosum is sexually transmitted in adults is indirect and is suggested by the fact that disease predominantly affects the genitalia, its higher prevalence in sexually active young adults and the frequent concurrence of other sexually transmitted diseases.[64,65]

Data from Britain and the US suggest that genital molluscum contagiosum has become more common in recent years.[66,67] This could partly reflect an increased awareness of this condition by patients and physicians alike. However, the trend is consistent with the sustained increase in other viral sexually transmitted infections such as genital herpes and genital warts over the same period.

Pathology and pathogenesis

Because MCV cannot be grown reproducibly in cell culture, the pathogenesis of the disease is not clear. Nevertheless, the pathological features of the lesions are characteristic and consist of focal areas of hyperplastic and hypertrophied epidermis surrounding a core of keratin and epithelial debris.[68] Ultrastructural studies of lesions have demonstrated MCV in all layers of the epidermis. Keratinocytes in the upper layers exhibit large intracytoplasmic inclusion bodies (molluscum bodies) which enlarge and compress the cell nucleus.[69] These molluscum bodies contain mature viral particles.

Cell-mediated immunity appears to play a role in the control of molluscum contagiosum infection.[68] The increased prevalence and dissemination of these lesions in AIDS patients is further evidence of the important role of cellular immunity in the pathogenesis of this condition.[70]

Clinical features and diagnosis

The majority of infections are caused by the major subtype MCV 1. The appearance and localization of lesions caused by the different subtypes are identical. In both children and adults, molluscum contagiosum lesions appear as small white or pink papules. The number of lesions varies from a single lesion to 20 or more. They may increase in size up to 3–5 mm and are often umbilicated. In children, the lesions are distributed widely but in adults, where sexual transmission is the likely cause of spread, lesions are commonly found on inner thighs, lower abdominal walls and external genitalia.[64] In HIV-infected individuals, molluscum contagiosum lesions may be extensive, large and disfiguring.[70] Unlike adults in general, lesions are commonly localized around the head and neck.

The most frequent complication of molluscum contagiosum is bacterial superinfection, usually from scratching of the lesions. Scratching and picking of lesions often contributes to their further spread and in adult males, a sycosis barbae-type picture may be seen. In a minority of cases, an eczematous reaction may arise around individual lesions. Lesions around the eyelids, particularly in HIV-infected individuals, can induce a chronic follicular or papillary conjunctivitis.[71]

The diagnosis of molluscum contagiosum is usually made on the basis of the characteristic clinical features of the lesions described above. They are most commonly confused with condyloma acuminata or verrucae vulgaris. Where necessary, the diagnosis can be readily confirmed on histological or electron microscopic studies of the biopsied lesions.[68,69]

Treatment and management

Treatment is simple and almost all modes of treatment that involve direct disruption of the lesions are effective. This is commonly done by piercing the lesion with a sharp needle or stick, enucleating the lesion and cauterizing the base with phenol, iodine or trichloroacetic acid. Curettage and cryotherapy are equally effective methods of destruction. Treatment may need to be repeated until all lesions are cleared. Treatment of molluscum contagiosum in immunocompromised hosts is less satisfactory due to the widespread nature of the lesions and high rate of recurrence.

Because of the high rates of concurrent STDs in adult patients, it is important that they and their sexual partners are screened for other sexually transmitted infections.

FOLLICULITIS, FURUNCULOSIS AND CARBUNCLES

Acute bacterial infection of the hair follicles can give rise to inflammatory lumps in the genital region. *Staphylococcus aureus* is the most frequently implicated organism and lesions commonly occur in areas subjected to minor trauma or friction, such as the groin or natal cleft. Folliculitis is a relatively trivial condition arising from inflammation, often infective, of the follicular ostium. Deeper infection of the hair follicle leads to follicular necrosis and abscess formation. This is manifest clinically as a furuncle. Lesions may be single or multiple. A carbuncle is a deep infection of a group of contiguous follicles and is a large lesion accompanied by intense inflammation of surrounding tissue. Healing of furuncles and carbuncles may be accompanied by scarring. Treatment of both conditions requires systemic antibiotics such as flucloxacillin.

BARTHOLIN'S GLAND ABSCESS

Bartholin's glands or the greater vestibular glands lie on either side of the introitus in the vaginal wall. Infection and blockage of the vestibular orifice of the duct may lead to the development of an abscess. This presents as an acute painful swelling at the introitus. In some instances purulent discharge may be visible from the duct orifice. Infection is commonly due to facultative aerobic and anaerobic organisms.[72] However, some cases are due to sexually acquired infection and *Neisseria gonorrhoeae* has been isolated from 6–28% of Bartholin's abscesses in various series.[73,74] *Chlamydia trachomatis* does not appear to play a significant aetiological role.[73]

Treatment of acute Bartholin's abscess requires incision and drainage with appropriate antibiotic therapy. Specimens of pus can be examined by microscopy and should be sent for full bacterial cultures. Patients should be screened for other sexually transmitted infections. Bartholin's cysts are chronic or subacute lesions that are less commonly associated with a sexually transmitted disease. If surgical treatment is indicated, they should be treated by marsupialization.

HIDRADENITIS SUPPURATIVA

Hidradenitis suppurativa is a chronic, recurrent, suppurative and cicatricial disease of apocrine glandular tissue. Cases show a female-to-male ratio of 10:1. Affected sites include the axilla, submammary and anogenital areas. The cause is uncertain but the onset of disease is rare before puberty and after middle age. The condition is sometimes associated with polycystic ovarian disease. Occlusion of apocrine glands leads to secondary infection and inflammation. Clinically early lesions form tender abscess-like swellings in apocrine areas which may rupture and heal by fibrosis. As the disease progresses and becomes chronic, extensive scarring, sinus and fistula formation may occur to urethra and anus.[75]

Minocycline is a reasonable choice as initial therapy, but some cases may require antiandrogens such as spironolactone or cyproterone acetate. Extensive or advanced disease may require wide surgical excision with healing by secondary intention.

CONDYLOMATA LATA

Condylomata lata are one of the most characteristic manifestations of secondary syphilis and occur predominantly in warm moist areas, such as around the anus, vulva, lateral aspect of the scrotum and inner thighs.[76,77] Condylomata lata are papules which are modified by their anatomical site and friction and appear as hypertrophic, dull red or pink fleshy wart-like lesions with a broad base and flat top. The surface is often eroded and exudate from lesions contains large numbers of *Treponema pallidum* and is highly infectious. Patients should be examined carefully for other mucocutaneous manifestations of syphilis and lymphadenopathy. The diagnosis is confirmed by dark-ground microscopy of exudate from the lesions and serological investigations. Treatment is usually procaine penicillin 1.2 mega units intramuscularly for 15 days unless the patient is penicillin sensitive.

SUMMARY

Papules or tumours occurring on the external genitalia are a common presentation. Genital warts, furuncles or molluscum contagiosum are by far the most frequent causes and will respond to appropriate therapy. However, the differential diagnoses include scabies, Bowenoid papulosis, condylomata lata and Buschke–Lowenstein tumours as well as abscesses and inflammatory lesions of the genitourinary tract. Biopsy should be undertaken for persistent lesions where there is any doubt as to the aetiology.

Rational management protocols are important in the treatment of genital warts as a significant minority of cases are refractory to treatment. Shared-care protocols between genitourinary and primary care physicians for genital warts are a logical development. Physicians should have a high index of suspicion for intraepithelial neoplastic lesions of the external genitalia and management should be in conjunction with the appropriate surgical specialist.

REFERENCES

1. Bafverstedt, B. (1967) Condylomata acuminata – past and present. *Acta Derm. Venereol.*, **47**, 376–381.
2. Burns, D.A. (1992) 'Warts and all' – the history and folklore of warts: a review. *J.R. Soc. Med.*, **85**, 37–40.
3. Sexually transmitted diseases in England and Wales: 1981–1990 (1992) *Communicable Disease Report*, **2**, R1–R7.
4. Quinn, T.C. and Cates, W. (1992) Epidemiology of sexually transmitted diseases in the 1990's, in *Sexually Transmitted Diseases*, (ed. T.C. Quinn), Raven Press, New York, pp. 15–18.
5. Dunn, A.E.G. and Ogilvie, M.M. (1968) Intranuclear virus particles in human genital wart tissue. *J. Ultrastruct. Res.*, **22**, 282–295.
6. Howley, P.M. and Schlegel, R. (1988) The human papillomaviruses – an overview. *Am. J. Med.*, **85**, (suppl 2A), 155–158.
7. Van Ranst, M.A., Tachezy, R., Delius, H. and Burk, R.D. (1993) Taxonomy of the human papillomaviruses. *Papillomavirus Rep.*, **4**, 61–65.
8. Reid, R. and Campion, M.J. (1988) The biology and significance of human papillomavirus infections in the genital tract. *Yale J. Biol. Med.*, **61**, 307–325.
9. Sugase, M., Moriyama, D. and Matsukura, T. (1991) Human papillomavirus in exophytic condylomatous lesions on different female genital regions. *J. Med. Virol.*, **34**, 1–6.
10. Oriel, J.D. (1971) Natural history of genital warts. *Br. J. Vener. Dis.*, **47**, 1–13.
11. Daling, J.R., Sherman, K.J. and Weiss, N.S. (1986) Risk factors for condyloma acuminatum in women. *Sexually Trans. Dis.*, **13**, 16–18.
12. De Villiers, E.M., Wagner, D., Schneider, A. *et al.* (1987) Human papillomavirus infections in women with and without abnormal cervical cytology. *Lancet*, **26**, 703–706.
13. Anon. (1988) Human papillomavirus: implications for clinical medicine. *Ann. Intern. Med.*, **108**, 628–630.
14. Bauer, H.M., Greer, C.E., Chambers, J.C. *et al.* (1991) Genital human papillomavirus infection in female university students as determined by a PCR-based method. *J. Am. Med. Assoc.*, **265**, 472–477.
15. Melkert, P.W.J., Hopman, E., van den Brule A.J.C. *et al.* (1993) Prevalence of HPV in cytomorphologically normal cervical smears, as determined by the polymerase chain reaction, is age dependent. *Int. J. Cancer*, **53**, 919–923.
16. Barrett, T.J., Silbar, J.D. and McGinley, J.P. (1954) Genital warts – a venereal disease. *J. Am. Med. Assoc.*, **154**, 333–334.
17. Barasso, R., de Brux, J., Croissant, O. and Orth, G. (1987) High prevalence of papillomavirus associated penile intraepithelial neoplasia in sexual partners of women with cervical intraepithelial neoplasia. *N. Engl. J. Med.*, **317**, 916–923.
18. Schneider, A., Kirchmayr, R., de Villiers, E.M. and Gissmann, L. (1988) Subclinical human papillomavirus infections in male sexual partners of female carriers. *J. Urol.*, **140**, 1431–1434.

19. Schneider, A. (1993) Pathogenesis of genital HPV infection. *Genitourinary Med.*, **69**, 165–173.
20. Vardy, D.A., Baadsgaard, Hansen, E.R. et al. (1990) The cellular immune response to human papillomavirus infection. *Int. J. Dermatol.*, **29**, 603–610.
21. Scott, G.M. and Csonka, G.W. (1979) Effects of injections of small doses of human fibroblast interferon into genital warts. A pilot study. *Br. J. Vener. Dis.*, **55**, 442–445.
22. Schonfield, A., Nitke, S., Schattner, A. et al. (1984) Intramuscular human interferon-beta injections in the treatment of condylomata acuminata. *Lancet*, **i**, 1038–1042.
23. Eron, L.J., Judson, F., Tucker, S. et al. (1986) Interferon therapy for condylomata acuminata. *N. Engl. J. Med.*, **315**, 1059–1064.
24. Keay, S., Teng, N., Eisenergg, M. et al. (1988) Topical interferon for treating condylomata acuminata in women. *J. Infect. Dis.*, **158**, 934–939.
25. Ferenczy, A., Mitao, M., Nagai, N. et al. (1985) Latent papillomavirus and recurring genital warts. *N. Engl. J. Med.*, **313**, 784–788.
26. Schneider, A. and Koutsky, L.A. (1992) Natural history and epidemiological features of genital HPV infection, in *The Epidemiology of Human Papillomavirus and Cervical Cancer*, (eds N. Munoz, F.X. Bosch, K.V. Shah and A. Meheus), International Agency for Research on Cancer, Lyon pp. 25–52.
27. Syrjanen, K.J. (1989) Histopathology, cytology, immunohistochemistry and HPV typing techniques, in *GPVI – Genitoanal Papilloma Virus Infection. A Survey for the Clinician*, (eds G. von Krogh and E. Rylandes), Conpharm AB, Karlstad, Sweden, pp. 33–67.
28. von Krogh, G. and Rylander, E. (1989) Clinical evaluation, in *GPVI A Survey for the Clinician*, (eds G. von Krogh and E. Rylander), Conpharm AB, Karlstad, Sweden, pp. 69–123.
29. Meisels, A., Fortin, R. and Roy, M. (1977) Condylomatous lesions of the cervix II. Cytologic, colposcopic and histopathologic study. *Acta Cytol.*, **21**, 379–390.
30. Vayrynen, M., Syrjanen, K. and Castren, O. (1985) Colposcopy in women with papillomavirus lesions of the uterine cervix. *Obstet. Gynaecol.*, **65**, 409–415.
31. Ferenczy, A., Richart, R.M. and Wright, T.C. (1991) Pearly penile papules: absence of human papillomavirus DNA by the polymerase chain reaction. *Obstet. Gynaecol.*, **78**, 118–122.
32. Thin, R.N. (1992) Meatoscopy: a simple technique to examine the distal anterior urethra in men. *Int. J. STD and AIDS*, **3**, 21–23.
33. Munoz, N. and Bosch, F.X. (1992) HPV and cervical neoplasia: a review of case control and cohort studies, in *The Epidemiology of Human Papillomavirus and Cervical Cancer*, (Eds N. Munoz, F.X. Bosch, K.V. Shah and A. Meheus), International Agency for Reseach on Cancer, Lyon, pp. 251–261.
34. Munoz, N., Bosch, F.X., de Sanjose, S. et al. (1993) The causal link between human papillomavirus and invasive cervical cancer: a population-based case control study in Columbia and Spain. *Int. J. Cancer*, **52**, 742–749.
35. Kataja, V., Syrjanen, S., Mantyjarvi, R. et al. (1992) Prognostic factors in cervical human papillomavirus infections. *Sexually Trans. Dis.*, **19**, 154–159.
36. Meijer, C.J.L., van den Brule, A.J.C. and Snijders, P.J.F. (1992) Detection of human papillomavirus in cervical scrapes by the polymerase chain reaction in relation to cytology: possible implication for cancer screening, in *The Epidemiology of Human Papillomavirus and Cervical Cancer*, (eds. N. Munoz, F.X. Bosch, K.V. Shah and A. Meheus), International Agency for Research on Cancer, Lyon, pp. 271–281.
37. Anon. (1991) Persistent anogenital warts. *Lancet*, **338**, 1114–1115.
38. Kinghorn, G.R. (1978) Genital warts: incidence of associated genital infections. *Br. J. Dermatol.*, **99**, 405–409.
39. Culp, O.S. and Kaplan, L.W. (1944) Condylomata acuminata. Two hundred cases treated with Podophyllin. *Ann. Surg.*, 251–256.
40. Wardropper, A. and Wooley, P. (1992) Treatment of anogenital warts in genitourinary clinics in England and Wales. *Int. J. STD and AIDS*, **3**, 439–441.
41. Loike, J.D. and Horwitz, S.B. (1976) Effects of podophyllotoxin and VP-16–213 on microtuble assembly *in vitro* and nucleotide transport in He La cells. *Biochemistry*, **15**, 5435–5443.
42. Simmons, P.D. (1981) Podophyllin 10% and 25% in the treatment of anogenital warts. A comparative double blind study. *Br. J. Vener. Dis.*, **57**, 208–209.
43. Marcus, J. and Camisa, C. (1990) Podophyllin therapy for condyloma acuminatum. *Int. J. Dermatol.*, **29**, 693–698.
44. Murphy, J. and Bloom, G.D. (1991) Podophyllin or podophyllotoxin as treatment for condylomata acuminata. *Papillomavirus Rep.*, **2**, 87–89.
45. Lassus, A. (1987) Comparison of podophyllotoxin and podophyllin in treatment of genital warts. *Lancet*, **ii**, 513.
46. Edwards, A., Atma-Ram, A. and Thin, R.N. (1988)

Podophyllotoxin 0.5% v podophyllin 20% to treat penile warts. *Genitourinary Med.*, **64**, 253–255.
47. Von Krogh, G. (1976) 5-Fluorouracil cream in the successful treatment of refractory condylomata acuminata of the urinary meatus. *Acta Derm. Venereol.*, **56**, 297.
48. Krebs, H.B. (1987) The use of topical 5-fluorouracil in the treatment of genital condylomata. *Obstet. Gynecol. Clin. North Am.*, **14**, 559–568.
49. Stone, K.M., Becker, T.M., Hadger, A. and Kraus, S.J. (1990) Treatment of external genital warts: a randomised clinical trial comparing podophyllin, cryotherapy and electrodessication. *Genitourinary Med.*, **66**, 16–19.
50. Kraus, S.J. and Stone, K.M. (1990) Management of genital infection caused by human papillomavirus. *Rev. Infect. Dis.*, **12 (suppl 6)**, 5620–5632.
51. Ling, M.R. (1992) Therapy of genital human papillomavirus infections. Part II: Methods of treatment. *Rev. Infect. Dis.*, **31**, 769–775.
52. Tyring, S.K. (1988) Treatment of condyloma acuminatum with interferon. *Semin. Oncol.*, **15 (suppl 5)**, 35–40.
53. Anon. (1988) Interferon and genital warts: much potential, modest progress. *J. Am. Med. Assoc.*, **259**, 570–572.
54. Patterson, J.W., Kao, G.J., Graham, J.H. and Helwig, E.B. (1986) Bowenoid papulosis. A clinicopathologic study with ultrastructural observations. *Cancer*, **57**, 827–836.
55. Lloyd, K. (1970) Multicentric pigmented Bowen's disease of the groin. *Arch. Dermatol.*, **101**, 48–51.
56. Ikenberg, H., Gissman, L., Gross, G. *et al.* (1983) Human papillomavirus type-16-related DNA in genital Bowen's disease and in bowenoid papulosis. *Int. J. Cancer*, **32**, 563–565.
57. Gross, G., Hagedorn, M., Ikenberg, H. *et al.* (1985) Bowenoid papulosis. Presence of human papillomavirus (HPV) structural antigens and of HPV-16-related DNA sequences in genital Bowen's disease and in Bowenoid papulosis. *Arch. Dermatol.*, **121**, 858–863.
58. Ananthakrishnan, N., Ravindran, R., Veliath, A.J. *et al.* (1981) Lowenstein–Buschke tumour of penis – a carcinomimic. Report of 124 cases with review of the literature. *Br. J. Urol.*, **53**, 460–465.
59. Judge, J.R. (1969) Giant condyloma acuminatum involving vulva and rectum. *Arch. Pathol.*, **88**, 46–48.
60. Postlethwaits, R. (1970) Molluscum contagiosum: a review. *Arch. Environ. Health*, **21**, 432–452.
61. Scholz, J., Rosen-Wolfe, A., Bugerl, J. *et al.* (1988) Molecular epidemiology of molluscum contagiosum. *J. Infect. Dis.*, **158**, 898–900.
62. Scholz, J., Rosen-Wolfe, A., Bugerl, J. *et al.* (1989) Epidemiology of molluscum contagiosum using genetic analysis of the viral DNA. *J. Med. Virol.*, **27**, 87–90.
63. Douglas, J.M. (1990) Molluscum contagiosum, In *Sexually Transmitted Diseases*, 2nd edn, (Eds K.K. Holmes, P. Marsh, P.J. Sparlure *et al.*), New York, pp. 443–447.
64. Wilkin, J.K. (1977) Molluscum contagiosum venereum in womens' outpatient clinics: a venereally transmitted disease. *Am. J. Obstet. Gynecol.*, **128**, 531–535.
65. Radcliffe, K.W., Daniels, D. and Evans, B.A. (1991) Molluscum contagiosum: a neglected sentinel infection. *Int. J. STD and AIDS.*, **2**, 416–418.
66. Lacey, C.J.N. (1993) Genital warts and molluscum contagiosum, in *Viral Infections*, The Medicine Group, Oxon, pp. 131–136.
67. Oriel, J.D. (1991) The increase of molluscum contagiosum. *Br. Med. J.*, **294**, 74.
68. Reed, R.J. and Parkinson, R.P. (1972) The histogenesis of molluscum contagiosum. *Am. J. Surg. Pathol.*, **2**, 161–166.
69. Stelley, W.B. and Burmeister, V. (1986) Demonstration of a unique viral structure: the molluscum viral colony sac. *Br. J. Dermatol.*, **115**, 557–562.
70. Schwartz, J.J. and Myskowski, P.L. (1992) Molluscum contagiosum in patients with human immunodeficiency virus infection. A review of twenty seven patients. *J. Am. Acad. Dermatol.*, **27**, 583–588.
71. Robinson, M.R., Udell, I.J., Garber, P.J. *et al.* (1992) Molluscum contagiosum of the eyelids in patients with acquired immune deficiency syndrome. *Ophthalmology*, **99**, 1745–1747.
72. Lee, Y.H., Rankin, J.S., Alpert, S. *et al.* (1977) Microbiological investigation of Bartholin's gland abscesses and cysts. *Am. J. Obstet. Gynecol.*, **129**, 150–156.
73. Bleker, O.P., Suralbraak, D.J.C. and Schulte, M.J. (1990) Bartholin's abscess: the role of *Chlamydia trachomatis*. *Genitourinary Med.*, **66**, 24–25.
74. Rees, E. (1967) Gonococcal bartholinitis. *Br. J. Vener. Dis.*, **43**, 150–156.
75. Radcliffe, K.W. (1991) Hidradenitis suppurativa. *Genitourinary Med.*, **67**, 58.
76. King, A., Nicol, C. and Rodin, P. (1990) Early acquired syphilis, in *Venereal Diseases*, 4th edn, Baillière Tindall, London, pp. 28–35.
77. (1980) Acquired syphilis: early stage, in *Clinical Practice and Sexually Transmitted Disease*, 2nd edn, (Eds. D.H.H. Robertson, A. McMillan and H. Young, Churchill Livingstone, Edinburgh, pp. 126–143.

12 The diagnosis and management of prostatitis

C. O'Mahony

INTRODUCTION

Prostatitis is a relatively uncommon condition seen in genitourinary medicine clinics. The diagnosis is often difficult to establish and the patients usually require a significant amount of time devoted to diagnosis and management. Because of the chronicity of some prostatic conditions, some patients become chronic attenders and provide a unique challenge to a service that by tradition has open access. Recent reviews include those by Colleen and Mardh,[1] Doble,[2] Meares,[3] Leigh,[4] Evans[5] and Chandiok.[6] From the genitourinary medicine viewpoint, little has changed since the comprehensive review by Thin[7,7a] in 1991 and 1997 and Luzzi[8] in 1996.

It is now largely accepted that prostatitis can be divided into acute and chronic bacterial prostatitis, chronic non-bacterial prostatitis and prostatodynia.[9] Clinically, bacterial and non-bacterial prostatitis have many similarities, but are distinguished by the fact that the causative organisms of non-bacterial prostatitis have not yet been elucidated. In prostatodynia inflammation of the prostate cannot be demonstrated.

ANATOMY OF THE PROSTATE GLAND

In order to understand prostate pathology and interpret the findings on examination or ultrasonography, it is important to review the anatomy. The prostate gland completely surrounds the prostatic urethra. It is made up of a central zone surrounding the urethra and a peripheral zone. It is the peripheral zone that can be palpated on rectal examination and that is largely involved in the commonest pathological processes, i.e. prostatitis and carcinoma of the prostate. The vas deferens fuses with the seminal vesicles to become the ejaculatory duct. This traverses the posterior wall of the prostate and opens into the prostatic urethra at the verumontanum.

The prostatic urethra is also studded with openings draining the prostate gland. These prostatic duct openings do not have sphincters and drain various sections of the prostate. Many of the ducts draining the peripheral zone have a tortuous passage through the prostate and enter the prostatic urethra at a perpendicular angle. Thus, these ducts are particularly likely to permit urine reflux into the prostate. This structure helps explain the often focal nature of infection within the prostate, particularly within the peripheral zone. On rectal examination, the normal prostate can readily be felt; it is about 4 cm long and 4 cm wide. The seminal vesicles are superior to the prostate and can generally only be felt if there is a pathological process causing swelling, i.e. infection or tumour.

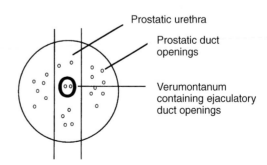

Fig. 12.1 The prostate gland.

BACTERIAL PROSTATITIS

Bacterial prostatitis can be divided into acute bacterial prostatitis (ABP) and chronic bacterial prostatitis (CBP). ABP is a rare complication of sexually transmitted urethral infections. However, historically gonorrhoea was a significant cause of acute prostatitis and prostatic abscesses. Nowadays, the commonest cause of ABP is urinary tract infection followed by penetration of the pathogens into the prostate, probably from urinary reflux. Thus, the commonest microbiological causes of acute prostatitis are bacteria which cause urinary tract infections in males. Table 12.1 contains a list of organisms that have been reported as causing acute or chronic bacterial prostatitis. However, the evidence implicating some of these organisms is controversial.

ABP presents with fever, rigors and symptoms of frequency, dysuria, urgency, haematuria and perineal pain. The infection can be preceded by urinary tract infection or be concomitant. If the prostate becomes significantly swollen there will be bladder outlet obstruction.

CBP, on the other hand, has a more subtle presentation with the patients usually complaining of a wide range of symptoms of varying severity. Although acute prostatitis can lead to chronic prostatitis, it is unusual to find such a clear precipitating event. There is often, however, a history of urinary tract infections or prior gonococcal urethritis. So, the patients typically present with a slowly developing symptomatology, largely characterized by pain. This pain can be suprapubic, perineal, testicular, radiating down the thighs, but especially focusing on the urethra and tip of the penis (Table 12.2). On occasions, the pain can be worsened or relieved by defaecation. Pain on ejaculation can also be a feature. There can be various urinary irritative complaints, e.g. frequency, dysuria, variation in stream, urgency, a sense of incomplete emptying of the bladder and haematuria and haematospermia. Sitting for prolonged periods generally worsens the condition. These symptoms are often coupled with a depressed nature, decreased libido and general lethargy. It is, of course, often difficult to assess whether it is the condition that caused the depression and lack of libido, etc. or whether these are features of the condition itself.

Table 12.1 Organisms implicated in prostatitis

Neisseria gonorrhoeae	++
Chlamydia trachomatis	+/–
Trichomonas vaginalis	+
Escherichia coli	++
Proteus species	++
Klebsiella	++
Salmonella	+
Staphylococci	+
Streptococcus faecalis	++
Candida (*albicans, glabrata, tropicalis*)	+/–
Cryptococcus neoformans	+
Mycobacterium tuberculosis	+
Ureaplasma urealyticum	+/–
Mycoplasma hominis	+/–

+/– indicating the relevant certainty of the association

Table 12.2 Symptoms of chronic prostatitis

Pain
   Slow insidious onset
   Perineal
   Testicular
   Suprapubic
   Urethra
   Tip of penis
   Radiating to thigh
   Worsened by prolonged sitting
   Pain on ejaculation
Urinary
   Variation in stream
   Frequency
   Dysuria
   Urgency
   Haematuria
   Haematospermia
Lethargy
Decreased libido
Depression

CHRONIC NON-BACTERIAL PROSTATITIS

Chronic non-bacterial prostatitis is much commoner than bacterial prostatitis and it must be said at the outset that these patients represent some of the 'heart sink' workload of genitourinary medicine. The often unremitting unresponsive chronic symptoms of the condition induce despair and exasperation in both the patient and the treating doctor. This situation is shared by our urology colleagues. In a comprehensive review in 1994, Doble[10] begins by stating 'The mere mention of chronic prostatitis sends the majority of practising urologists diving for cover.' A consultant opinion should be sought at a very early stage in the management of this condition, so that the patient can feel there is some continuity of his care. His symptoms are likely to last longer than the contract of the junior hospital doctor making the initial diagnosis! Repeated courses of antibiotics with follow-up appointments in three months' time (i.e. when the doctor has changed posts) generate far more problems in the long run than time spent in the early stages on investigation, diagnosis and explanation.

Patients with chronic non-bacterial prostatitis usually have signs and symptoms similar to patients with chronic bacterial prostatitis, except they generally do not have a predisposing history of urinary tract infections or urethritis. Despite repeated investigations, organisms are never found in the urinary tract or in prostatic secretions. Interpretation of laboratory investigations can be difficult in these patients as they have generally had multiple courses of antibiotics from various sources before attending for a definitive examination.

PROSTATODYNIA

Patients with this condition have many of the symptoms of prostatis but physical findings, laboratory investigations and cultures are all normal. Video-urodynamic studies can show some dysfunction of the bladder neck and prostatic urethra. This has been referred to by Meares[11] as bladder neck urethral spasm syndrome, now termed bladder neck dyssynergia. A combination of spasm of the internal urinary sphincter and the perineal muscles serves to increase the prostatic urethral pressure, encouraging pressure and possibly backflow into the prostate. Backflow has already been demonstrated by voiding cystourethrography.[12] Kirby et al.[13] also proved urinary reflux into the prostate gland using carbon particles instilled into the bladder prior to transurethral prostatic resection. It has even been suggested that urate itself may be an irritant chemical. Persson[14] found a correlation between urate concentrations in prostate secretions and pain levels in 56 patients with non-bacterial prostatitis.

AETIOLOGY OF PROSTATITIS

Of the organisms listed in Table 12.1, there is little doubt that gonorrhoea and the common urinary tract pathogens cause bacterial prostatitis. Infected urine gets refluxed under pressure into the prostate and sets up foci of infection.

Neisseria gonorrhoeae can cause urethritis and epididymo-orchitis and has also been isolated from prostatic abscesses. In a study in 1931, 75% of 42 reviewed cases of prostatic abscess were found to be caused by *Neisseria gonorrhoeae*.[15] *Trichomonas vaginalis* infection has been described, but must be uncommon.[16]

Chlamydia trachomatis has also been shown to cause urethritis and epididymoorchitis and might be expected thus to cause a prostatitis on occasions. Definitive proof, however, is still awaited. In one study that used transrectal ultrasound to guide transperineal aspiration biopsies in 50 men with non-bacterial prostatitis, chlamydia could not be demonstrated by culture techniques or immunofluorescence.[17] Mardh et al..[18] could not demonstrate chlamydia in any of 28 specimens of prostatic fluid in a similar group of patients. Others have examined the local immune response in the prostate and have used locally produced IgG and IgA to aid identification of possible pathogens. These studies have been reviewed by Meares in 1992,[3] leading him to conclude that 'Chlamydia appears to play an insignificant role in the aetiology of prostatitis.'

The main support for *Ureaplasma urealyticum* causing prostatitis comes from a study done in 1983,[19] when Brunner et al. examined 597 patients with non-bacterial prostatitis. Using the segmental specimen technique, they noticed increased cultures of ureaplasma in prostatic secretions compared to urethral specimens in 13.7% of this group. Ohkawa et al.[20] isolated *Ureaplasma urealyticum* from 18 of 143 prostatitis patients. Treatment with either minocycline or ofloxacin eradicated *Ureaplasma urealyticum* in 14 patients complying with treatment. Ten had resolution of symptoms and three showed complete resolution with clearance of leucocytes from expressed prostatic fluid. The major difficulty, of course, is differentiating urethral contamination from prostatic secretions, particularly when dealing with organisms that are known to be present in the male urethra, even in the absence of urethritis. In 1989 Doble[21] failed to find either *Ureaplasma urealyticum* or *Mycoplasma hominis* by direct sampling of the prostate in prostatitis, using ultrasound-guided biopsies.

The role of prostatic calculi is uncertain, as calculi are a common finding in the normal prostate. About 75% of middle-aged men can be shown to have small prostatic calculi by transrectal ultrasonography.[22] This figure rises to 100% in elderly men. The prostatic calculi can be shown to be made of urinary constituents rather than constituents of prostatic secretions.[23] It does, however, seem likely that if calculi are present and an infective prostatitis develops, the infected calculi can act as a focus of continued infections. These patients may require surgical intervention for eradication of the condition.

Haematogenous spread has been implicated in prostatitis caused by *Staphylococcus aureus* and *Mycobacterium tuberculosis*. Even fungi have been implicated in some cases and a recent case report[24] describes infection with *Cryptococcus neoformans*, successfully treated with fluconazole. However, parasitic, mycotic and granulomatous prostatitis are rare, with the latter

INVESTIGATION OF ACUTE BACTERIAL PROSTATITIS

being of histological importance, as it can be confused with carcinoma of the prostate.

The acute presentation and severe symptoms in acute prostatitis usually point to the diagnosis. The key investigation is urine culture to identify the organism and obtain sensitivities. A routine STD screen should first be performed, i.e. swabs for chlamydia and gonorrhoea from the urethra, a two-glass urine test should be done and a midstream urine sample should be sent immediately to the laboratory for analysis and culture. It must be emphasized to the laboratory that they must pursue culture and sensitivity on any organisms that are isolated. The request should also indicate that culture should be extended for fastidious organisms. This ensures that effort has been made to culture organisms like Neisseria, β-haemolytic streptococcus and anaerobes that might not routinely be detected on standard urine culture. In addition to routine culture, our laboratory centrifuges these urines at 3000 rpm for five minutes and inoculates the deposit on to blood and chocolate agar plates. The chocolate agar plate is incubated in CO_2 and the blood agar plate incubated anaerobically at 37°C for 18–24 h.

Many laboratories may not proceed to identification and sensitivity testing if a 'mixed growth' is found. The urine is often assumed to be contaminated and discarded. However, in the exceptional situation of acute prostatitis, all organisms should be identified and sensitivity testing done and the results interpreted in the light of symptoms and response to treatment. Mixed infections can be significant in prostatitis. A full blood count (looking for raised white cells) and plasma viscosity should be requested, as should a serological chlamydia antibody test (chlamydia microimmunofluorescence test), to obtain baseline titres. The prostate is excruciatingly tender in acute prostatitis. Prostatic massage should never be performed when an acute infection is suspected, as this would cause severe pain and could precipitate bacteraemia, a seminal vesiculitis or an epididymitis.

A plain radiograph of the pelvic area can show calculi in the bladder, occasionally prostatic calculi can be identified and, on very rare occasions, lost foreign bodies can be identified.

In the case outlined in Figure 12.2, self-instrumentation with soldering iron wire had resulted in the wire inadvertently disappearing into the bladder, coiling up and becoming calcified over a two-year period. This culminated in the precipitation of acute prostatitis with urethral discharge, secondary to urinary tract infection,

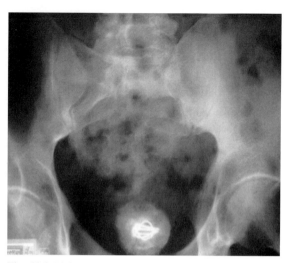

Fig. 12.2 Plain pelvic film revealed the presence of a 5 × 5 cm bladder calculus with a length of wire embedded at its centre. (Courtesy of the *British Journal of Sexual Medicine*.)

leading to presentation at a GU clinic.[25] If prostatic abscesses are suspected, CT or MRI scanning of the prostate would be preferable to transrectal ultrasound in the acute phase of the condition.

INVESTIGATION OF CHRONIC PROSTATITIS

The main distinguishing feature of chronic bacterial and non-bacterial prostatitis is the fact that organisms are successfully identified in the former, but not in the latter. The investigations for both conditions are largely identical. A full STD screen to include Gram stain from the urethra, swabs for gonorrhoea and chlamydia, a two-glass urine test and a midstream urine sample should be performed. Blood should be taken for prostate-specific antigen to exclude any underlying malignancy. Blood should also be taken for chlamydia serology and processed for syphilis and hepatitis B serology and, of course, as usual, the patient should be offered HIV testing with informed consent. An HIV test should be discussed, as genital symptoms can be caused by unspoken anxiety relating to often unfounded worries about having contracted HIV.

A plain radiograph of the pelvis should also be organized. It is probably ideal to rebook the patient for further examination, including prostatic massage, for a specific slot during a quiet clinic time when this examination can be explained to the patient and performed without haste. This also allows the examination to be undertaken in the confidence of knowing there is no urinary tract infection present, i.e. the second of the

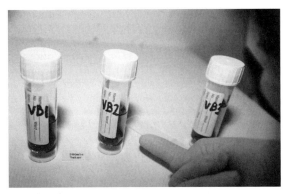

Fig 12.3 Segmental investigations. First catch urine (VB1), mid-stream urine (VB2), prostatic massage then final urine (VB3)

two-glass test is clear and midstream urine sample is negative for growth. The pivotal investigation is, of course, analysis of expressed prostatic secretions. The time-honoured method of Meares and Stamey[26] entails segmental investigations that allow localization of infection to urethra, prostate or bladder.

The usual STD and Gram-stain swabs are taken from the urethra, unless done previously. The patient then passes 10 ml of urine into a sterile urine container – this is labelled VB1 (first voided bladder urine) (Figure 12.3).

The patient then urinates a further approximate 200 ml into the toilet, then collects a further 10 ml specimen. This is labelled VB2 and is the equivalent of a midstream urine sample. Prostatic massage is then undertaken. With a lubricated index finger the prostate gland is gently massaged from side to side towards the centre for approximately one minute. Expressed prostatic secretion (EPS) usually appears in a droplet at the penile tip and this is collected directly onto a plain slide for Gram staining and a counting chamber for white cell count and swabs are also taken for culture and analysis (gonorrhoea, chlamydia, other organisms).

It is often necessary to 'strip' the urethra. Gentle pressure is applied from as far back on the urethra as possible, i.e. gentle pressure through the scrotum, along the urethra, squeezing towards the tip and expressing as much prostatic fluid as possible. The patient then urinates a further 10 ml into a sterile container and this is labelled VB3.

Dilutions of each of the urine samples and EPS are made and colony counts are assessed and in this way some estimation can be made of where the site of infection is likely to be. In CBP colony numbers from the prostatic secretions and VB3 are usually 10-fold greater than those from VB1 and VB2. The presence of numerous pus cells in the expressed prostatic secretion is considered by some authors to be indicative of prostatitis. Clumping of large aggregates of pus cells or casts of pus cells are also considered to be highly significant. Macrophages containing fat (oval bodies) are also considered to be highly indicative of prostatitis. The exact number of total leucocytes per high-power field in prostatic secretion that can be considered abnormal is still controversial. Ten white blood cells per high-power field is often quoted as the upper limit of normal. However, it is well known that the white cell count can vary between individuals and even within the same individual, depending on factors such as timing of last ejaculate. Recent ejaculation prior to prostatic massage can give white cell counts of >10 per high-power field.[27] On this basis, it has been suggested that no ejaculation should take place for 3–5 days prior to prostatic massage.

There is controversy over the relevance of semen culture. Results often do not correlate with expressed prostatic secretion microscopy and culture. This is probably due to the fact that infected foci in the prostate gland may have blocked ducts and contribute little to the ejaculate which is therefore largely made up of normal prostatic secretions and secretions from other glands. However, if the patient has complained of haematospermia, it may be worth requesting.

In circumstances where further investigation is considered important, e.g. suspected prostatic abscesses or calculi, then transrectal ultrasonography (TRUS), CT or MRI scanning can be useful. There is now considerable experience with transrectal ultrasonography.[28] It is a useful investigation but rarely conclusive. A recent attempt by de la Rosette[29] to quantify the abnormal features by automated analysis of ultrasonographic prostate images (AUDEX) showed a high sensitivity (90%) but poor specificity (64%). Recently colour Doppler ultrasonographic scanning showed that colour intensity matched the severity of symptoms in patients with prostatic syndromes.[30] MRI is certainly useful in distinguishing other pelvic pathology that may be masquerading as prostatitis.[31]

Prostatic biopsy, either through the perineum or transrectally, has been employed in the past but in view of the focal nature of prostatic disease, results were difficult to interpret. Ultrasound-guided transrectal biopsies are more specific and are employed when a focal abnormality can be demonstrated. Transrectal biopsy does, however, carry a risk of infection and one recent report found nine of 110 patients reattending with infectious symptomatology after biopsy.[32] *Pseudomonas cepacia* caused the majority of the infections and contaminated ultrasound transmission gel was implicated as the source of infection. Aus *et al.*[33] showed a marked reduction in infection rates by using norfloxacin 400 mg bd for one week as prophylaxis, prior to biopsy. However, interpretation of culture results can then be difficult.

MANAGEMENT

In practical terms, if organisms are implicated, appropriate antibiotic therapy is used. If no organisms are found on investigation but clinical suspicion of infection remains, empirical therapy is often instituted anyway, in the presence of objective evidence of prostatitis.

Table 12.3 gives a list of appropriate drugs in bacterial prostatitis. Although co-trimoxazole has been used extensively in the past, the Committee on Safety of Medicines (CSM) has recently recommended caution in its use in view of the serious reported reactions involving blood and generalized skin disorders, particularly in elderly patients. This recent CSM recommendation is included in current editions of the *British National Formulary* under 'sulphonamides and trimethoprim'. The aminoquinolones are currently becoming drugs of first choice. Because of their excellent absorption following oral administration, they can even be used orally in acute prostatitis. In acute prostatitis the inflammation of the prostate does allow easy diffusion of antibiotics. There is usually a shift in pH from slightly acidic to a more alkaline milieu in the prostate. This facilitates the diffusion of antibiotics with the appropriate pharmacokinetics, i.e. lipid soluble and weakly acidic. However, Nickel et al.,[34] using a rat model, compared levels of norfloxacin in uninflamed and inflamed prostate glands. Norfloxacin levels did not change significantly in the uninfected and infected glands. Naber et al.[35] found ofloxacin levels in prostatic fluid to be twice that in the corresponding plasma with even higher levels in seminal fluid. Ofloxacin would currently be a personal choice because of its wide spectrum of activity which includes *Chlamydia trachomatis* and *Neisseria gonorrhoea*.

If there are symptoms of urinary tract obstruction suprapubic catheterization may be necessary. Urethral catheterization should never be performed in prostatitis. Any predisposing cause of urinary tract infection, i.e. urinary calculi, would also have to be dealt with to prevent recurrences. Oral therapies must be continued for at least 4–6 weeks to ensure eradication of the organisms and might possibly prevent acute prostatitis becoming chronic. Cure rates of up to 100% have been claimed for aminoquinolones in prostatitis.[36] In established chronic prostatitis, antibiotic courses of up to three months are sometimes indicated depending on symptoms and signs of inflammation. Patients should be reviewed after one week of therapy. Further reviews are dictated by symptoms and progress. Improvement can often be objectively measured as a decrease in the number of white cells found in expressed prostatic secretion. Surgical management is rarely employed, unless infected prostatic calculi are noted. Microwave treatment in the form of local hyperthermia, either transrectal[37] or transurethral,[38] gives symptomatic relief in some patients. An extract of pollen, Cernilton, has antiinflammatory and smooth muscle relaxant properties and has been used successfully in chronic prostatitis and prostatodynia.[39]

In prostatodynia, where pain in the absence of inflammation is the main feature, three main approaches are used.

1. Simple pain relief in the form of antiinflammatory agents, given either intermittently or long term in the form of ibuprofen 400 mg tds or any of the enteric-coated range.
2. If there are signs and symptoms of intermittent ureteric dysfunction, e.g. hesitancy, intermittent poor stream, pulse voiding, etc., then therapies aimed at relieving the internal sphincter spasm can help. This sphincter tone is maintained by sympathetic stimulation mediated via α-1 receptors. Selective α-1 blockers can inhibit the tone of the smooth muscle and allow relaxation. Prazosin or indoramin can be used, but must be commenced in low dosage and increased to therapeutic levels that give relief. Typical schedules are prazosin 0.5 mg bd for one week, increasing the second week to 1 mg

Table 12.3 Drugs used in the treatment of prostatitis

Ofloxacin 400 mg bd
Ciprofloxacin 500 mg bd
Norfloxacin 400 mg bd
Gentamicin IV 4 mg per kg per day and ampicillin 2 g six-hourly
Doxycycline 100 mg bd
Minocycline 100 mg daily
Erythromycin 500 mg bd
(New macrolides, i.e. azithromycin/clarithromycin to be evaluated)
Ibuprofen 400 mg tds
Prazosin 0.5 mg bd, increasing to 2 mg bd
Indoramin 5 mg bd, increasing to 20 mg bd
Alfuzosin hydrochloride 2.5 mg × 1 nocte, increasing to × 1 tds
Diazepam 2 mg tds

High-dose antibiotic therapy should continue for four weeks in acute prostatitis. A minimum of six weeks therapy is required for chronic prostatitis, possibly using lower drug dosage to minimize side effects bd and finally to maintenance dosage of 2 mg bd. Maintenance dosage with indoramin is 20 mg bd. Postural hypotension, micturition syncope, drowsiness, dry mouth, blurred vision and ejaculatory delay can all be problems with this therapy. Alfuzosin is a new α-1 blocker that possibly has a more selective affinity for the α-1 receptors of the genitourinary tract rather than those of the systemic vasculature. *In vitro* studies have shown specificity for α-1 adrenoreceptors located in the trigone of the urinary bladder, urethra and prostate. However, all of the α receptor blockers have some hypotensive effect. Patients should be warned to lie down if symptoms of dizziness or fainting occur, particularly with the first dose. Alfuzosin is given as 2.5 mg in the evening initially, then morning and evening, increasing gradually to tds.

3. For cases of prostatodynia who are felt to have tension myalgia of the pelvic floor, muscle relaxants in the form of diazepam are a useful treatment. The usual dosage is 2 mg tds. Relief of symptoms for a prolonged period may allow decrease of this dosage eventually without much symptom recurrence. These treatments can be coupled with pelvic floor retraining.[40] It is of interest that in one prospective trial diazepam was as effective as minocycline in relieving symptoms of chronic prostatitis.[41]

Finally, it is important that patients with chronic prostatitis are managed in a situation where continuity of care is likely. Regular communication with the patient's general practitioner is essential to avoid duplication of therapies, exclusion of serious disease and background anxieties relating to sexually transmitted infections. These anxieties should all be dealt with early on and the patient reassured as much as possible. Once a sexually transmitted infection has been ruled out or eradicated there is no restriction whatsoever on frequency of intercourse and the patient should be thoroughly reassured that they cannot pass anything on to their partner when this is the case. Despite intensive antibiotic therapy, pain relief, etc., many patients will continue to have symptoms and need to feel their condition is understood and need help to cope with the chronicity of the condition. It is also important that patients are reassured that all investigations which need to be done have been done, to avoid them shopping around different GUM clinics or urologists.

URINARY TRACT INFECTION

Urinary tract infections are an important cause of problems which affect patients attending GUM clinics. Patients may present with symptoms typical of a cystitis or even pyelonephritis, thinking they may indeed have acquired a sexually transmitted infection. However, other patients often present with these symptoms knowing they will be dealt with without delay and also get immediate treatment. These are often patients who have previously attended the clinic for another reason or know from partner/friends about the rapid service available through GUM clinics.

It is usually easy to differentiate between anterior or posterior urethritis, cystitis and pyelonephritis on history alone. Dysuria is, of course, common to all infections, but frequency, suprapubic pain and loin tenderness are not generally features of more commonly presenting urethritis.[42]

The usual routine GUM clinic investigations, i.e. urethral Gram stain, culture for *Neisseria gonorrhoeae* and chlamydia, as outlined in previous chapters, should be done. The two-glass urine test is useful in differentiating a cystitis or nephritis from a urethritis. The first 10–20 ml of urine collected is compared to a second sample (which can act as a midstream urine sample). In the presence of cystitis, there are usually threads and clouding of the urine in both specimens. Routine dipstick analysis generally reveals haematuria and proteinuria in the midstream urine sample. Newer urinary dipsticks contain the leucocyte esterase test which, when combined with the nitrate test, is a good predictor of urinary tract infection. Direct microscopy of centrifuged or uncentrifuged fresh urine is also useful in experienced hands by showing increased number of white cells and motile bacteria. However, in most clinics, clinical assessment combined with dipstick urinalysis would form the initial basis for treatment.

The vast majority of patients attending GUM clinics who turn out to have urinary tract infection tend to be young and this is often their first infection. This often poses the dilemma of whether further investigations should be done, who should do them and who should oversee follow-up.

Who to investigate?

The addition of a midstream urine sample for culture and analysis to routine genital tract tests already performed is of little inconvenience to either the patient, the clinic or the laboratory. While awaiting culture report, antibiotic therapy will usually be initiated if there is a clinical suspicion of cystitis or pyelonephritis. The patient can be told to take the antibiotics as directed, rest at home, take increased fluids, phone in 2–3 days for discussion of preliminary results or, if the condition has failed to respond or has worsened, reattend immediately. At the second visit, if symptoms have cleared and culture is also back indicating UTI with organisms sensitive to the initiated antibiotic, patients should be given a further follow-up appointment for about three weeks later. A follow-up MSU sample should be sent to ensure continued clearance of the infection and absence of red cells and protein. At this point it is also vital to get the patient's cooperation to allow communication of this episode of urinary tract infection to his GP. It is important that there is a central record documenting a patient's urinary tract infections and the best place for this is in the patient's own GP notes. The GP needs to know that infection was documented, what the organism was, the reported sensitivities and response to treatment. This information is necessary for the GP who may wish to pursue further investigations at a hospital of his choice rather than have further investigations done through the GUM service.

If the patient reattends with worsening of symptoms or other complications, admission may be necessary for referral to urology and intravenous antibiotic therapy.

The following groups need further investigation:

- first urinary tract infection in men;
- recurrent urinary tract infection in women;
- acute pyelonephritis;
- frank haematuria;
- culture of more pathogenic organisms, i.e. proteus, klebsiella, pseudomonas, candida, etc.

There is some discussion over whether the first investigation should be intravenous pyelogram or renal tract ultrasound. Much depends on the clinical suspicions, i.e. renal ultrasound is excellent at detecting hydronephrosis, renal abscesses, obstructed ureters and other complicating pelvic pathology, i.e. pelvic abscesses, etc. It can be used in patients where renal function is uncertain and the risks of intravenous pyelography might be increased. Cost, ease of access and speed of reporting of service would also be prime considerations. Again, it is crucial that the patient's GP be involved at an early stage or it may result in some duplication of investigations or difficulty in recovering cost of investigations from the relevant body. It is our practice that once a urinary tract infection is confirmed and it is felt further investigation is warranted, immediate communication with the GP takes place. The findings to date are detailed and it is clearly stated that a referral is to be made to either intravenous pyelography or ultrasound on the GP's behalf for a particular date, and if the GP does not wish this appointment to go ahead the onus lies with the practice to cancel it. In the future, difficulties over funding for further investigations for non-STD conditions in attending via GUM departments may mean patients being immediately referred back to general practice with no further investigations being organized by GUM. This would, however, mean a delay in patient management and would be a retrograde step to good medical practice.

The more aggressive urological investigations should probably be initiated by the renal physicians or urologists who would have to deal with any subsequent pathology found. Further investigations could include a micturating cystogram or a radionuclide cystogram using technetium-99 m labelled glycoprotein and ^{99m}Tc dimercapto-succinic acid (DMSA renal scans).

Likely pathogens

Obviously, in the event of an STD being confirmed, i.e. infection due to gonorrhoea or chlamydia, elimination of these may eradicate the symptoms but, more commonly, the standard urinary tract pathogens would be isolated, i.e. coliforms, proteus, etc. The vast majority of infections seen in GUM clinics would be community acquired and, as such, would be relatively sensitive to standard antibiotic treatment. A knowledge of local organisms and sensitivities is of great value in deciding therapy. For instance, in our locality, the PHLS published six-monthly reviews of common urinary tract organisms isolated and percentage sensitivities (Table 12.4).

Table 12.4 Urinary tract infection

Type of organisms	% of total isolates
Coliforms (*E. coli*, Klebsiella)	81.0%
Proteus species	7.7%
Enterococci	5.9%
Staphylococci	2.4%
Pseudomonas species	1.9%
Streptococci	1.1%

Sensitivity patterns of the common isolates

Type of organism	Coliforms	% of total isolates Coliform and proteus species
Amoxycillin	57%	59%
Co-amoxiclav	94%	94%
Cephradine	96%	96%
Trimethoprim	81%	80%
Nitrofurantoin	95%	87%
Ciprofloxacin	99%	99%

Mycobacterium tuberculosis is not a common isolate but should be looked for when there is persistent pyuria and, of course, in cases where tuberculosis is more likely, i.e. HIV disease. Short courses of antibiotic therapy, i.e. stat doses of three-day or five-day courses, have been shown to be very effective in simple urinary tract infections. In our clinic, sensitivity patterns indicate excellent sensitivity to aminoquinoloines and familiarity with their use in dealing with gonorrhoea has led us to employ them as first-line therapies for suspected UTIs. Ciprofloxacin 250 mg bd for five days or ofloxacin 400 mg daily for five days or norfloxacin 200 mg for three days affords good compliance. Obviously, ofloxacin has a potential added advantage of being effective against both gonorrhoea and chlamydia, should these be subsequently shown to be the causative agent or a coincidental infection.

HONEYMOON CYSTITIS

This is a term used to describe cystitis associated with intercourse. It is a common finding in the age group we deal with in GU medicine and some women often have their lives ruined by inevitable cystitis 12–36 h after intercourse. Simple hygiene before intercourse by both partners and after intercourse by the female, coupled with voiding urine within half an hour of intercourse, can be useful in decreasing the frequency of the syndrome but more often than not, long-term antibiotic prophylaxis is needed. This can be given as a low-dose antibiotic every night or in some cases can simply be confined to a low-dose antibiotic taken after intercourse. It is also important to use lubrication during intercourse even if both partners feel this is unnecessary. Artificial lubrication does mean there is less trauma to the delicate introital tissues with less oedema and likelihood of ascending infection. Traditional lubricants such as KY Jelly can be used but this is often cold to the touch and dries relatively quickly. Newer lubricants like Replens or the latest addition, Senselle, are all available over the counter in chemists and they are very useful as self-help in the management of recurrent cystitis. Nitrofurantoin, particularly in the non-nauseating form of macrodantin 100 mg after intercourse, is a very useful agent to stop this distressing condition occurring. Patients must be warned, however, that if they do get cystitis even though they are following this prophylactic course, they must take a urine specimen to their general practice or local clinic for analysis, as it is possible that nitrofurantoin prophylaxis may select out a proteus type of urinary tract infection which would need urgent and aggressive treatment. Other antibiotics have been used, i.e. one tablet co-trimoxazole and latterly norfloxacin have all been shown to be useful.[43] It is far better to have successful prophylaxis with just an occasional breakthrough than to be continually treating repeated urinary tract infections. As well as reducing the amount of damage, the amount of antibiotic eventually taken in the year is usually less with prophylaxis than in treating true episodes of full-blown cystitis or pyelonephritis.

REFERENCES

1. Colleen, S. and Mardh, P.A. (1990) Prostatitis, in *Sexually Transmitted Diseases*, 2nd edn, (eds K.K. Holmes *et al.*) McGraw-Hill, New York, pp. 653–661.
2. Doble, A. (1991) Prostatitis, in *Recent Advances in Sexually Transmitted Diseases and AIDS*, (eds J.R.W. Harris and S.M. Foster), Churchill Livingstone, Edinburgh, pp. 129–157.
3. Meares, E.M. (1992) Prostatitis and related disorders, in *Campbell's Urology*, 6th edn W.B. Saunders Philadelphia, pp. 807–822.
4. Leigh, D.A. (1993) Prostatitis – an increasing clinical problem for diagnosis and management. *J. Antimicrob. Chemother.*, **31**, 1–9.
5. Chandiok, S., Fisk, P.G. and Riley, V.C. (1992) Prostatitis – clinical and bacterial studies. *Int. J. STD and AIDS*, **3**, 188–190.
6. Evans, D.T.P. (1994) Treatment of chronic abacterial prostatitis: overview. *Int. J. STD and AIDS*, **5**, 157–164.
7. Thin, R.N. (1991) The diagnosis of prostatitis: a review. *Genitourinary Med.*, **67**, 279–283.
7a. Thin, R.N. (1997) Diagnosis of chronic prostatitis: a review and update. *Int. J. STD and AIDS*, **8**, 475–481.
8. Luzzi, G. (1996) The prostatitis syndromes. *Int. J. STD and AIDS*, **7**, 471–479.
9. Drach, G.W., Fair, W.R., Meares, E.M., Jr and Stamey, T.A. (1978) Classification of benign diseases associated with prostatic pain: prostatitis or prostatodynia? *J. Urol.*, **120**, 266.
10. Doble, A. (1994) Chronic prostatitis. A review. *Br. J. Urol.*, **74**, 537–541.
11. Meares, E.M. Jr (1986) Prostatodynia: clinical findings and rationale for treatment in *Therapy of Prostatitis*, (eds W. Weidner, H. Brunner, W. Krause and C.F. Ruthauge), Zuckswerdt Verlag, Munich W, pp. 207–212.
12. Hellstrom, W.K.G., Schmidt, R.A., Lue, T.F. *et al.* (1987) Neuromuscular dysfunction in non-bacterial prostatitis. *Urology*, **30**, 183–188.

13. Kirby, R.S., Lowe, D., Baltitude, M.I. and Shuttleworth, K.E.D. (1982) Intraprostatic urinary reflux: an aetiological factor in abacterial prostatitis. *Br. J. Urol.*, **54**, 729–731.
14. Persson, B.E. and Ronquist, G. (1996) Evidence for a mechanistic association between nonbacterial prostatitis and levels of urate and creatinine in expressed prostatic secretion. *J. Urol.*, **155**(3), 958–960.
15. Sargent, J.C. and Irwin, R. (1931) Prostatic abscess: clinical study of 42 cases. *Am. J. Surg.*, **11**, 334–337.
16. Van Laarhoven, P.H. (1987) *Trichomonas vaginalis*, a pathogen of prostatitis. *Netherlands J. Surg.*, **19**, 263–273.
17. Doble, A., Thomas, B.J., Walker, M.M. *et al.* (1989) The role of *Chlamydia trachomatis* in chronic abacterial prostatitis: a study using ultrasound guided biopsy. *J. Urol.*, **141**, 332–333.
18. Mardh, P.A., Ripa, K.T., Colleen, S. *et al.* (1978) Role of *Chlamydia trachomatis* in non-acute prostatitis. *Br. J. Vener. Dis.*, **54**, 330–334.
19. Brunner, H., Weidner, W. and Schiefer, H.G. (1983) Studies of the role of *Ureaplasma urealyticum* and *Mycoplasma hominis* in prostatitis. *J. Infect. Dis.*, **147**, 807–813.
20. Ohkawa, M., Yamaguchi, K., Tokunaga, S., Nakashima, T. and Shoda, R. (1993) Antimicrobial treatment for chronic prostatitis as a means of defining the role of ureaplasma urealyticum. *Urol. Int.*, **51**(3), 129–132.
21. Doble, A., Thomas, B.J., Furr, P.M. *et al.* (1989) A search for infectious agents in chronic abacterial prostatitis using ultrasound guided biopsy. *Br. J. Urol.*, **64**, 297–301.
22. Peeling, W.B. and Griffiths, G.J. (1984) Imaging of the prostate by ultrasound. *J. Urol.*, **147**, 217–224.
23. Rameriz, C.T., Ruiz, J.A., Gomez, A.Z. *et al.* (1980) A crystallographic study of prostatic calculi. *J. Urol.*, **124**, 840.
24. Fuse, H., Ohkawa, M., Yamaguchi, K., Hirata, A. and Matsubara, F. (1995) Cryptococcal prostatitis in a patient with Behçet's disease treated with fluconazole. *Mycopathologia*, **130**(3), 147–150.
25. O'Mahony, C., Scott, G. and Kinghorn, G.R. (1988) Solderin' the bladder – unusual foreign body embedded in a bladder calculus. *Br. J. Sexual Med.*, **15**, 88–90.
26. Meares, E.M. and Stamey, T.A. (1968) Bacteriologic localization patterns in bacterial prostatitis and urethritis. *Invest. Urol.*, **5**, 492–518.
27. Jameson, R.M. (1967) Sexual activity and the variations of the white cell content of the prostatic secretion. *Invest. Urol.*, **5**, 297–302.
28. Ludwig, M., Weidner, W., Schroeder-Printzen, I., Zimmermann, O. and Ringert, R.H. (1994) Transrectal prostatic sonography as a useful diagnostic means for patients with chronic prostatitis or prostatodynia. *Br. J. Urol.*, **73**(6), 664–668.
29. De la Rosette, J.J., Giesen, R.J., Huynen, A.L., Aarnink, R.G. and van Iersel, M.P. (1995) Automated analysis and interpretation of transrectal ultrasonography images in patients with prostatitis. *Eur. Urol.*, **27**(1), 47–53.
30. Veneziano, S., Pavlica, P. and Mannini, D. (1995) Color Doppler ultrasonographic scanning in prostatitis: clinical correlation. *Eur. Urol.*, **28**(1), 6–9.
31. Fowler, J.E., Peters, J.J. and Hamrick-Turner, J. (1995) Mullerian duct cyst masquerading as chronic prostatitis: diagnosis with magnetic resonance imaging using a phased array surface coil. *Urology*, **5**(4), 676–678.
32. Keizur, J.J., Lavin, B. and Leidich, R.B. (1993) Iatrogenic urinary tract infection with *Pseudomonas cepacia* after transrectal ultrasound guided needle biopsy of the prostate. *J. Urol.*, **149**(3), 523–526.
33. Aus, G., Ahlgren, G., Bergdahl, S. and Hugosson, J. (1996) Infection after transrectal core biopsies of the prostate – risk factors and antibiotic prophylaxis. *Br. J. Urol.*, **77**(6), 851–855.
34. Nickel, J.C., Downey, J., Clark, J., Ceri, H. and Olson, M. (1995) Antibiotic pharmacokinetics in the inflamed prostate. *J. Urol.*, **153**(2), 527–529.
35. Naber, K.G., Kinzig, M., Sorgel, F. and Weigel, D. (1993) Penetration of ofloxacin into prostatic fluid, ejaculate and seminal fluid. *Infection*, **21**(2), 98–100.
36. Andriole, V.T. (1991) Use of quinolones in treatment of prostatitis and lower urinary tract infections. *Eur. J. Clin. Microbiol. Infect. Dis.*, **10**, 342–350.
37. Montorsi, F., Guazzoni, G., Bergamaschi, F. *et al.* (1993) Is there a role for transrectal microwave hyperthermia of the prostate in the treatment of abacterial prostatitis and prostatodynia? *Prostate*, **22**(2), 139–146.
38. Nickel, J.C. and Sorensen, R. (1996) Transurethral thermotherapy for nonbacterial prostatitis: a randomised double-blind sham controlled study using new prostatitis specific assessment questionnaires. *J. Urol.*, **155**(6), 1950–1954.
39. Rugendorff, E.W., Weidner, W., Ebeling, L. and Buck, A.C. (1993) Results of treatment with pollen extract (Cernilton® N) in chronic prostatitis and prostatodynia. *Br. J. Urol.*, **71**, 433–438.
40. Segura, J.W., Opitz, J.L. and Greene, L.F. (1979) Prostatitis and pelvic floor tension myalgia. *J. Urol.*, **122**, 168–169.

41. Simmons, P.D. and Thin, R.N. (1985) Minocycline in chronic abacterial prostatitis: a double blind prospective trial. *Br. J. Urol.*, **57**, 43–45.
42. Wilkie, M.E., Almond, M.K. and Marsh, F.P. (1992) Diagnosis and management of urinary tract infection in adults. *Br. Med. J.* **305**, 1137–1141.
43. Nicole, I.F., Harding, G.K.M., Thompson, M. *et al.* (1989) Prospective randomized placebo-controlled trial of norfloxacin for the prophylaxis of recurrent urinary tract infection in women. *Antimicrob. Agents Chemother.*, **33**, 1032–1035.

13 The diagnosis and management of pelvic inflammatory disease

S.N. Mann, J.R. Smith and S.E. Barton

Pelvic inflammatory disease (PID) is a complex and poorly understood condition which causes considerable morbidity in the female population worldwide and consumes vast resources in the management of its consequences. It is estimated that more than one million women per year in the USA will suffer from an episode of acute PID;[1] a proportion of these will also suffer the chronic sequelae of infertility, ectopic pregnancy, pelvic pain and menstrual disorders.

In the UK, women with symptoms of acute PID will present to a diversity of clinical settings such as the general practitioner, accident and emergency department, family planning clinic or genitourinary medicine clinic. In many countries there is also an array of private doctors, particularly gynaecologists, whom they may first consult. This results in great variation in how extensively available diagnostic methods are used and, often, whether treatment is adequately prescribed.

An additional management problem lies in the changing pattern of acute infection. As *Chlamydia trachomatis* is the main pathogen responsible for PID in the Western world,[2] symptoms can be mild and women are increasingly likely to be managed as outpatients. Indeed, patients may have such mild symptoms that they do not actually present acutely for treatment, but only many years later with the effects of chronic tubal damage.[3]

The best opportunities for control of acute PID and its longer term effects lie largely in primary prevention and early detection and treatment of lower genital tract infection by screening. Scholes *et al.* (1996)[4] have shown that active screening and treatment of high-risk asymptomatic women can reduce the incidence of acute PID by 56%, in the USA.

There is also scope, however, for improving the way in which upper genital tract infection can be managed when it does present clinically. By applying the best principles of both gynaecology and genitourinary medicine, management protocols can be developed for use in any setting, in association with appropriate training of staff. It may then be possible to ensure safe and adequate treatment and follow-up to ensure resolution or referral, where appropriate.

One such protocol has been piloted for use in an accident and emergency department[5] and has demonstrated that dramatic improvements in history taking, microbiological sampling and treatments prescribed can be made (see Appendix).

This chapter aims to outline the current knowledge of the epidemiology, aetiology, diagnosis and treatment of PID which can form the basis for standardization of management strategies and improved clinical care.

DEFINITION

PID is classically defined as inflammation of the upper genital tract. It is, nonetheless, difficult clearly to distinguish between lower and upper genital tract infection since it has been shown that 40% of those with asymptomatic and presumed lower genital tract chlamydial infection already have a plasma cell endometritis.[6] Given that, in other studies, endometritis shows 85% concordance with salpingitis,[7] it may be more accurate to reflect upon a disease continuum where these clear boundaries cannot be assumed.

The disease is most commonly caused by the ascent of microorganisms from the female lower genital tract (vagina and endocervix) to the endometrium and fallopian tubes, causing inflammation. It is usually initiated by a sexually transmitted disease (STD) which may lead to damage in a previously normal pelvis. This can predispose it to further colonization by a variety of organisms, both aerobic and anaerobic and opportunistic as well as pathogenic, giving rise to the polymicrobial aetiology which is now well recognized. The risk of microbial ascent is increased by factors that may further breach the normal host genital tract defences, such as childbirth, miscarriage or surgical procedures, and rates of acute infection are higher in these groups.

EPIDEMIOLOGY

In the UK, there was a 50% increase in women between the ages of 20 and 24 diagnosed with PID between 1975 and 1985[8] and there has been no apparent decrease since then. This increase paralleled an increase in the incidence of *Chlamydia trachomatis* infection reflected in the KC60 returns from genitourinary medicine clinics[9] and further supports the accumulated evidence for this organism as a leading cause of acute PID in the UK. In Sweden over the same period since the introduction of active case finding and primary prevention of STDs, a decline in incidence has been mirrored by a fall in hospital admissions for acute PID with no apparent concomitant increase in community diagnosis.[10] This supports the evidence suggesting early diagnosis and prevention as the way forward in management of this difficult condition.

RISK FACTORS

Associated factors have been identified that relate not only to acquisition of STDs but also to progression to PID. These can be a useful diagnostic adjunct to substantiate or refute clinical indicators. They may also facilitate the targeting of so-called 'at-risk' populations in the quest to improve diagnosis of asymptomatic or clinically mild disease.

Age, ethnicity and sexual activity

A sexually transmitted agent can be identified in up to 60% of cases of proven cases of PID.[11] Risk indicators can be divided into those of sexually transmitted infection and those of upper genital tract infection. Young age, black race and early coitarche[12] have all been associated with an increased risk of STDs, along with a high frequency of sexual intercourse and change of sexual partners.[13] The relationship with young age is felt to be due to both behavioural and physiological factors. Teenagers tend to show less regular use of barrier contraceptives, change sexual partner more frequently and have higher rates of infection in the partner pool. However, they also show higher rates of cervical ectopy and greater penetrability of cervical mucus which in turn may increase susceptibility to infection.[14]

Sexual behaviour plays a key role in the development of PID, emphasizing the importance of detailed sexual history taking as a routine part of the consultation.[15]

Contraception

Oral contraception
In spite of the fact that the oral contraceptive pill has been associated with an increased risk of cervical chlamydial infection which in turn may be related to an increased prevalence of cervical ectopy,[16] there is an apparent protective effect against PID. The incidence is reduced, disease is milder when assessed laparoscopically and tubal infertility is reduced to one-third in pill takers relative to controls who were not using oral contraception at the time of recorded infection.[17]

Intrauterine contraceptive device (IUCD)
Use of the IUCD has been limited in the past by its negative associations with PID. More recently, studies have confirmed that the increase in risk is confined to the first 20 days following insertion and thereafter is related solely to background risk of STD.[18] This has practical implications for screening for infection prior to insertion of IUCD and also suggests that devices should be left *in situ* for their maximum lifespan if there are no strong indications for early removal. However, removal of IUCD is in most cases advisable in the presence of acute PID, along with appropriate advice to the patient to avoid subsequent conception.

Bacterial vaginosis

The association of PID and bacterial vaginosis is controversial. Soper *et al.*[19] identified vaginal microorganisms associated with bacterial vaginosis in 61.8%

of women with laparoscopically confirmed PID. This was not confirmed by the findings of Faro et al.[20] who found that the vaginal bacterial flora seen was not typical of bacterial vaginosis as classified by microbiological criteria, although microscopy was not used.

Other risk factors

Factors that have been implicated aetiologically in PID but that are less well substantiated include vaginal douching, smoking, drug use and alcohol.[21,22] It is unclear whether any links here are direct or indirect.

DIAGNOSIS OF PID

Diagnostic sensitivity is maximized by combining microbiological and, if indicated, laparoscopic findings with clinical indicators. Clinical manifestations of disease vary widely in both severity and presentation of symptoms, making diagnosis on clinical grounds alone both difficult and inaccurate. In a study of women presenting to a genitourinary medicine clinic with abdominal pain and who had laparoscopic diagnosis, those with PID were clinically indistinguishable from those with other diagnoses or no obvious cause for their pain.[23] Clinical diagnosis alone is therefore a delicate balance between missing the diagnosis, and therefore subjecting the individual to the risks of no treatment or delayed treatment and the consequent long-term sequelae of infection, versus overdiagnosis. Most clinicians would rightly initiate treatment with a low index of suspicion. However, in this approach there is the inherent risk of failing to make an alternative important diagnosis plus the added anxiety attached to the label of PID and the future diagnoses of recurrent disease that may be made as a result.

On the basis of this, many clinicians would advocate the use of laparoscopy routinely in diagnosis, whilst the more conservative may reserve this for cases presenting diagnostic difficulty or unresponsive to antimicrobial therapy.

Clinical diagnosis

Different clinical criteria have been applied for the diagnosis of PID. Jacobsen and Westrom[24] used criteria of abdominal pain plus two or more of the following: vaginal discharge, fever, vomiting, menstrual irregularity, urinary symptoms, tenderness on bimanual examination, adnexal swelling and erythrocyte sedimentation rate (ESR) greater than 15 mm/h. They were the first to verify diagnosis using laparoscopy and observed a 65% concordance.

Many variations on this theme have been used in an attempt to standardize clinical diagnosis, without significant improvement in sensitivity. Westrom and Mardh[25] defined the basic necessary criteria of lower abdominal pain, cervical motion pain and adnexal tenderness and demonstrated an increased sensitivity of diagnosis between 70% and 96% with between one and three of the following additional features: ESR greater than 15 mm/h, fever and adnexal mass. In most studies, these same basic criteria are now used with some variation in the additional criteria as defined by different authors. There is no standard, however, where sensitivity and specificity are considered satisfactory.

Supplementary non-specific laboratory tests may aid diagnosis to some extent. A raised ESR and white blood cell count are frequently helpful. The ESR (or C-reactive protein) is raised according to severity of disease and can aid decisions about treatment as well as monitoring response to therapy.[26] More recently, measurement of CA125 has been used as a marker of pelvic inflammation[27] although this is unlikely to influence diagnostic precision.

Microbiological diagnosis

Microbiological sampling is essential if a diagnosis of PID is being considered. Both lower and upper genital tracts may be sampled although there is poor correlation between the two. Increased diagnostic yield is achieved by testing a maximum number of sites, particularly for detection of chlamydial infection,[28] although endometrial sampling has not been shown to improve yield of organisms over endocervical sampling alone.[29] It may, however, have a role in the histological diagnosis of PID.

Laparoscopic diagnosis

Laparoscopy is routinely employed for diagnosis in Scandinavia and the USA, whereas in the UK its use remains confined to selected cases.[30] Costs and the risks of the procedure are two factors limiting its use. Clearly there are a number of situations in which it is a requirement for diagnosis: in the older woman, the shocked and very ill woman, if there is diagnostic doubt or if there is no history of sexual contact. Beyond this, threshold for use may vary.

Since use of laparoscopy is only available to the gynaecologist, it is essential that good channels of communication exist from genitourinary medicine and the general practitioner to increase its role in diagnosis.

In favour of extending its use, one can argue that it not only increases diagnostic accuracy but also provides a therapeutic opportunity for possible pelvic lavage and division of adhesions and an opportunity for upper genital tract microbiological sampling. In addition, there is evidence that the severity of the inflammatory reaction of the fallopian tubes as visualized at laparoscopy is more important as a prognostic indicator for future fertility than both aetiology or treatment.[17] This is not necessarily correlated with severity of symptoms.[31]

Laparoscopy is traditionally regarded as the 'gold standard' in diagnosis. Minimum criteria used are oedema, tubal hyperaemia and exudate.[24] However, this probably represents a relatively late stage in disease and may miss a proportion of cases if the investigation is performed early in the process. It is also becoming increasingly recognized that there may be an early stage or mild form of disease where there is histological endosalpingitis shown by fimbrial microbiopsy[32] or positive chlamydial tube isolates[33] without macroscopic evidence of disease. The prognostic significance of this is not fully elucidated but the efficacy of laparoscopy as a 'gold standard' for diagnosis is thrown into question and the importance of detailed microbiological investigation is highlighted.

MICROBIAL AETIOLOGY OF PID

Neisseria gonorrhoeae and *Chlamydia trachomatis* are the primary pathogens that cause PID. Between 1960 and 1980, isolation of the gonococcus from the genital tract fell from 50% to 10% of cases of PID, in association with its declining prevalence in industrialized countries.[34] However, aetiology is reflected in local prevalences and in the developing world gonococcal sepsis continues to be an important cause of morbidity.

In the developed world, *Chlamydia trachomatis* has been isolated from the Fallopian tubes in up to 30% of cases of PID[35] with chlamydial antibodies identified in up to two-thirds.[36] In the most recent microbiological study, *C. trachomatis* was identified in the genital tract of 38.5% of 104 women and *N. gonorrhoeae* in 14.4%.[37] Stacey et al. confirmed similar microbiological findings, with *C. trachomatis* identified in 48% and *N. gonorrhoeae* in 30%.[38]

It is common to see both organisms together. Coinfection probably occurs by a mutual facilitation[11] although pelvic infection may occur in the absence of either organism in 25–50% of cases.[39] Other organisms are implicated in the polymicrobial aetiology; most important of the aerobic bacteria are *Escherichia coli* and *Haemophilus influenzae* and anaerobes are also commonly identified. The role of the mycoplasmas remains ill defined. There is some serological evidence to imply a role in a proportion of women with PID but they have rarely been isolated from Fallopian tubes.

The mechanism of infection is thought to be a primary infection by a sexually transmitted organism providing an initial insult to the female genital tract and rendering it susceptible to invasion by organisms that are normally non-pathogenic. The isolation of anaerobes tends to be associated with disease of increased severity and they are invariably present in tuboovarian abscess formation.[40] The significance of anaerobes in mild disease remains unclear.

PATHOGENESIS OF PID

Organisms ascend from the cervix via the endometrium to the Fallopian tubes. Evidence for this includes the demonstration of *N. gonorrhoeae* and *C. trachomatis* in the epithelial linings of all three sites.[41]

Spontaneous ascent is thought to occur in about 20% of women although the factors responsible are largely unknown.[42] This proportion is significantly increased by iatrogenic introduction of organisms during procedures involving transcervical instrumentation, such as insertion of IUCDs and termination of pregnancy. Here the rate has been quoted as high as 63% where there is preexisting cervical chlamydial infection.[43] *C. trachomatis* and *N. gonorrhoeae* give rise to characteristically different clinical pictures, although rate of development and severity of infection are also determined by the host response. *N. gonorrhoeae* initiates damage by invasion of epithelial cells, predisposing the then compromised genital tract to invasion by opportunistic aerobic and anaerobic bacteria.[44] It is more virulent than *C. trachomatis* and more frequently causes symptoms and pyrexia. It is therefore likely to be more highly represented in hospital populations. It is unusual to isolate *N. gonorrhoeae* from the upper genital tract,[45] which is in line with a proposed role as primary invader giving way to organisms of lower pathogenicity which invade the upper genital tract and produce the classic picture of polymicrobial disease.

The picture seen with *C. trachomatis*, in contrast, is generally more benign. It is likely to present later in disease and be underrepresented in the hospital population.[46] The organism itself shows little direct cytotoxicity and severity of sequelae are out of proportion to the clinical disease. There is now increasing evidence that a host cell-mediated immune response to a chlamydial heat-shock protein may be implicated in chronic disease, where recurrent exposure to chlamydial antigens is likely to provoke tubal damage.[47,48] An alternative theory is that asymptomatic or subclinical infection may persist as a chronic tubal infection and predispose to ectopic pregnancy and tubal infertility.[49]

TREATMENT OF PID

Although evaluation of most therapeutic regimes is based on immediate clinical response, it is important also to consider efficacy in terms of prevention of long-term morbidity. Follow-up of a cohort of women with a documented episode of PID showed a 10-fold increase in admissions for abdominal pain and ectopic pregnancy, an eight-fold increase in hysterectomy and a six-fold increase in occurrence of endometriosis, relative to control subjects.[50]

Whether treatment is initiated on an inpatient or ambulatory basis is largely related to clinical severity. Alternatively, a patient may be hospitalized during treatment if clinical response to outpatient therapy has been poor. The logic of this is questionable since the severity spectrum of clinical disease is poorly correlated with degree of tubal inflammation or extent of long-term damage. It has been suggested that outpatient management may be associated with more frequent adverse outcome although there are fewer data available than for hospitalized patients.[51]

Related to this is whether patients actually comply well with oral medication out of hospital.[52] A recent study showed only 31% compliance with doxycycline which could inevitably be improved by inpatient management, while also providing the opportunity for early laparoscopic intervention. However, in reality, to hospitalize or not is an individually based decision, setting expense and inconvenience against presumed therapeutic advantage.

Early treatment has been shown to improve outcome. Women who delay care for more than three days after the onset of symptoms are three times as likely to develop tubal infertility or ectopic pregnancy; these women are more likely to have recently undergone gynaecological procedures and thus attribute symptoms accordingly.[53] It is therefore important to be particularly vigilant in these cases.

Antimicrobial chemotherapy

Combination antibiotic therapy is the mainstay of treatment although there are efforts underway to develop a single antibiotic agent that is fully effective. In a statistical analysis of 101 clinical trials, the following conclusions were drawn.[54]

- There is no difference in cure rates between the regimens including cover for *C. trachomatis* and those that do not. In spite of these findings, it is still essential to cover for chlamydial infection to prevent long-term sequelae which are not quantified in these studies. Also, there are high levels of infection within the community which require identification and treatment.
- There is a significant therapeutic advantage with regimens that employ good anaerobic cover compared with those that do not. Severe disease is invariably polymicrobial and includes anaerobes. The role of antianaerobe therapy in mild disease is less clear.[40] Nevertheless, most would advocate its routine use.
- Higher cure rates are seen with the newer regimens, mainly the second- and third-generation cephalosporins and newer penicillins, compared with the older regimens, mainly tetracyclines and penicillins. The authors currently use doxycycline 100 mg bd in combination with augmentin one tablet tds. If treatment is given intravenously, cefuroxime 750 mg tds and metronidazole 500 mg tds are given along with doxycycline. For those patients known to have been in contact with *N. gonorrhoeae*, or where the organism is isolated, ciprofloxacin 250 mg stat is added.

Non-antibiotic therapy

More invasive therapeutic techniques that have been used in severe polymicrobial disease include transvaginal aspiration of tuboovarian abscess under ultrasound guidance[55] and laparoscopic aspiration of abscesses with adhesiolysis.[56] Both are associated with improved short- and long-term prognosis in the small series studied.

Simple measures such as bed rest and non-steroidal antiinflammatories[57] do not produce long-term benefit although they have immediate and subjective benefits which are important. Physiotherapy in the form of balneotherapy has been shown to reduce frequency of abdominal pain, but has no effect on tubal occlusion or adhesion formation.[58]

Treatment of contacts

PID should be considered as a disease of both women and their sexual partners. If partners are not treated, reinfection will inevitably occur. Most repeat episodes occur within two years and 20% of women will have a repeat infection.[17]

Partners are often asymptomatic of infection and therefore will not spontaneously present for treatment. In one study, 59.7% of male partners of women with PID were found to have *N. gonorrhoeae*, *C. trachomatis* or non-specific urethritis (NSU).[59] Only 32% of those with *N. gonorrhoeae* or *C. trachomatis* and 8.5% with

NSU were symptomatic. Active initiation of follow-up and treatment therefore needs to be pursued. Epidemiological treatment also serves to reduce the infection pool in the community.

Consultation with the woman and her partner should also be used as an opportunity to provide information and education about preventive measures. Treatment issues such as compliance and abstinence from sexual contact during therapy should be fully addressed.

PREVENTION

The Swedish example has clearly shown how an active approach to surveillance and prevention may cause a significant reduction in morbidity from PID and ultimately the long-term sequelae of infertility, ectopic pregnancy and pelvic pain. The evidence clearly confirms that treatment is more effective prior to upper genital tract involvement[60] and many cases will be asymptomatic or undiagnosed and therefore remain untreated.

Awareness of STDs and how to avoid exposure is information that requires wide dissemination if prevention is to be effective. Since the majority of infections are asymptomatic, active case finding in the form of screening should be considered in certain situations. Also, accurate and early diagnosis of suspicious symptoms is required to prevent spread of infection from the lower to upper genital tract.

Population screening across the board is the ideal but is limited largely by cost. Buhaug et al., in an epidemiological model, have estimated that routine testing of all women between the ages of 18 and 24 having a gynaecological examination for any reason would be cost-effective if tests were at least two years apart.[61]

It is important to know local prevalences of infection within any given community as these vary and will affect cost-effectiveness of a screening programme. Another factor is the sensitivity of test employed. This has limited chlamydia screening in the past but with the introduction of new molecular biological techniques (LCR and PCR), this is less likely to be a problem provided that the costs of the tests allow their use as a screening tool.

Targeted screening for high-risk groups is currently employed to some extent. Screening of women prior to transcervical instrumentation, e.g. before termination of pregnancy or insertion of IUCD, is recommended.[62] Particularly in the former, the evidence clearly points to a reduction in rates of post-operative infection as well as cost benefits.[43,62,63] It may also be appropriate to target 'at-risk' women on the basis of epidemiological data and sexual history taking.

The success of screening for *C. trachomatis* in reducing disease prevalence will partly depend upon assiduous follow-up and treatment of sexual contacts. Facilities for this need to be in place and will often involve collaboration between different specialties (genitourinary medicine, gynaecology, family planning, general practice).

CONCLUSION

There are a number of areas where there is scope for improvement in detection and management of PID and its antecedents. The results of a large longitudinal study of women in Sweden with confirmed PID identify areas where efforts should be concentrated in order to minimize the impact of infertility.[17] The recommendations include improved surveillance of PID-producing STDs, focus on under-25-year-olds, prompt symptom evaluation, diagnosis and treatment and consideration of couples rather than individuals. They also suggest encouragement towards combined oral contraceptive use which may improve PID outcome.

These strategies are now to be evaluated in the UK, following a report from the Chief Medical Officer's expert advisory group.[64]

APPENDIX PROTOCOL USED IN A & E FOR MANAGEMENT OF WOMEN WITH POSSIBLE PID[5]

Are you going to see a woman presenting with ABDOMINAL PAIN?
Please make sure you ask the following questions.

Presenting Symptoms
  PV discharge
  Deep dyspareunia
  Urinary symptoms
  LMP
  Contraception
  Last sexual partner (regular/casual)

Past gynaecological history
  PID
  Ectopic pregnancy
  IUCD
  Laparoscopy – when and why?
  Pelvic surgery – when and why?

Past genitourinary history
  STDs

Past obstetric history
  Gravida   para
  Miscarriages(s)/abortion(s)

On examination, please check for these findings
  Temperature
  Abdominal tenderness/guarding/rebound
  Genital examination:
    Discharge (colour/smell/amount)
    Bleeding
  Speculum examination:
    Appearance of cervix

REMEMBER TO TAKE SWABS

  Bimanual examination:
    Uterine size/mobility/adnexal mass/tenderness

Investigations
  Urinalysis (MSU if indicated)
  Pregnancy test
  FBC
  Endocervical swabs for chlamydia and gonorrhoea
  Vaginal swabs for candida/TV/BV
  bHCG

What is your diagnosis?
  ?Ectopic pregnancy   Refer to gynae STAT
  ?Severe PID   Refer to gynae NOW
  ?Not sure   Discuss with gynae

Clinical PID which does not require hospital admission
  Start on appropriate antibiotics and analgesia
  Refer to GU clinic: same day/next morning
                      patient and sexual contacts

Treatment
  Mild/moderate PID

        Doxycycline 200 mg stat
                   +
        Doxycycline 100 mg bd 10 days
                   +
        Metronidazole 400 mg bd 10 days
        or Augmentin 375 mg tds 10 days

  & if GC a possibility
        Ciprofloxacin 500 mg stat
                   or
        Ampicillin 3 g, probenicid 1 g

Checklist
  Adequate history
  Proper examination
  Relevant tests
  Reliable diagnosis
  Appropriate antibiotics

REFERRAL

REFERENCES

1. Rolfs, R.T., Galaid, E.I. and Zaidi, A.A. (1992) Pelvic inflammatory disease: trends in hospitalisations and office visits 1979 through 1988. *Am. J. Obstet. Gynecol.*, **166**, 983–990.
2. Gjonnes, H., Dalaker, K., Anstad, G. et al. (1982) Pelvic inflammatory disease: etiologic studies with emphasis on *Chlamydia trachomatis. Obstet. Gynecol.*, **59**, 550–555.
3. Buchan, H. and Vessey, M. (1989) Epidemiology and trends in hospital discharges for pelvic inflammatory disease in England 1975–1985. *Br. J. Obstet. Gynaecol.*, **396**, 1219–1223.
4. Scholes, D., Stergachis, A., Heidrich, F.E. et al. (1996) Prevention of pelvic inflammatory disease by screening for cervical chlamydial infection. *N. Engl. J. Med.*, **334**(21), 1363–1366.
5. Wales, N.M., Barton, S.E., Boag, F.C., Booth, S.J. and Smith, J.R. (1997) An audit of the management of pelvic inflammatory disease. *Int. J. STD and AIDS*, **8**, 409–411.
6. Paavonen, J., Kiviat, N., Brunham, R. et al. (1985) Prevalence and manifestations of endometritis among women with cervicitis. *Am. J. Obstet. Gynecol.*, **152**, 280–286.
7. Paavonen, J., Aine, R., Teisala, K., Heinonen, P.K. and Punnonen, R. (1985) Comparison of endometrial biopsy and peritoneal cytologic fluid testing with laparoscopy in the diagnosis of acute pelvic inflammatory disease. *Am. J. Obstet. Gynecol.*, **151**, 645–650.
8. Catchpole, M. (1992) Sexually transmitted diseases in England & Wales: 1981–91. *CDR Rev.*, **2**(1), 1–7.
9. Anon. (1995) Sexually transmitted diseases quarterly report: genital infection with *Chlamydia trachomatis* in England and Wales. *CDR Rev.*, **5**(26), 122–123.
10. Westrom, L. (1988) Decrease in incidence of women treated in hospitals for acute salpingitis in Sweden. *Genitourinary Med.*, **62**, 59–63.
11. Mardh, P-A. (1986) An overview of the infectious agents of salpingitis, their biology and recent advances in methods of detection. *Am. J. Obstet. Gynecol.*, **138**, 933–951.
12. Evans, B., Tasker, T. and MacCrae, K.D. (1993) Risk profiles for genital infection on women. *Genitourinary Med.*, **69**(4), 257–261.
13. Grodstein, F. and Rothman, K.J. (1994) Epidemiology of pelvic inflammatory disease (review). *Epidemiology*, **5**(2), 234–242.
14. Washington, A.E., Cates, W. and Wasserheit, J. (1991) Preventing pelvic inflammatory disease. *J. Am. Med. Assoc.*, **266**(18), 2575–2580.

15. Bevan, C. and Ridgway, G.L. (1992) Pelvic inflammatory disease. *Br. J. Obstet. Gynaecol.*, **99**, 944–945.
16. Bontis, J., Varilis, D., Panidis, D. *et al.* (1994) Detection of *Chlamydia trachomatis* in asymptomatic women: relationship to history of contraception and cervicitis. *Adv. Contracep.*, **10**(4), 309–315.
17. Westrom, L. (1994) Sexually transmitted diseases and infertility. *Sexually Trans. Dis.*, **21**(2)(suppl), 32–37.
18. Farley, T.M., Rosenberg, M.J., Rowe, P.J., Chen, J.H. and Meink, O. (1992) Intra-uterine devices and pelvic inflammatory disease: an international perspective. *Lancet*, **339**, 785–788.
19. Soper, D.E., Brockwell, N.J., Dalton, H.P. and Johnson, D. (1994) Observations concerning the microbial aetiology of acute salpingitis. *Am. J. Obstet. Gynecol.*, **170**(4), 1008–1014.
20. Faro, S., Martens, M., Maccato, M., Hammill, H. and Pearlman, M. (1993) Vaginal flora and pelvic inflammatory disease. *Am. J. Obstet. Gynecol.*, **169**(2), 470–474.
21. Scholes, D., Daling, J.R., Stergachis, A. *et al.* (1993) Vaginal douching as a risk factor for acute pelvic inflammatory disease. *Obstet. Gynaecol.*, **81**(4), 601–606.
22. Mueller, A., Daling, J.R., Weiss, N.S. *et al.* (1990) Recreational drug use and the risk of primary infertility. *Epidemiology*, **1**, 195–200.
23. Stacey, C.M. and Munday, P.M. (1994) Abdominal pain in women attending a genitourinary medicine clinic: who has PID? *Int. J. STD and AIDS*, **5**, 338–342.
24. Jacobsen, L. and Westrom, L. (1969) Objectivized diagnosis of acute pelvic inflammatory disease. *Am. J. Obstet. Gynecol.*, **105**, 1088–1098.
25. Westrom, L. and Mardh, P-A. (1990) Acute pelvic inflammatory disease, in *Sexually Transmitted Diseases*, 2nd edn, (eds K.K. Holmes, P-A. Mardh, P.F. Sparling *et al.*), McGraw-Hill, New York, pp. 593–613.
26. Miettenen, A.K., Heinonen, P.K., Laippala, P. and Paavonen, J. (1994) Test performance of erythrocyte sedimentation rate and C-reactive protein in assessing the severity of acute pelvic inflammatory disease. *Int. J. Gynaecol. Obstet.*, **44**(1), 53–57.
27. Mozas, J., Castilla, J.A., Jimen, P. *et al.* (1994) Serum CA125 in diagnosis of acute pelvic inflammatory disease. *Am. J. Obstet. Gynecol.*, **171**(7), 102–110.
28. Hay, P.E., Thomas, B.J., Horner, P.J. *et al.* (1994) *Chlamydia trachomatis* in women: the more you look, the more you find. *Genitourinary Med.*, **70**(2), 97–100.
29. Fish, A.N.J., Fairweather, D.V.I., Oriel, J.D. and Ridgway, G.L. (1988) Isolation of *Chlamydia trachomatis* from the endometriums of women with and without symptoms. *Genitourinary Med.*, **64**, 75–77.
30. Johal, B., Ridgway, G.L. and Siddle, N.C. (1990) Management of pelvic inflammatory disease. *Int. J. STD and AIDS*, **1**, 401–404.
31. Patton, D.L., Moore, D.E., Spadoni, L.R. *et al.* (1989) A comparison of the Fallopian tube's response to overt and silent salpingitis. *J. Obstet. Gynaecol.*, **73**, 622–630.
32. Sellors, J., Mahoney, J., Goldsmith, C. *et al.* (1991) The accuracy of clinical findings and laparoscopy in pelvic inflammatory disease. *Am. J. Obstet. Gynecol.*, **164**(1), 113–120.
33. Stacey, C., Munday, P., Thomas, B. *et al.* (1990) *Chlamydia trachomatis* in the fallopian tubes of women without laparoscopic evidence of salpingitis. *Lancet*, **336**, 960–963.
34. Mardh, P-A. (1980) An overview of infectious agents of salpingitis, their biology and recent advances in methods of detection. *Am. J. Obstet. Gynecol.*, **138**, 933–951.
35. Mardh, P-A., Ripa, K.D., Svensson, I. and Westrom, L. (1977) *Chlamydia trachomatis* infection in patients with acute salpingitis. *N. Engl. J. Med.*, **296**, 1377–1379.
36. Trehorne, J.D., Ripa, K.D., Mardh, P-A., Svensson, L. and Westrom, L. (1979) Antibodies to *Chlamydia trachomatis* in acute salpingitis. *Br. J. Vener. Dis.*, **55**, 26–29.
37. Bevan, C.D., Johal, B.J., Mumtaz, G., Ridgway, G.L. and Siddle, N.C. (1995) Clinical, laparoscopic and microbiological findings in acute salpingitis: report on a UK cohort. *Br. J. Obstet. Gynaecol.*, **102**, 407–414.
38. Stacey, C.M. *et al.* (1992) A longitudinal study of pelvic inflammatory disease. *Br. J. Obstet. Gynaecol.*, **99**(12), 992–997.
39. Rice, G. and Schachter, J. (1991) Pathogenesis of pelvic inflammatory disease: what are the questions? *J. Am. Med. Assoc.*, **266**(18), 2587–2593.
40. Heinonen, P.K. and Miettenen, A. (1994) Laparoscopic study of the microbiology and severity of acute pelvic inflammatory disease. *Eur. J. Obstet. Gynecol. Reprod. Biol.*, **57**(2), 85–89.
41. Paavonen, J., Teisala, K., Heinonen, P.K. *et al.* (1987) Microbiological and histopathological findings in acute pelvic inflammatory disease. *Br. J. Obstet. Gynaecol.*, **94**, 454–460.
42. Gilbert, G.C. and Weisberg, E. (1993) Infertility as an infectious disease – epidemiology and prevention. *Baillière's Clin. Obstet. Gynaecol.*, **7**, 159–181.

43. Blackwell, A., Thomas, P.D., Wareham, K. and Emery, S.J. (1993) Health gains from screening for infection of the lower genital tract in women attending for termination of pregnancy. *Lancet*, **342**, 207–210.
44. Paavonen, J., Valtonen, V.V., Kasper, D.L., Malkamaki, M. and Makela, P.K. (1989) Serological evidence for the role of *Bacteroides fragilis* and enterobacteriaceae in the pathogenesis of acute PID. *Lancet*, **i**, 293–295.
45. Eschenbach, D.A., Buchanan, T.M., Pollock, H.M. et al. (1975) Polymicrobial aetiology of acute pelvic inflammatory disease. *N. Engl. J. Med.*, **293**, 166–171.
46. Svensson, L., Westrom, L., Ripa, R.T. and Mardh, P-A. (1980) Differences in some clinical and laboratory parameters in acute salpingitis related to culture and serologic findings. *Am. J. Obstet. Gynecol.*, **138**, 1017–1021.
47. Wagar, E.A., Schachter, J., Bavoil, P. and Stephen, R.S. (1990) Differential human serologic response to two 60 000 molecular weight *Chlamydia trachomatis* antigens. *J. Infect. Dis.*, **162**, 922–927.
48. Toye, B., Laferriere, C., Claman, P., Jessamine, P. and Peeling, R. (1993) Association between antibody to the chlamydial heat shock protein and tubal infertility. *J. Infect. Dis.*, **168**, 1236–1240.
49. Henry-Suchet, J., Utzmann, C., Debrux, J., Ardoin, P. and Catalan, F. (1987) Microbiologic study of chronic inflammation associated with tubal factor infertility: role of *Chlamydia trachomatis*. *Fertil. Steril.*, **47**(2), 274–277.
50. Buchan, H., Vessey, M., Goldacre, M. and Fairweather, J. (1993) Morbidity following pelvic inflammatory disease. *Br. J. Obstet. Gynaecol.*, **100**(6), 558–562.
51. Peterson, H.B., Walker, C.K., Kahn, J.G. et al. (1991) Pelvic inflammatory disease: key treatment issues and options. *J. Am. Med. Assoc.*, **266**(18), 2605–2611.
52. Brookoff, D. (1994) Compliance with doxycycline therapy for outpatient treatment of pelvic inflammatory disease. *South. Med. J.*, **87**(11), 1088–1091.
53. Hillis, S.D., Joesoef, R., Marchbanks, P.A. et al. (1995) Delayed care of PID as a risk factor for impaired fertility. *Am. J. Obstet. Gynecol.*, **168**(5), 1503–1509.
54. Dodson, M.G. (1994) Antibiotic regimes for treating acute PID: an evaluation. *J. Reprod. Med.*, **39**(4), 285–296.
55. Cacciatore, B., Leminen, A., Ingman-Freiberg, S., Ylostalo, P. and Paavonen, J. (1992) Transvaginal sonographic findings in ambulatory patients with suspected pelvic inflammatory disease. *Obstet. Gynaecol.*, **80**(6), 912–916.
56. Henry-Suchet, J. and Tesquier, L. (1994) Role of laparoscopy in the management of pelvic adhesions and pelvic sepsis. *Baillière's Clin. Obstet. Gynaecol.*, **8**(4), 759–772.
57. Landers, D.V., Sung, M.L., Bottles, K. and Schachter, J. (1993) Does addition of anti-inflammatory agents to antimicrobial therapy reduce infertility after murine chlamydial salpingitis? *Sexually Trans. Dis.*, **20**(3), 121–125.
58. Gerber, B., Wilkin, H., Zachanas, K., Barten, G. and Splitt, G. (1992) Treatment of acute salpingitis with tetracycline/metronidazole with or without additional balneotherapy; a second look laparoscopy study. *Geburshilfe Frauenheilk*, **52**(3), 165–167.
59. Kamwendo, F., Johansson, E., Moi, H., Forslin, L. and Danielsson, D. (1993) Gonorrhoea, genital chlamydial infection and non-specific urethritis in male partners of women hospitalized for acute PID. *Sexually Transm. Dis.*, **20**(3), 143–146.
60. Cates, W.C., Rolfs, R.T. and Aral, S.O. (1990) STDs, PID and infertility: an epidemiologic update. *Epidemiol. Rev.*, **12**, 119–120.
61. Buhaug, M., Skejeldestad, F.E., Backe, B. et al. (1989) Cost-effectiveness of testing for chlamydial infections in asymptomatic women. *Med. Care*, **27**, 833–841.
62. Qvigstad, E., Skaug, K., Jerve, F. et al. (1983) Pelvic inflammatory disease associated with *Chlamydia trachomatis* in first trimester abortion. *Br. J. Vener. Dis.*, **59**, 189–192.
63. Skejeldestad, F.E., Tuveng, J., Solberg, A.G. et al. (1988) Induced abortion – *Chlamydia trachomatis* and post-abortal complications: a cost-benefit analysis. *Acta Obstet. Gynecol. Scand.*, **67**, 525–529.
64. Chief Medical Officer's expert advisory group (1998) *Chlamydia trachomatis*. London.

14 The diagnosis and management of psychosexual problems in genitourinary medicine

P. Woolley

INTRODUCTION

At first sight it might appear that people who attend a GUM clinic should not have sexual problems, apart from any discomfort from the disease. Indeed, few individuals are sent to, or attend, GUM clinics primarily because of overt sexual difficulties. Unfortunately, due to the high pressure of work in clinics, this has meant that frequently there is insufficient time to delve into the possibility of psychosexual problems at the patient's first attendance.

The treatment of sexual dysfunction has never been a high priority in health service planning, with sex therapy services generally being left to district health authorities. It is difficult to estimate the need for such services as the level of sexual dysfunction in the population cannot be accurately gauged. However, what studies have been performed show a high level of morbidity in the general community[1] and in patients attending GPs' surgeries.[2]

Any person attending a doctor has some degree of apprehension and this applies to those who seek advice regarding genital tract infections, particularly as it involves discussions of the person's sexual activity. Indeed, the estimated prevalence rates of psychological and psychiatric morbidity amongst GUM clinic attenders may be artificially high, with reports of rates of up to 31% using the Hospital Anxiety Depression Scale,[3] particularly as several studies using the General Health Questionnaire (one additionally using the Crown-Crisp Experimental Index and Illness Concern Questionnaire) have reported that one quarter were misclassified and in many instances, psychological distress was related to the presenting problem.[4,5]

Persons attending a genitourinary clinic primarily because of psychosexual difficulties form only a minority of new attendances. In a problem-orientated categorization of 'other conditions' seen in a genitourinary clinic, these accounted for only 0.2% of attendances.[6] The prevalence of self-reported psychosexual problems amongst patients attending a genitourinary clinic has been reported to be between 10% and 20% for men and 9% and 25% for women.[7,8]

The main psychosexual complaint seen in a London genitourinary medicine clinic was shown to differ markedly from those conditions seen in a community psychosexual clinic (Table 14.1).[9] It is difficult to compare these results with others, as in an Oxford genitourinary medicine clinic, the most common male problem was premature ejaculation and coital-orgasmic dysfunction was the most common indication in females, but problems with relationships or fears of infection were not included in the questionnaires used.[7]

Table 14.1 Differences in psychosexual problems between genitourinary medicine and community psychosexual sessions

	Genitourinary clinic	Community clinic
Male problems		
Premature ejaculation	14.1%	41.7%
Fear of infection	26.6%	–
Erectile failure	7.8%	25.0%
Relationships	32.8%	8.3%
Others*	15.6%	25.0%
None	3.1%	–
Female problems		
Vaginismus	16.7%	12.2%
Low libido	5.6%	51.0%
Fear of infection	19.4%	–
Anorgasmia	2.8%	12.2%
Relationships	38.9%	18.4%
Other**	11.1%	6.1%
None	5.6%	–

* Includes non-ejaculation, gender orientation.
** Includes vaginal pain, postsexual assault.

PROBLEMS IN MEN

Few men pass through life without experiencing erectile inadequacy on at least one occasion, but it is only when this happens on more than an occasional basis that it has significance. It is difficult to ascertain the incidence of erectile dysfunction in the general population, but it may be as high as one in 10 men across all ages. In a questionnaire study conducted in the United States aimed to determine the frequency of sexual dysfunction in predominantly Caucasian, well-educated and 'happily married' couples, 40% reported erectile or ejaculatory dysfunction; difficulty in achieving an erection was reported by 7%; difficulty in maintaining an erection was reported by 9%.[10]

Premature ejaculation

Premature ejaculation is a condition in which a man is unable to exert voluntary control over his ejaculatory reflex with the result that, once sexually aroused, he reaches orgasm sooner than he would like. Often this leaves his partner 'unsatisfied', to the extent that the couple feel that their sexual life is being spoiled by the man's too rapid ejaculation. In biological terms most primates, as well as lower species, ejaculate within seconds and only man's desires for a 'pleasure phase' have evolved as sex acquires functions beyond that of reproduction.

In the most comprehensive study of premature ejaculation, it became apparent that men with this problem fall into two types. The first group are generally younger, have good erections but, from the start of their erectile awareness, complain of rapid ejaculation. The other group are mainly older men who usually have periods of normal ejaculation, but who find themselves ejaculating quickly in association with erectile loss.[11] The former exhibit true premature ejaculation, while the response of the latter is really an expression of impotence; their rapid response often alternates with erection loss without ejaculation.

The main causes of premature ejaculation can be divided into either physical or psychological/emotional. It is essential that a thorough examination of both the neurological and urological systems be undertaken in a person who has had previous good control of ejaculation.

Much of the literature has focused on psychological and emotional causes. Analytic ideas are based on the belief that such men are hostile to or frightened of women and that their rapid ejaculation is one way in which they can 'punish them' by prematurely terminating coitus. Similar dynamics have been applied to men who are unable to ejaculate into the vagina, this being linked with their unconscious fears of soiling from an early age.

Early conditioning theories view the first sexual contacts as important.[12] Learning control is not possible until sensory feedback is perceived, e.g. a young child has to learn what a full bladder feels like before he or she can learn to control the bladder reflex. Causes for not learning control of the ejaculatory reflex are infrequently identified, but examples of such men consorting with prostitutes or having sex in the back seat of a car and other scenarios in which speed of response is important set their pattern of rapid response which remains with them.

When physical causes have been excluded, treatment programmes are based on the assumption that sensory feedback or the sensation of preorgasmic arousal will result in ejaculatory control. The man is instructed to focus his attention on prolonged, intense levels of arousal so that he can learn to perceive the intense preorgasmic sensations. The programmes begin with a ban on sexual intercourse and are based upon sensate focus (Table 14.2). The couple are initially instructed on non-demand, non-genital touching, in which the emphasis is on improving their communication skills. When this has been achieved, stage 2 is discussed. This is non-demand, genital touching, the emphasis being on erotic physical feelings and allowing them to be perceived. The next step is to continue the programme

Table 14.2 Sensate focus

Stage 1	Non-demand, non-genital touching
Stage 2	Non-demand, genital touching
Stage 3	Instruction in activity related to the particular dysfunction, e.g. squeeze technique for premature ejaculation
Stage 4	Sexual intercourse

Table 14.3 Organic causes of erectile dysfunction

Endocrine causes	Erectile dysfunction may result from androgen deficiency caused by pituitary or testicular disease. This leads to a reduced plasma testosterone level
Neurological disease	Nerve damaged either through demyelinating conditions such as multiple sclerosis or autonomic disease or as a result of abdominal/pelvic surgery may be causative factors
Diabetes mellitus	Diabetic neuropathy, either involving the sacral nerves which mediate erection or central autonomic disease, may cause erectile dysfunction, sometimes as the first symptoms
Drugs	Alcohol, opiates, antihypertensive agents, antidepressants and oestrogen may lead to erectile loss
Vascular disease	Obstruction of the abdominal aorta at its bifurcation into the common iliac arteries may cause claudication of the lower limbs and/or erectile dysfunction. Venous leaks due to prolonged ischaemia of the corpora cavernosa result in blood entering the penis but not being trapped by the subcorporeal venous plexus; consequently erection does not occur
Peyronie's disease	A painful plaque in the corpora cavernosum causing fibrosis of the sinusoidal spaces or corporal artery occlusion may lead to penile curvature and decreased rigidity
Embarrassment or fear	e.g. persons with ileostomies, colostomies, postmyocardial infarction, cerebrovascular accident
Other causes	Penile trauma, phimosis, priapism

but the partner is specifically instructed to heighten sexual erotic feelings using a start/stop technique, whereby genital stimulation produces intense pre-orgasmic arousal to a point where orgasm is almost inevitable, at which point the man signals to stop. When the sensations have subsided, restimulation takes place.

When this part of the programme is proceeding well, penetration can be attempted, initially in the female superior position. Vaginal containment with female movement follows using the start/stop technique at the moment the man feels his ejaculation is inevitable. It is suggested that this should be repeated every 2–3 weeks to maintain progress, progressing to male thrusting and position variation.

The 'squeeze' technique is commonly used as an adjunct to the start/stop technique, in which the woman firmly squeezes the penis at the point where the glans meets the shaft which reduce sensations more quickly.

Drug therapy may also be helpful as an adjunct but is rarely successful when used alone. Clomipramine, a tricyclic antidepressant, works by increasing the arousal latency and delays ejaculation. It can be given in a single daily dose of 10 mg, increasing by 10 mg at weekly intervals until a total daily dose of 50–75 mg is achieved.[13] However, success is frequently lost once treatment is withdrawn.

Erectile dysfunction

Erectile dysfunction may be defined as the failure to produce penile erection of sufficient rigidity to achieve sexual intercourse. Long-term erectile dysfunction is thought to affect at least 15% of men of middle age, 20% of men over 60 and 65% of those in their 70s. Until recently, most cases were assumed to be psychogenic in origin but it is now realized that there may be an organic component in up to 50% of cases (Table 14.3).[14]

An organic cause should be suspected when, following a previous uninterrupted period of normal sexual activity, erectile dysfunction occurs under all circumstances not associated with a psychiatric disorder. With physical disorders, the onset tends to be insidious, whereas with psychological disorders, it is fairly acute and frequently intermittent or only occurs in certain situations. A full physical examination with laboratory tests should be performed.

Psychogenic erectile dysfunction can be distinguished from organic dysfunction by monitoring nocturnal penile tumescence. Spontaneous erections during periods of rapid eye movement (REM) sleep are normal in the former but may be absent or impaired in the latter. However, nocturnal penile tumescence is not diagnostic and increases in penile circumference are not necessarily accompanied by penile rigidity.[15] A computer-linked system that records and analyses change in both penile circumference and rigidity has been developed which can be used to assess nocturnal penile activity or penile responses to erotic stimuli. However, it is not generally available in clinical practice.

Significant advances in the diagnosis and management of erectile dysfunction have been made in recent years. Injecting drugs into the corpus cavernosum to induce erection, often referred to as PIPE (pharmacologically induced penile erection), CUMRI (cavernosal unstriated muscle relaxant injection) or ICP (intracorporeal injection of papaverine), has been used in the management of erectile dysfunction for the last decade.[16] This is especially appropriate for patients with diabetic neuropathy, other neuropathies or other evident neurological cause but is less successful in those whose dysfunction is attributed to peripheral vascular disease.

Papaverine is the most commonly injected agent which can be self-administered and exerts muscle-relaxing activity by direct inhibition of phosphodiesterase. When ineffective alone, it can be combined with the α-adrenoreceptor antagonist phentolamine mesylate.[14] The dose needed to give an erection of the desired duration, i.e. at least 40 min but not more than four hours, varies from patient to patient but is usually between 12 mg and 80 mg. Side effects include burning or bruising at the site of injection, fibrotic changes in the corpus cavernosa which may lead to a Peyronie's-like erectile distortion and painful prolonged erection which should be considered a medical emergency if it persists for more than four hours.[17] The aspiration of blood through a large-bore needle is undoubtedly the initial treatment of choice in drug-induced priapism, since it is relatively safe and a prerequisite for the injection of an α-adrenoreceptor agonist if it fails. If these measures fail, patients may require a shunt procedure.[18]

Intrapenile injection of vasodilators such as phentolamine and vasoactive intestinal peptide (VIP) as well as prostaglandin E1 can also be used to induce erection.[19,20] The latter does not produce prolonged erections and beneficial effects appear to be long lasting (months)[21] (Table 14.4).

Failure to sustain an erection following an adequate dose of papaverine or prostaglandin E1 suggests a venous leakage, especially if erections subside very quickly. Those with neuropathic erectile dysfunction tend to have an exaggerated response and may require lower doses.

Dynamic infusion cavernosometry can be used to diagnose venous leaks which do not become smaller or disappear on erection and can be treated by surgical ligation. In the short term, the results are encouraging but the problem may recur in other veins. Arterial occlusive disease is common and, if it affects the iliac arteries, can be a cause of erectile dysfunction. Diagnosis frequently rests on finding arterial stenoses on angiography and treatment can be surgical, by laser therapy or balloon dilatation.

The use of the implantable penile prosthetic splint has become more frequent in those with organic erectile dysfunction, particularly if there is no evidence of response to intracorporeal injections. The simplest type is the semirigid or malleable rod implanted into each corpus cavernosum. The second main group consists of two cylinders with self-contained fluid-filled reservoirs. At each end of the device, situated beneath the glans, there is a pump which is activated by pressure on the glans, causing transfer of fluid from the terminal reservoirs into the longitudinal cylinders, producing rigidity and simulating an erection. However, the best functional result is produced by the inflatable penile prosthesis in which a pair of inflatable silicone cylinders are placed into each corpus cavernosum. A fluid-filled reservoir is buried beneath the lower ends of the rectus abdominis muscle and a pump is implanted in the scrotum. The three components are connected by silicone tubing and the prosthesis is activated by repeated pressure on the scrotal pump which transfers fluid from the reservoir into the cylinders.

Table 14.4 Commonly used treatment for erectile dysfunction

Compound	Action	Efficacy	Rapidity	Side effects
Alprostadil (Caverject)	Strong, smooth muscle relaxant	75%	5–20 min	Dizziness, penile pain, hypotension
Papaverine plus phentolamine	Strong, smooth muscle relaxant. Blocks α-adrenoreceptors	65%	10–20 min	Fibrosis, hypotension, bruising
Yohimbine (oral treatment)	Blocks presynaptic α_2-adrenergic receptors	Unknown	Unknown	CNS symptoms, e.g. sweating, tremor, nausea, tachycardia

The indole alkaloid yohimbine is an antagonist of α_{-2}-adrenoreceptors and 5-HT and has been shown to restore erectile function in a small group of men with organic or psychogenic impotence.[22] At present, it is not registered in the United Kingdom but is available on a named patient basis.

Summary
The management of psychogenic erectile dysfunction is extremely difficult and may require prolonged counselling sessions to determine the precipitating cause. Sometimes showing the patient that an erection is possible following the injection of papaverine may be sufficient to overcome his problem.

Retarded ejaculation

Retarded ejaculation is defined as a specific inhibition of the ejaculatory reflex, whereby a man finds it difficult or is unable to ejaculate, although he has a great desire to do so and the stimulation he receives should be adequate for this. The severity of the problem varies from occasional intermittent episodes to such gross inhibition that orgasm cannot be achieved. In all cases other sexual feelings such as libido, desire and arousal are normal.

Occasionally the problem arises because of an unconscious fear of the possible negative consequences of ejaculation, such that he avoids depositing semen in his partner's vagina since to do so would produce great anxiety for him and, as such, is a learned response. More commonly, the problem is caused by destructive communication within the relationship, i.e. it may be the final act left to him in a marital power struggle.

Treatment is aimed at looking at the effect of negative feelings on ejaculation and trying to loosen the association between the two. Non-demand, non-genital touching followed, if successful, by non-demand, genital touching (allowing orgasm through manual stimulation) immediately takes pressure off the sexual system and leads to increased sexual arousal and thereby, in the man, produces an association between being sexual with his partner and ejaculation without unconscious negative feelings. The man can then be instructed to manually stimulate to impending orgasm and then enter the vagina as he is about to ejaculate. Gradually the man is able to enter the vagina at lower levels of arousal, progressively more removed from ejaculation.

PROBLEMS IN WOMEN

The role of endogenous testosterone in female sexual desire and orgasm is poorly understood, but it appears from studies that coital frequency increases at the time of ovulation which coincides with an increase in testosterone level.[23,24] As, in clinical practice, exogenous androgen usually takes two weeks to affect behaviour, it is possible that this midcycle surge is responsible for the premenstrual rise in desire felt by many women.

Vaginismus

Vaginismus is a conditioned response that most likely results from an association of pain and fear consisting of a phobia of penetration and involuntary spasm of the muscle surrounding the lower third of the vagina. Although it can occasionally be associated with general inhibition or orgasmic dysfunction, most sufferers remain sexually responsive. Efforts at penetration lead to physical pain, fear, humiliation and frustration that frequently cause feelings of inadequacy and fear of abandonment.

It is important to take a detailed sexual history to determine whether the sufferer has ever experienced vaginal penetration (primary vaginismus) or has experienced vaginal penetration without problems in the past but vaginismus has developed subsequently (secondary vaginismus).

Primary vaginismus frequently stems from unresolved sexual conflicts in childhood or early adolescence. Often such women are emotionally immature and may have been subject to sexual abuse in childhood or were brought up to believe that sex was 'dirty' or 'bad'.

Secondary vaginismus can occur as a result of painful lesions of the genitalia, e.g. vulvovaginal infections, inflammatory dermatological lesions of the vulva, Bartholin's abscess or cyst, postpartum tears or episiotomy, endometriosis or pelvic pathology which can produce discomfort during penetrative intercourse. Even though the initial problem may have been a physical one, psychological factors are important in the development of the condition. In most cases a specific cause or initiating event for the vaginismus cannot be found and the problem is assumed to be multifactorial in origin.

All couples follow a basic programme tailored to meet individual needs which progresses at a pace determined by the woman (Table 14.5).

Initially, it is important to spend time on basic sex education which provides the opportunity for the couple to talk about sexual attitudes and expectations. If the couple have any degree of physical contact, a programme of non-demand, non-genital touching progressing, if all goes well, to non-demand, genital touching can be undertaken. If the couple have had no physical contact then communication exercises will

Table 14.5 Treatment programmes for the management of vaginismus

1. Education regarding sexual function
2. Exerting control of vaginal muscles
3. Visual exploration of the vulva
4. Vaginal insertions (non-penile)
5. Conjoint examination with the partner
6. Transfer of vaginal insertion to partner (non-penile)
7. Insertion of penis with woman in control
8. Transfer of control of insertion of penis to partner

have to be commenced on a more basic level. Many women withdraw from all physical contact because they feel that any contact automatically leads to sexual intercourse. Once they have discovered that this is not the case, they can develop trust in their partner and allow further, increasingly intimate physical contact.

While the above programme is running, a second programme for the woman can be run alongside, attempting to overcome the woman's fear of penetration. The first part of the programme is to teach the woman to relax her vaginal muscles by instructing her to tighten and relax the muscle group regularly for 10 min twice daily, focusing on relaxation. Once control of the vaginal muscles has been achieved, the woman is instructed to look at her vulva and perineum and, when she is able to do this without feeling threatened, she is instructed to touch them gently.

The next stage is to start vaginal insertion by relaxing the vaginal muscles, applying plenty of lubrication and entering the vagina, gently moving only when the muscles are relaxed. Quite often a woman is happy to commence with a finger then, once this has been achieved, she can move to the insertion of two or ever three fingers in the same way. Some women, however, may prefer to use graded vaginal trainers, the size being gradually increased.

Frequently the programme is helped by the partner being involved. Instructing the couple to look at each other's genital area can be very useful, especially if accompanied by discussion about their feelings for each other.

Control of penetration can gradually be given to the partner, either inserting his finger or a vaginal trainer. Women who have achieved insertion of a range of trainers are instructed to consider the penis as a kind of trainer, using the female superior position in which she relaxes her vaginal muscles and inserts the penis using plenty of lubrication. Once the woman has managed to insert the penis she can attempt movement and when this has been successfully achieved, movement can be attempted by the man. Finally control of insertion of the penis can be returned to the man. Following this, different positions can be used.

Other help can be offered to the couple if the programme gets 'stuck' but this is probably the province of the experienced psychosexual counsellor. Systematic desensitization to eliminate irrational fears using imagery is sometimes helpful, while in other situations, hypnosis can be tried.

Support of the vaginismus sufferer during therapy is vitally important, as is the understanding that she will probably experience some degree of anxiety and will find the programme hard work.

Orgasmic dysfunction

For men, orgasm is an essential part of the sexual act whilst some women appear to be quite content without it. However, women can feel under pressure to experience orgasm, often stemming from the male perception that his partner cannot enjoy sexual satisfaction without it, and this may lead to disenchantment with her sexuality if she fails to achieve it.

Orgasmic dysfunction or anorgasm is the involuntary inhibition of the orgasmic reflex. Women frequently have a strong sexual desire and enjoy intercourse but 'hold back' even though the stimulus would normally be sufficient for orgasm. This fear, either conscious or unconscious, is of losing control over feelings and/or behaviour and the resultant reaction of 'holding back' can be powerful. Orgasmic dysfunction can be divided into absolute anorgasmia, in which the woman is anorgasmic in both coital and clitoral stimulation under any circumstances, and situational anorgasmia in which orgasm can be reached in low-tension situations but fails when there is any form of anxiety.

The aim of therapy is to shift sexual experience from 'achievement' of orgasm to the giving and receiving of pleasure while being distracted from inhibitory 'over control.' The treatment programme should include both partners and concentrate on correcting misconceptions of sexual roles, enhancing communication, education, particularly relating to the need for clitoral stimulation, and effective techniques to improve female sexual response. This is best achieved in a sensate focus programme involving non-demand, non-genital touching in which both partners explore their own and each other's genital area finding out what are the most pleasant sensations. Success at this stage can be followed by non-demand, genital touching during which simple exercises can be introduced to heighten female sexual arousal, such as stop/start clitoral stimulation techniques, trying different types of stimulation, e.g. sex toys, pelvic rocking and tightening and relaxing vaginal muscles. If a woman still finds that she is unable to achieve orgasm, she can be encouraged to explore sexuality alone and, once orgasm is reached, can involve her partner. Once orgasm has been achieved in this way, the couple are instructed to have intercourse using the female superior position, since this allows maximum clitoral stimulation while enjoying the pleasurable sensations rather than attempting orgasm. However, once her partner has ejaculated, he should continue clitoral stimulation manually to orgasm. Gradually, orgasm for the woman is achieved at lower levels of stimulation, concluding with orgasm during intercourse on a regular basis.

The majority of women will succeed in their quest for orgasm, but some will not and these tend to be women who have an unconscious fear of loss of control which can also be seen in other areas of their lives. However, even these women should have received considerable reassurance in lovemaking, leading to the development of a more satisfying sexual relationship.

Dyspareunia and pelvic pain

It is rare for sexually active women to go through life without experiencing dyspareunia on at least one occasion, while for some, introital pain may be a persistent or recurring problem resulting in severe strain on their sexual relationship. In a community-based survey in Oxfordshire, 8% of women reported pain or discomfort during intercourse and 17% vaginal dryness causing difficulties more than half of the time during the previous three months.[25] However, few women attending medical practitioners report coital pain as their major complaint unless specifically asked.[26]

When vulvovaginal pathology or infection have been excluded, the commonest causes of superficial dyspareunia are inadequate arousal and vaginismus, both of which are amenable to education and/or specific therapy.[27] A condition commonly seen in GUM clinics is when a physical cause of dyspareunia such as pelvic inflammatory disease or severe candidal vulvovaginitis is complicated by a secondary psychological reaction consisting of a fear of pain and sexual avoidance. A vicious circle is established, whereby fear or pain inhibits vaginal lubrication giving rise to vaginal dryness, superficial dyspareunia on attempted intercourse, more physical pain and a secondary vaginismus.

When symptoms of vaginal dryness occur at the expected time of the menopause, they can easily be attributed to oestrogen deficiency. However, in younger women, vaginal dryness may signal the forthcoming onset of premature menopause or hyperprolactinaemia. As might be expected, dyspareunia due to vaginal dryness is associated with decreased frequency of intercourse, lower orgasmic experience and reduced coital enjoyment, all of which can be managed by the addition of a suitable lubricant gel (remembering that oil-based lubricants have a deleterious effect on latex condoms), topical oestrogen creams or hormone replacement therapy.

Deep dyspareunia during intercourse is usually due to a mechanical problem, the commonest cause being the upward displacement of the uterus by the penis with stretching of the richly innervated supporting ligaments. Occasionally, if the penis negotiates one of the lateral fornices, especially if the ovary is enlarged and tender around the time of ovulation, ovarian compression can result in pain. Such problems can usually be overcome by varying sexual positions. Deep dyspareunia is not uncommon after severe pelvic inflammatory disease, presumably due to adhesions, or after pelvic surgery. Pelvic ultrasound is recommended to exclude pelvic pathology, particularly ovarian cysts.

Women with non-organic pelvic pain present to a wide spectrum of medical specialties as the pain, as far as the woman is concerned, is a real entity and is of sufficient severity to warrant seeking medical attention. The pain is frequently of more than six months duration, is felt in the lower abdomen, often localized to one side or one iliac fossa and radiates to the vagina, lower back or down the thighs. The upright position and movement of the pelvic organs during running, jumping and deep penetration during intercourse frequently aggravate the condition. Premenstrual exacerbation with relief after the onset of menstruation and prolongation of the Mittelschmerz have been described. Women with non-organic pelvic pain are usually of reproductive age and the syndrome seldom occurs in the perimenopause and never after the menopause.[28]

Many women are polysymptomatic and, besides pelvic pain, also complain of bowel and bladder disturbances, vaginal discharge, menstrual difficulties and vague bodily discomfort reflecting somatization of anxiodepressive states.[29] Examination of sites where pain is experienced, particularly in the lower abdomen, may reveal defined areas of hyperaesthesia.[30] Gynaecological assessment frequently elicits marked tenderness when the uterus, adnexae and posterior parametrium are palpated. A past history of an organic source of pelvic pathology, such as pelvic inflammatory disease or trauma following childbirth, enhances the likelihood of expression of pelvic pain.

Often women with non-organic pelvic pain are found to have had an insecure family life during their childhood and are emotionally immature. They have difficulties in accepting their feminine roles and lack the ability to perform satisfactorily both as sexual partners and as mothers.[31] Interpersonal relationships are often shallow and a tendency towards hysteria and anxiodepression is common.[32,33] Furthermore, a history of incest in over a third of women has been found in one study, suggesting that this may contribute to their problems with dependency and trust.[34] Expression of pain in the pelvic organs is a logical extension of the psychopathology of such women.

Appreciation of the multifactorial nature and complexity of non-organic pelvic pain is essential, as is the realization that some women may not necessarily be seeking a cure and a diligent search for an organic explanation may serve only to drive them to express pain at another bodily site. A detailed history and thorough examination should be conducted, as this illustrates the intent to take the woman's complaint and symptoms seriously. Laparoscopic examination will exclude an organic cause, although it is difficult to ascertain the emphasis which should be placed on pelvic varicosities, pelvic congestion, peritoneal fluid in the pouch of Douglas or minor pathology. Although each has been proposed as a possible cause, detailed studies have not revealed such pathology in every case.[28,35] Following laparoscopic examination, women should be reassured that no major disease process is evident. Psychological reactance is avoided by indicating that since pain still exists, the search should continue to discover its causation. This explanation may modify the woman's focus or belief.[35]

THE ASSESSMENT OF SEXUAL DYSFUNCTION

Sexual problems are frequently complex, with physical, psychological and relationship aspects all playing a role. Additionally, the majority of sufferers tend to postpone seeking help longer than those with other disorders.

First interviews

It is important to clarify that both partners are willing and involved before inviting both to an interview. A patient may have initiated a request for psychosexual counselling without the knowledge of a partner who may not be aware that a problem exists. In addition, the partner involved in the problem may not be the one the patient is living with or to whom he or she is married.

It is advisable that both partners are seen separately as well as together. In the separate single interview, the patient or partner is free to discuss things that they might not wish to bring up in front of each other and to express their point of view. When they are seen together, it is possible to see how the two interact: whether they are 'at war' whether one dominates the other and to determine the language to be used when discussing sexual matters.

At an early stage in the joint interview it is helpful to include a small anatomy and physiology lesson; a simple diagram is sufficient to help the conversation flow and a light-hearted rather than overserious approach is more likely to put the couple at ease. Different terminologies for anatomical sites and sexual acts are commonly used by different sexes, social groups and age groups, while there are also regional variations.

History

Having put the couple at their ease, a thorough history should be taken, not only of the sexual problem but also of the background to the problem, family history and social history (Table 14.6). If a sexual problem has arisen after a period of normal sexual functioning, then it is important to establish events preceding that dysfunction, both physical and emotional.

Many medicinal agents may affect sexual performance by their action on the central nervous system, hormonal systems or by their sympathetic and/or parasympathetic action on the autonomic nervous system. In particular, antihypertensive agents, hormone therapies and drugs used for the treatment of psychiatric conditions may have side effects which are responsible for the dysfunction.

Table 14.6 Taking a sexual history from a patient

Medication	Antihypertensive agents
	Hormone therapy
Family history	Diabetes
	Hypertension
	Depressive illness
Past injury/ illness	Spinal cord injuries/operations
	Pelvic injury/operations
	Conditions likely to produce neuropathy
	Drug taking
	History of previous emotional trauma
	Delay in sexual or developmental milestones
	Diseases affecting the genitalia/ reproductive system
Social history	Development of childhood relationships
	Development of adult relationships
	Previous sexual partners
Personal history	Smoking
	Drinking and drug habits
	Diet and exercise

Family history may reveal inherited disease or familial tendencies to diabetes, hypertension or depressive illness.

Additionally, the sexual attitudes of parents and key figures may have influenced the person's thinking and contributed to the psychosexual problem. Any serious illness, injury or operation that might affect the pelvic anatomy, genitalia or nerve supply to the region should be identified, as should any systemic illness likely to contribute to the problem. Occasionally, emotional trauma such as rape, indecent assault, sexual abuse and worries about genital mutilation following childbirth may be elicited.

Social history may be particularly relevant to sexual dysfunction and attention should be paid not only to the development of present and previous partnerships but also to the way in which childhood relationships were formed – all this may reveal habitual patterns of reaction relevant to the present problem.

Details about smoking (excessive nicotine can produce a neuropathy), drinking and drug habits (including tranquillizer and any illicit drug use), diet and exercise may all be helpful in assessing the patient's present condition. A detailed sexual history is then required to establish the present sexual complaint and how long it has been present.

Table 14.7 Routine tests performed in patients with sexual dysfunction

Routine tests
Urinalysis:
   Diabetes mellitus
Full blood count:
   Iron deficiency anaemia
   Pernicious anaemia
   Leukaemia
ESR:
   Systemic inflammatory process
   Systemic infective process
Serological tests for syphilis

Tests in males
Testosterone level
Prolactin level

Tests in females
Follicle-stimulating hormone level
Luteinizing hormone level
Prolactin level

Optional tests
Thyroid function tests
Vitamin B_{12} and folate levels
Fasting blood sugar

Physical examination

A general physical examination is required, looking particularly for evidence of diseases such as diabetes, hypertension and any condition that may cause a neuropathy or alteration in one of the endocrine systems, as well as a local examination of the genitalia not only to eliminate physical disease but also to provide reassurance to both partners.

Examination of the male should include testing for abdominal, cremasteric and anal reflexes. The size, shape and consistency of the testes, presence of any abnormality within the scrotal sac, the presence of hernias and any inguinal glands should be noted. The penis itself should be examined, paying particular attention to the shape and size in the flaccid state, any plaques or hardened tissue suggesting Peyronie's disease, whether the foreskin (if present) retracts fully and whether there is any evidence of urethritis. Occasionally it may be necessary to examine the penis in the erect position to observe bowing or distortions. The injection of papaverine into the corpora cavernosa to establish whether filling is possible is a specialist technique and should not be considered as part of the routine examination.

A thorough examination of the female genitalia should include assessing the normality or otherwise of the labia, clitoris, introitus and any signs to suggest vaginitis or cervicitis. An internal examination will allow the assessment of dyspareunia spasm of the pubococcygeal muscles or palpable ovarian cysts.

Initial tests (Table 14.7)

It is usual to screen for hormonal levels; normally this implies a serum testosterone level in men and serum follicle-stimulating hormone (FSH) and luteinizing hormone (LH) in women. However, these latter measurements are probably unnecessary in a woman who is menstruating normally and regularly but are indicated when this is not the case. Serum prolactin levels are useful in both men and women, especially if the menstrual cycle is disturbed. Hyperprolactinaemia secondary to a pituitary adenoma is an uncommon but important condition which can present early with secondary sexual dysfunction. The results of serum prolactin levels must be interpreted with caution, however, since they can be raised in anxiety, sexual arousal and drugs (particularly those used for the treatment of psychiatric conditions). A lateral skull X-ray of the pituitary fossa and scanning techniques should be employed if pituitary pathology is suspected. For those with hyperprolactinaemia, in the absence of an identifiable cause, bromocriptine may be effective in restoring sexual function.

Specific tests

More specific tests may be indicated for a small number of patients, but most are specialist techniques outside the range of the average genitourinary clinic.

Penile plethysmography

This is a useful aid for distinguishing between the physiological and psychological causes of erectile dysfunction. The device consists of a mercury-filled strain gauge looped around the penis and attached to a recording drum which runs overnight while the patient is asleep. Those with organic disease do not exhibit the normal phases of penile tumescence associated with rapid eye movement (REM) sleep which occur in bursts of about 10 min or more during the night. Men with anxiety-induced erectile dysfunction are likely to have a normal nocturnal tumescence pattern.

The same process occurs in females, but the monitoring of intravaginal or clitoral responses poses problems and the technique is only available at a research level.

Radiological investigations

Radiology may be used to look for causes of vasculogenic erectile dysfunction and frequently involves the use of angiography when preceding investigations have suggested a vascular lesion. Advances in contrast media and catheter design have made angiography safer and less traumatic, being performed under local anaesthesia with or without sedation. The femoral artery is punctured and a flexible guidewire inserted and passed up into the lower aorta. A catheter can be threaded over the wire (which is then removed) and the contrast medium is injected to determine any vascular abnormality. There are few contraindications to angiography other than the presence of uncontrolled hypertension or coagulation defects.

In cases of vasculogenic erectile inadequacy due to arterial stenosis, angioplasty is the logical extension of diagnostic angiography in which, when a suitable stenosis is identified, a guidewire is manoeuvred through the obstruction and a special catheter is passed over the wire. At the end of the catheter is a non-elastic balloon which can be distended with fluid to a preset diameter, thus dilating the structure and restoring blood flow. Not all occlusions are suitable for this treatment, but its scope is increasing as more experience is gained.

Neurological investigations

When there are indications of a neurological disorder, either in the history or by the elicitation of abnormal signs, then special investigations are indicated. Lesions affecting the frontal lobes can produce disinhibition with either an increase or decrease in sexual behaviour. Hypersexual activity can be found after encephalitis, particularly if affecting the limbic system. Computed tomography (CT) scanning of the brain will exclude any focal cerebral lesion.

Epilepsy can occur during sexual intercourse, either as a result of hyperventilation or because orgasm can trigger reflex epilepsy. Postictal sexual phenomena include disinhibition and masturbation. An electroencephalogram (EEG) is essential and, if symptoms occur for the first time over the age of 20 years (late-onset epilepsy), a CT scan is indicated.

Spinal cord lesions are the commonest neurological cause of sexual dysfunction, producing erectile impotence in men and genital insensitivity in women. The most common aetiology is multiple sclerosis and sexual difficulties may be the earliest manifestation. Weak ejaculation and sexual dysfunction are often associated with bladder, sensory and sweating problems. In women, sexual dysfunction occurs in half the cases, frequently consisting of a combination of anorgasmia, dyspareunia, lack of vaginal lubrication and general paraesthesia.

The three most helpful investigations are visual evoked response (VER), examination of the cerebrospinal fluid (either increased lymphocytes or raised protein) and magnetic resonance imaging (MRI). Even in the absence of a history of optic neuritis, the majority of patients with multiple sclerosis have a delayed VER. MRI is not widely available, but in multiple sclerosis it may reveal demyelinated plaques which may have been clinically silent.

Peripheral neuropathy is characterized by sensory and motor disturbances in the distal limb areas and by areflexia, but it is the autonomic involvement which is mainly responsible for failure of orgasm in the female and impotence in the male.

Electrophysiological studies of the peripheral nerves will reveal delayed conduction times and there are now more direct methods for measuring autonomic dysfunction.

SPECIAL CIRCUMSTANCES

Psychosexual difficulties can occur secondary to a physical condition or may even be caused by its therapeutic treatment.

SEXUALLY RELATED HEADACHES

The classification of sexually related headaches is essentially a descriptive one, particularly as several mechanisms can occur in the same clinical setting. Headaches related to sexual activity include:

- intracranial space-occupying lesions
- muscle contraction (tension headache)
- benign orgasmic cephalalgia
- postcoital migraine
- postexertional migraine.

Sexual arousal and physical exertion raise cerebrospinal fluid pressure and, in the presence of an intracranial lesion, may produce traction on sensitive structures giving an abrupt, brief and well-localized headache. Causes include cerebral tumour, cerebral aneurysm, subdural haematoma, hydrocephalus and chronic infections of the nervous system. Non-neurological diseases may also produce exertional headache, e.g. dental disease, sinusitis, acute anaemia. Persons with exertion-related headache need to be investigated thoroughly to exclude any underlying serious pathology.

Benign orgasmic cephalalgia is severe, usually acute in onset and exclusive to sexual activity. It is usually bilateral and occipital but can be generalized and even radiate to the shoulders. It lasts from 10 min to a few hours and the pain is described as pulsatile and independent of muscular contraction. Up to half are migraine sufferers and it is presumed to be vascular in origin, but why it is intermittent is unknown.[37]

Postcoital onset of typical migraine is well recognized but the migraine follows sexual intercourse and does not occur during it. Essentially a disorder of the young, it can occur at any age but care should be taken in diagnosing it for the first time in anyone over the age of 50 years.

Reassuring the patient that sexually related headaches are benign helps him or her to cope with the problem which is particularly important where treatment is not entirely effective.

Posttraumatic stress disorder

Posttraumatic stress disorder consists of a collection of affective and autonomic disturbances, the features of which include reexperience of a traumatic event as nightmares and flashbacks. Sexual dysfunction can be directly related to the trauma, e.g. childbirth, sexual abuse, rape. The diagnostic criteria include:

- presence of a recognizable stressor;
- intrusive reexperience of the trauma;
- numbing of responsiveness to the external world;
- two of the following: sleep disturbance, survivor guilt, poor concentration, intensification of symptoms, exaggerated startle response.

Treatment consists of helping the patient to talk about the traumatic experience frequently and assisting them to ventilate their feelings and emotions about the event.[38]

Sex and ageing

There is a wide variation in what can be considered as 'normal' sexual behaviour in old age. Studies of healthy elderly men have shown a decrease in the frequency and degree of nocturnal penile tumescence, while morning erections and nocturnal emissions also become less frequent. More importantly, erection takes longer to develop and requires direct tactile stimulation of the penis as psychological stimulation no longer produces spontaneous erections. Ejaculation becomes less powerful and more difficult to control, with a longer refractory period.[39]

The physiological changes in female sexual function are more clearly related to the menopause and include reduced libido and vaginal atrophy. Hormone replacement therapy enhances an older woman's sexuality by increasing general feelings of well-being and a sense of attractiveness, but it has been disappointing when used as a treatment for loss of libido.

Sex after a CVA

In general, the quality of sexual adjustment for the post-CVA patient is influenced by two factors: the quality of the prestroke relationship and the partner's attitude to continuing sexual activity. It is usual to advise patients to abstain from sexual activity for six weeks following a stroke, but this does not appear to be founded on any scientific evidence. It is unusual for there to be a direct effect on the neurophysiological aspect of sexual functioning, although damage to the higher centres of the autonomic nervous system may result in impairment of the sexual response. Muscle weakness may make intercourse difficult, while loss of speech and sphincter control may cause considerable embarrassment to both patient and partner. Advice on different coital positions may be necessary. Counselling aimed primarily at promoting a positive body image and enhancing feelings of self-worth is frequently helpful in promoting successful sexual readjustment.

Chronic lumbar pain

Low back pain is a common problem which can lead to sexual dysfunction. It may be either a direct consequence of interruption to the nerves or blood supply leading to erectile inadequacy in the male and anorgasmia in the female, possibly due to a lack of lubrication or restriction of sexual activity because of constant pain. Additionally, referred pain may give rise to unusual sensations in the genital area.

At least one-third of men with low back pain cease sexual activity. Those who continue do so on a reduced basis and frequently experience pain during and after intercourse for several hours to days afterwards.[41]

Physical treatments range from alteration of coital positions, reducing pain-provoking movements, enhancing stimulation and alternatives to coitus. Psychosexual counselling should be a routine adjunct to physical treatment in such cases.

Sex following myocardial infarction

As with any serious illness those who recover from a myocardial infarction often find they have a reduced libido and, in men, erectile problems. The severity of the infarct and the extent of cardiac decompensation are much less important causes of sexual debility than the psychological condition of the patient. Depression and anxiety frequently follow a myocardial infarction and may also have an adverse effect on sexual desire and functioning. It is important that sexual advice is given to all patients early in their rehabilitation programme, with the aim of restoring the couple to approximately their preinfarct level of sexual activity.

Hypertension

Several studies have shown that between 17% and 20% of men with untreated hypertension suffer from erectile dysfunction, compared with only 7% of age-matched normotensive controls[42,43], while ejaculatory failure and reduced libido were also common. Consequently, patients attending with sexual dysfunction should be screened for hypertension. The effect of untreated hypertension in female arousal and orgasmic function has not been extensively studied, although anorgasmia may be common.[44]

Many of the therapeutic agents used to treat hypertension increase the risk of sexual difficulties and these are an important cause of non-compliance with prescribed medication for the condition[45] (Table 14.8).

Psychosexual problems associated with STDs

The presence of a sexually transmitted disease, especially when this is chronic, recurrent or a condition infective to others, can cause psychosexual problems. Many persons with genital herpes, particularly when recurrences are frequent, develop psychosexual problems and may even avoid sexual activity despite counselling and reassurance.[46] Loss of attractiveness, low self-esteem and anxiety, significant negative changes in sexual responsiveness and feelings of trauma, depression and vulnerability have been found in women attending follow-up clinics after an abnormal cervical cytology, colposcopy or laser therapy to the cervix.[47,48] Explanation, education and an open discussion at the time of examination/referral may be beneficial.

Table 14.8 Effects on sexual function of commonly used antihypertensive agents

Therapeutic agent	Possible effect on sexual function
Ganglion blockers, e.g. methyl dopa	Reduced sexual drive
Diuretics:	
Bendrofluazide	Impairment of erectile function
Spironolactone	Loss of sex drive, erectile dysfunction
β-blockers:	
Propanolol	Erectile dysfunction, reduced sex drive, reduced arousal
Acetutolol, atenolol	As above, but incidence less
Calcium channel blockers:	
Verapamil	Erectile dysfunction
Nifedipine	Ejaculatory failure
ACE inhibitors, e.g. captopril, enalapril	Low propensity for sexual difficulties

REFERENCES

1. Osborn, M., Hawton, K. and Gath, D. (1988) Sexual dysfunction among middle-aged women in the community. *Br. Med. J.*, **296**, 959–962.
2. Rust, J., Golombok, S. and Pickard, C. (1987) Marital problems in general practice. *Sex. Marit. Ther.*, **2**, 127–130.
3. Barczak, P., Kane, N., Andrews, S. *et al.* (1988) Patterns of psychiatric morbidity in a genitourinary clinic: a validation of the Hospital Anxiety Depression Scale. *Br. J. Psychiat.*, **152**, 698–700.
4. Tarnoplsky, A., Hand, D.J., McLean, E.K. *et al.* (1979) Value and uses of a screening questionnaire (General Health Questionnaire). *Br. J. Psychiat.*, **134**, 508–515.
5. Fitzpatrick, R., Frost, D. and Ikkos, G. (1986) Survey of psychological disturbances in patients attending an STD clinic. *Genitourinary Med.*, **62**, 111–115.

6. Pattman, R.S. and Schofield, C.B.S. (1986) Problem-orientated categorisation of 'other conditions' seen in a genitourinary medicine clinic. *Br. J. Vener. Dis.*, **59**, 63–65.
7. Catalan, J., Bradley, M., Gallway, J. and Hawton, K. (1981) Sexual dysfunction and psychiatric morbidity in patients attending a clinic for sexually transmitted diseases. *Br. J. Psychiat.*, **138**, 292–296.
8. Slatford, K. and Currie, C. (1984) Prevalence of psychosexual problems in patients attending a genitourinary clinic. *Br. J. Vener. Dis.*, **60**, 398–401.
9. Jones, A.J. and Thin, R.N. (1991) Contrasting psychosexual problems of patients attending a genitourinary medicine clinic and a community-based clinic. *Int. J. STD and AIDS*, **2**, 124–127.
10. Frank, E., Anderson, C. and Rubenstein, D. (1987) Frequency of sexual dysfunction in 'normal' couples. *N. Engl. J. Med.*, **299**, 111–115.
11. Shapiro, B. (1943) Premature ejaculation: review of 1,130 cases. *J. Urol.*, **50**, 374–379.
12. Masters, W.H. and Johnson, V.E. (1970) *Human Sexual Inadequacy*, J. and A. Churchill, London.
13. Eaton, H. (1973) Clomipramine (Anafranil) in the treatment of premature ejaculation. *J. Int. Med. Res.*, **1**, 432–434.
14. Williams, G. (1990) Management of impotence – a urologist's perspective. *Prescribers J.*, **30**, 71–77.
15. Meisler, A.W. and Carey, M.P. (1990) A critical review of nocturnal penile tumescence monitoring in the diagnosis of erectile dysfunction. *J. Nerv. Ment. Dis.*, **178**, 78–89.
16. Virag, R. (1982) Intracavernous injection of papaverine for erectile failure. *Lancet*, **ii**, 938–940.
17. Levine, S.B., Althof, S.E., Turner, T.A. *et al.* (1989) Side effects of self-administration of intracavernous papaverine and phentolamine for the treatment of impotence. *J. Urol.*, **141**, 54–57.
18. Halstead, D.S., Weigel, J.W., Noble, M.J. and Mebust, I.N.K. (1986) Papaverine-induced priapism. *J. Urol.*, **136**, 109–110.
19. Ottesen, B., Wagner, G., Virag, R. and Fahrenkrug, J. (1984) Penile erection: possible role for vasoactive intestinal polypeptide as neurotransmitters. *Br. Med. J.*, **288**, 9–12.
20. Stackel, W., Hason, R. and Marberger, M. (1988) Intracavernous injection of prostaglandin E1 in impotent men. *J. Urol.*, **140**, 66–68.
21. Virag, R. and Adaikan, P.G. (1987) Effects of prostaglandin E1 on penile erection and erectile failure. *J. Urol.*, **137**, 1010–1013.
22. Reid, K., Morales, A., Harris, C. *et al.* (1987) Double-blind trial of yohimbine in the treatment of psychogenic impotence. *Lancet*, **i**, 421–423.
23. Persky, H., Lief, H.I., Strauss, D. *et al.* (1978) Plasma testosterone level and sexual behaviour of couples. *Arch. Sex. Behav.*, **3**, 157–173.
24. Morris, N.M., Udry, J.R., Khan-Dawood, F. and Dawood, M.Y. (1987) Marital sex frequency and mid-cycle female testosterone. *Arch. Sex. Behav.*, **16**, 27–37.
25. Osborn, M., Hawton, G. and Gath, D. (1988) Sexual dysfunction among middle-aged women in the community. *Br. Med. J.*, **296**, 959–962.
26. Semmens, J.P. and Semmens, J.F. (1974) Dyspareunia: brief guide to office counselling. *Med. Aspects Human Sexual.*, **8**, 85–86.
27. Black, J.S. (1976) Sexuality and the gynaecologist. *Patient Management*, **9**, 57–61.
28. Renaer, M. (1980) Chronic pelvic pain without obvious pathology in women. *Eur. J. Obstet. Gynaecol. Reprod. Biol.*, **10**, 455–463.
29. Reading, A.E. (1982) A critical analysis of psychological factors in the management and treatment of chronic pelvic pain. *Int. J. Psychiat. Med.*, **12**, 129–139.
30. Slocumb, J.C. (1984) Neurological factors in chronic pelvic pain: trigger points and the abdominal pelvic pain syndrome. *Am. J. Obstet. Gynecol.*, **149**, 536–543.
31. Duncan, C.H. and Taylor, H.C. (1952) A psychosomatic study of pelvic congestion. *Am. J. Obstet. Gynecol.*, **64**, 1–12.
32. Beard, R.W., Belsey, E.N. and Lieberman, J.C.M. (1977) Pelvic pain in women. *Am. J. Obstet. Gynecol.*, **128**, 566–570.
33. Magni, G., Salmi, A., de Leo, D. and Ceola, A. (1984) Chronic pelvic pain and depression. *Psychopathology*, **17**, 132–136.
34. Gross, R.J., Doerr, H., Caldirola, D. *et al.* (1981) Borderline syndrome and incest in chronic pelvic pain. *Int. J. Psychiat. Med.*, **10**, 79–86.
35. Rosenthal, R.H., Ling, F.W., Rosenthal, J.L. and McNeeley, S.G. (1984) Chronic pelvic pain: psychological features and laparoscopic findings. *Psychosomatics*, **25**, 833–841.
36. Watts, F.N., Austin, S. and Powell, G.E. (1973) The modification of abnormal beliefs. *Br. J. Med. Psychol.*, **46**, 359–363.
37. Paulson, G.W. and Klawans, H.C. (1974) Benign orgasmic cephalalgia. *Headache*, **13**, 181–187.
38. Davidson, J., Walker, J.I. and Kitts, C. (1987) A pilot study of phenelzine in the treatment of post-traumatic stress disorder. *Br. J. Psychiat.*, **150**, 252–255.
39. Schiavi, R.C., Schreiner-Engel, P. and Medneli, J. (1990) Healthy ageing and male sexual function. *Am. J. Psychiat.*, **147**, 776.
40. Leiblum, S., Bachmann, G., Kemmann, E. *et al.*

(1983) Vaginal atrophy in the post-menopausal woman: the importance of sexual activity and hormones. *J. Am. Med. Assoc.*, **249**, 2195–2198.
41. Coates, R. and Ferroni, P. (1991) Sexual dysfunction and marital disharmony as a consequence of chronic lumbar pain. *Sex. Marit. Ther.*, **6**, 65–69.
42. Bulpitt, C.J., Dollery, C.T. and Carne, S. (1976) Changes in symptoms of hypertensive patients after referral to hospital clinic. *Br. Heart J.*, **38**, 121–128.
43. Bauer, G.E., Hunuor, S.N., Baker, J. and Marshall, P. (1981) Clinical side effects during antihypertensive therapy: a placebo-controlled, double-blind study. *Postgrad. Med. Commun.*, **14**, 49–54.
44. Riley, A.J., Steiner, J.A., Cooper, R. and McPherson, C.K. (1987) The prevalence of sexual difficulties in male and female hypertensive patients. *Sex. Marit. Ther.*, **2**, 131–138.
45. Medical Research Council (1981) Report of the Working Party on Mild to Moderate Hypertension. Adverse reactions to bendrofluazide and propanolol for the treatment of mild hypertension. *Lancet*, **ii**, 539–543.
46. Goldmeier, D., Johnson, A., Byrne, M. and Barton, S. (1988) Psychosocial implications of recurrent genital herpes simplex virus infection. *Genitourinary Med.*, **64**, 327–330.
47. Campion, M.J., Brown, J.R., McCance, D.J. *et al.* (1988) Psychosexual trauma of an abnormal cervical smear. *Br. J. Obstet. Gynaecol.*, **95**, 175–181.
48. Boag, F.C., Dillon, A.M., Catalan, J. *et al.* (1991) Assessment of psychiatric morbidity in patients attending a colposcopy clinic situated in a genitourinary medicine clinic. *Genitourinary Med.*, **67**, 481–484.

15 Management of asymptomatic and early symptomatic patients with HIV infection
(with special reference to antiretroviral therapy and viral resistance patterns)

D.A. Hawkins, G. Moyle and M.S. Youle

INTRODUCTION

The natural history of human immunodeficiency virus (HIV) infection has become clearer since it was first isolated in 1983.[1] The time course from primary infection to symptomatic illness, the acquired immunodeficiency syndrome (AIDS) and death appears dependent on a number of factors, which are interlinked. These include severity of seroconversion illness,[2,11] phenotype of virus acquired,[3] state of the host immune function and therapeutic and lifestyle interventions.

In the past 10 years the number of people known to be carrying HIV infection has escalated dramatically. The visible epidemic has spread from gay men and injecting drug users into mainstream sexual communities, especially within the heterosexual milieu in developing societies where there is little access to information concerning HIV or protection from sexual transmission. Meanwhile perinatal transmission has become commoner as the numbers of women affected has risen. In contrast, those infected by blood products have remained a small proportion of the total, reflecting changes, at least in the developed world, in blood product screening and manufacture. At the outset of the epidemic patients presenting with HIV infection were mainly those who had already developed clinical illness and often had severe symptomatic disease or AIDS.[4] However, as awareness of HIV infection rose and testing for HIV antibodies became more widely available, increasing numbers of people were identified who were asymptomatic or at an early stage in their disease course. It is the management of this group which will be discussed in this chapter.

One of the central problems of HIV infection is the difficulty in predicting progression from well and infected to HIV-related illness. The use of clinical markers in HIV disease is fraught with difficulties since by the time reliable markers of progression occur, e.g. oral hairy leucoplakia, multidermatomal herpes zoster or recurrent oral candidiasis,[5,6] there has often been considerable damage to the patient's immune function which may be difficult or impossible to reverse.

Many of the earliest signs of HIV disease are conditions that occur commonly in the general population and may be associated with transient immune dysfunction due to other processes, for instance chest infections, or may be due to concomitant medication, such as oral candidosis in those using inhaled steroids. This results in a poor predictive value for many conditions which occur in the first 5–10 years of HIV infection.

Laboratory tests of immune deficiency have been used to assess the damage caused to the patient with HIV since well before the virus had been identified. It was the discovery of disordered T-cell lymphocyte populations in those presenting with AIDS that led to the recognition of the syndrome as a distinct entity and it is clear from the pathogenesis that a progressive decline in the numbers and function of the CD4+ lymphocytes is one of the hallmarks of the condition. The use of CD4 counts as a marker of disease progression is now central to the care of most patients with HIV[7] since the test has become readily available in most hospitals or HIV treatment facilities. Concerns exist, however, over the reliability of this as a single marker of HIV progression, due to wide inter- and intrapatient variability, normal fluctuations that occur in early disease and variations in results between laboratories.[8] Many other tests have been used to try and quantify the progression occurring in the early stages of HIV infection, including other lymphocyte subpopulations, e.g. CD8+,[9] markers of immunological activation, e.g. serum neopterin and β_2-microglobulin, and virologic markers such as HIV p24 antigenaemia and viral DNA or RNA as measured by polymerase chain reaction (PCR). Measures of the latter (viral load) are now the best single test to predict progression and will be discussed in more detail later.

Undoubtedly we are at an early stage in the development of tests to characterize progression of HIV and clearly the repertoire of markers and their quantification will greatly improve the ability of the clinician to predict deterioration in individual patients. For any of these tests to be useful they must be reliable, reproducible, easy to carry out and above all else cheap. If these criteria are not met they are unlikely to gain acceptance. The use of combinations of tests has already been attempted to further assess HIV-related immunological damage and appears to improve predictive value to some degree.[10]

In this chapter we will review the clinical presentation and management of the asymptomatic and early symptomatic HIV-infected patient with special reference to psychological issues and monitoring of disease. Following this will be a review of therapy for HIV itself in this group of individuals, including the importance of clinical trials, especially in the face of the emergence of antiretroviral drug resistance, with a further discussion of the possible treatments which may be available in the future.

CLINICAL PRESENTATION

Primary HIV infection

Symptomatic primary HIV infection is probably quite common but is usually not reported by patients. This is because seroconversion illnesses mimic other acute infections and the diagnosis may not be thought of by the patient or his/her doctor should they seek medical attention. Although often described as a glandular fever type illness there are a number of distinguishing features which differentiate primary HIV infection from EBV mononucleosis[11,12] (Table 15.1).

Symptoms may include an acute fever, sore throat and lymphadenopathy. There may be quite marked lethargy and arthralgia and myalgia are common. A skin rash frequently occurs and is typically erythematous, non-pruritic and sparsely maculopapular. As well as involving the face and trunk, it can also be found on the extremities including the palms and soles (cf. secondary syphilis). HIV is a neurotropic virus and a mild aseptic meningoencephalitis may present with headache, photophobia and retroorbital pain. Less commonly, myelopathy, peripheral neuropathy, facial palsy, brachioneuritis and Guillain–Barré syndrome have been described. Finally, gastrointestinal symptoms may be evident with, in particular, mucocutaneous ulceration, nausea, vomiting and diarrhoea (Fig. 15.1).

Table 15.1 Differentiation of primary HIV infection from Epstein–Barr virus (EBV) mononucleosis

Feature	EBV infection	HIV infection
Onset	Gradual	Acute
Tonsillar hypertrophy	Common	Mild enlargement
Exudative pharyngitis	Common	Rare
Skin rash	Rare	Common
Mucocutaneous ulcers	Rare	Common
Liver dysfunction	Jaundice 8%	Mild abnormalities occur
Atypical lymphocyte	Occurs 80–90%	Occurs <50%
Diarrhoea	Unknown	Occurs

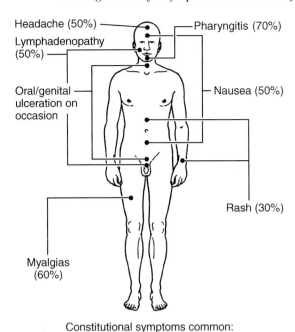

Fig. 15.1 Common features of primary HIV-1 infection.

Laboratory tests confirm the presence of generalized viraemia with positive P24 antigenaemia but negative or equivocal HIV antibody tests.[13] Tests for HIV viral load confirm high levels of viral replication which is associated with a high degree of infectiousness of an individual by any means of transmission and explains, in part, the rapid increase in HIV incidence in newly exposed populations by sexual transmission where many people are seroconverting.

There may be a modest or occasionally precipitous drop in CD4 count, leading to oral candida or even opportunistic infections such as PCP or candidal oesophagitis, but immune function and helper cells usually recover rapidly. Liver function tests have also been abnormal in patients presenting with severe symptomatic seroconversion.[14] There is evidence that progression to AIDS may be dependent on the severity of the initial primary illness (fever, duration of symptoms) and also the degree of viral load.[15] A number of studies are currently evaluating the effects of antiretroviral therapy at this early stage (see p. 194).

Asymptomatic infection

Following primary infection and a possible seroconversion illness the patient usually enters a period of variable length where they remain asymptomatic. If they were unaware of their infection it is during this period that they may be tested for reasons other than those resulting from clinical illness. These are the patients that are found to be HIV antibody positive at routine screening of blood donors or following medical examinations for insurance or employment purposes. They may be the partners of patients who have developed clinical disease or may be screened for HIV as part of an STD examination. Women may be found to be carrying HIV during antenatal screening although an antibody test is not taken by all pregnant women at present. In view of the success of antiretroviral intervention (ACTG 076 trial)[16] in reducing vertical transmission, HIV testing should be offered routinely and be the norm, particularly in areas of high prevalence.

In most of these situations the patient is likely to be surprised and shocked at the result although in some countries where HIV testing is not freely available some people use the blood transfusion service as an HIV testing facility. These are a group who need much in the way of post-test counselling, since they are frequently not from the traditional high-risk groups and therefore have less in the way of peer information and support. As the result is unexpected, it is often even more than usually difficult to come to terms with and a programme of ongoing support which encompasses formal counselling or referral to other agencies should be implemented. It is important that these patients be clear about their physical condition whilst dealing with the psychological issues.

The asymptomatic patient who is known to be infected should be followed up regularly to assess his/her clinical condition, including the measurement of laboratory tests, to monitor physical and mental well-being and to discuss advances in the care and treatment of HIV disease. The sexual health of these individuals is also important since the occurrence of other pathogens in the HIV person may both affect the progression of HIV disease[17,18] and render them more likely to pass on the virus to their sexual partners.[19] Usually patients would be seen every 3–6 months depending on their immune function. It is very important to differentiate between those asymptomatic patients with well-preserved immune parameters and those who, although developing no clinical disease, have evidence of markedly abnormal immunology. The likelihood of dying with HIV disease is quite clearly related to CD4 function[20] so it is always important to measure CD4+ lymphocyte counts and other surrogate markers at least twice in a given period before attempting to determine what stage a patient has reached (see below, p. 193). The fact that immunosuppression and CD4 decline are important in HIV disease

CD4+ T-cell categories	Clinical categories		
	(A) Asymptomatic or PGL	(B) Symptomatic, not (A) or (C) conditions	(C) AIDS-indicator conditions
1 ≥500/mm^3	A1	B1	C1
2 200–499/mm^3	A2	B2	C2
3 <200/mm^3	A3	B3	C3

NB Categories are defined by both CD4 count and clinical presentation.
Shaded area indicates AIDS
Where there is an overlap of conditions, (C) takes precedence over (B), which takes precedence over (A). For classification purposes, once a category B condition has occurred, the subject will remain in category B. The same goes for progression to category C.

Category A
- Asymptomatic HIV infection
- Persistent generalized lymphadenopathy
- Acute (primary) HIV infection with accompanying illness or history of acute HIV infection

Category B
- Bacillary angiomatosis
- Candidiasis, oropharyngeal (thrush)
- Candidiasis, vulvovaginal; persistent, frequent or poorly responsive to therapy
- Cervical dysplasia (moderate or severe)/cervical carcinoma in situ
- Constitutional symptoms, such as fever (38.5°C) or diarrhoea lasting >1 month
- Hairy leucoplakia, oral
- Herpes zoster (shingles), involving at least two distinct episodes or more than one dermatome
- Idiopathic thrombocytopenic purpura
- Listeriosis
- Pelvic inflammatory disease, particularly if complicated by tuboovarian abscess
- Peripheral neuropathy

Category C
- Candidiasis of bronchi, trachea or lungs
- Candidiasis, oesophageal
- Cervical cancer, invasive
- Coccidioidomycosis, disseminated or extrapulmonary
- Cryptococcosis, extrapulmonary
- Cytomegalovirus disease (other than liver, spleen or nodes)
- Cytomegalovirus retinitis (with loss of vision)
- Encephalopathy, HIV-related
- Herpes simplex: chronic ulcer(s) (>1 month duration); or bronchitis, pneumonitis or oesophagitis
- Histoplasmosis, disseminated or extrapulmonary
- Isosporiasis, chronic intestinal (>1 month duration)
- Kaposi's sarcoma
- Lymphoma, Burkitt (or equivalent term)
- Lymphoma, primary, of brain
- *Mycobacterium avium* complex or *M. kansaii*, disseminated or extrapulmonary
- *Mycobacterium tuberculosis*, any site (pulmonary or extrapulmonary)
- *Mycobacterium*, other species or unidentified species, disseminated or extrapulmonary
- *Pneumocystis carinii* pneumonia
- Pneumonia, recurrent
- Progressive multifocal leucoencephalopathy
- *Salmonella* septicaemia, recurrent
- Toxoplasmosis of brain
- Wasting syndrome due to HIV

Fig. 15.2 Current CDC classifications of HIV, 1993.

has been formally recognized in the 1993 AIDS definition which utilizes CD4 levels as well as clinical parameters to stage HIV-infected patients (Fig. 15.2). Infections and HIV-related malignancies usually occur at specific ranges of immune dysfunction with increasing incidence thereafter as greater immune dysfunction occurs. However, the type of conditions that occur are reliant on the prior exposure of the patient to the condition, for example, cytomegalovirus or toxoplasmosis, often geographically determined, such as histoplasmosis (found predominantly in South and Central America), or due to as yet poorly defined mechanisms, e.g. Kaposi's sarcoma. However, a new human herpes virus (HHV 8) infection has recently been implicated.[21]

What is clear is that patients should be aware of the importance of monitoring their immunological health in tandem with their clinical well-being since the former will have as much, if not more, effect on their survival as the latter. As the availability of quantitative markers of virologic function increases, these may be added to the battery of tests used to determine the progression and or stage of HIV disease in an individual and be used for rational determination of the most appropriate time to institute or change therapy.

Early symptomatic infection

Often the first signs of HIV disease are minor. If there has not been a clear seroconversion illness, and diagnosis at that time, then the patient may present in a number of ways and in various settings. They may be seen first by their general practitioner or in hospital casualty departments and as the numbers of HIV-infected individuals in a community rises so do the experience and levels of suspicion concerning the diagnosis.

Persistent skin problems, especially eczema and warts; recurrent bacterial chest infections; oral or vaginal candidosis for no apparent reason; night sweats and unexplained generalized lymphadenopathy should all raise suspicion. The patient may present to a variety of hospital outpatient consultants, especially the dermatologist, oncologist, gastroenterologist, neurologist or rheumatologist. Anyone who has an unexplained clinical condition, especially with bizarre features, should now be considered possibly to have HIV disease. As syphilis was the great mimic in the Victorian era, so HIV is in the late 20th century.

It is often difficult to raise the issue of HIV testing when concerns exist that this may be the underlying diagnosis, especially in an unexpected setting. However, a clear and adequate discussion of the possibility along with supportive pretest counselling is a vital part of good clinical care.

Patients should be quite aware of all issues around testing prior to it being done since much psychological morbidity may otherwise result. Patients should not be pressurized into taking an HIV test nor should testing be deferred when HIV is considered to be an important part of the differential diagnosis and knowledge of the result will impact on treatment.

Once again, the sexual habits of this group, as with the asymptomatic, are important and regular screening should be offered along with vaccination against hepatitis B as appropriate. Thus, all patients should be screened for hepatitis B surface antigen (HBsAg), antibodies to HBs Ag and antibodies to hepatitis B core antigen. Those without antibody or antigen are susceptible to hepatitis B and should be given a standard course of vaccination and antibody response checked two months after the last vaccination. Should there be no antibody response, it may be worth giving one additional booster.

Discussion concerning safer sex and the possibility of infection in sexual partners is perhaps easier in a screening clinic of well HIV patients compared to when they have developed more serious disease and are justifiably focusing on this as the major issue. The psychological effects of an HIV diagnosis on someone's sexual behaviour should not be underestimated and maintenance of safer sex may be difficult. Clearly, this is an important issue but not one which a doctor may want to bring up regularly with his patient who may have many clinical symptoms to be dealt with. In addition, the patient may feel reluctant to let the doctor down by admitting to their difficulties or the fact that they have had unsafe sex. In view of this, it is important overtly to recognize these difficulties and we organize behavioural advice and support to be readily available outside the routine HIV clinic consultations.

Some of the conditions or symptoms that are seen early in HIV disease are simple to treat whilst others, such as night sweats, are more problematic. Most conditions occur as relapsing and remitting conditions that vary in frequency, duration and severity of episode, often related to the stage of immune depression. Thus folliculitis at a high CD4 cell count may be mild and transient but as the immune parameters drop lower, the episodes become more severe and difficult to treat. Several exceptions to this rule exist. Night sweats are often very heavy and resistant to therapy in early disease and psoriasis is often worse in HIV when the CD4 cell count is high. This may be in part due to the pathogenesis of these conditions, being related to the early activation that occurs in the B-cell limb of the immune system rather than T-cell decline.

CONSTITUTIONAL SYMPTOMS

Fatigue, malaise and night sweats are common early signs of HIV disease. Fevers are also seen but if these persist there is usually an underlying cause. The specific treatment of such constitutional symptoms is for the most part ineffective. Rest and adequate nutrition (with or without added vitamins or other dietary manipulation) may be important and the use of regular antiinflammatory medication, especially non-steroidals, may help alleviate these symptoms.

SKIN CONDITIONS

Seborrhoeic eczema

This form of dermatitis is common in HIV-infected individuals. Although also frequently occurring in the uninfected general population, it is more severe and persistent in HIV disease. Usually it presents as flaky red areas of skin especially over the forehead, in the nasolabial folds and in the hair-bearing areas of the body. If it occurs at the angle of the mouth painful cracking, called angular cheilitis, may result (differential diagnosis includes oral candidiasis). The arms and chest are common sites although it can occur in any site. The skin is frequently dry and the rash may be itchy, which results in excoriation and possible secondary infection. Thought to be due in part to an overgrowth of fungal elements or yeasts in the dermis, the mainstay of therapy is adequate moisturizing and combination creams or ointments containing low-dose corticosteroids and antifungal drugs such as Daktacort or Canesten HC. More severe seborrhoeic dermatitis may be treated with stronger steroid creams combined with antifungal agents.

Folliculitis

Intensely itchy papular lesions, often in crops, in the hairy areas of the body can be problematic. Excoriation due to scratching is common and treatment is often difficult. Emollients and the use of topical or systemic antihistamines is usually first-line therapy. Antifungal agents and broad-spectrum antibiotics are indicated for more persistent episodes. Psoralens and ultraviolet light (PUVA) have also been used successfully.

Rosacea and acneiform rashes

These are relatively common as facial rashes in HIV disease and can be treated effectively with long-term tetracycline therapy. Sometimes acne may be associated with concomitant drug therapy (e.g. zidovudine) and withdrawal may be necessary. The use of retinoic acid derivatives is indicated for persistent acne.

Psoriasis

This is seen in patients who previously have not manifested the condition as well as those who have had prior psoriatic episodes. Psoriasis can be severe and difficult to control in the HIV-infected patient. Treatment options are the same as for the uninfected subject but difficult cases may require regular review by an experienced dermatologist. Concurrent therapy with antiretrovirals may be associated with an improvement in the psoriasis whilst, conversely, decreasing immune function often results in a worsening of disease (although not always – see above, p. 189).

Table 15.2 Drugs that have been shown to induce frequent reactions in HIV-infected patients

Aminopenicillines
Atovaquone
Clindamycine
Dapsone-trimethoprim
Ketoconazole
Pentamidine
Rifampicin
Sulphadiazine-pyrimethamine
Sulphamethoxazole-trimethoprim
Thiacetazone
Delavirdine
Nevirapine

Drug-related allergic rashes

Allergic skin rashes caused by treatment for or prophylaxis against infections in HIV disease are a major problem. They limit the use potential of many agents and result in significant morbidity. The drugs which are commonly associated with allergic rashes are shown in Table 15.2. The rashes seen are usually maculopapular and intensely itchy, some are photosensitive (e.g. dapsone), whilst a few may result in the Stevens–Johnson syndrome or exfoliative dermatitis. It is clear that the incidence of allergic reactions to drugs is much higher in the HIV-infected subject and that the likelihood of developing a rash to a particular agent increases with the degree of immunodeficiency and possibly the dose level.[22]

Early diagnosis, stopping the agent and prompt intervention with antihistamines and occasionally steroids, for severe reactions, may be beneficial. Rechallenge is only useful when the culprit is unclear due to multiple medications. With some drugs and modest reactions, it may be possible to 'treat through' with tolerance developing to the drug. Some success has been achieved with desensitization regimens for trimethoprim-sulphamethoxazole, which is important since this drug appears to be the best agent for prophylaxis of both PCP and toxoplasmosis.

Fungal skin conditions

Fungal overgrowth becomes increasingly common on the skin in all bodily sites as HIV disease progresses. Skin problems due to ringworm and other dermatophytes are an early manifestation of the condition and tend to be both more florid and difficult to treat than in the uninfected individual. Spread to nails with dystrophia and whitening is common and until recently has been difficult to treat.

Skin lesions should be treated with mild steroid and antifungal combination creams. More persistent lesions or those affecting the nail beds can be given courses of terbinafine orally for one month, in the case of skin lesions alone, or for at least three months when the nails are also involved. This agent appears to be effective and less toxic than griseofulvin. Isolated nail infections may be painted with tioconazole nail solution.

Warts

As immune deficiency progresses so does the incidence of viral warts. Although mainly genital, facial, hand and occasionally intraoral warts do occur. They may spread rapidly and grow to a large size. Since infection with some human papillomavirus (HPV) subtypes can result in dysplastic changes, leading to increased risks of cervical and anal carcinoma, treatment is extremely important. Immunosuppressed women who have HIV infection should have regular colposcopic examinations since cytology smears alone fail to detect all premalignant cervical changes. Cervical carcinoma in HIV-infected women is an AIDS diagnosis.

Destruction has, until recently, been the mainstay of therapy for viral warts. Chemical agents (podophyllin, trichloroacetic acid) and physical destruction (cryotherapy, hyfrecation) are usually effective. For resistant or massive genital warts surgical excision is often required but without adequate follow-up with standard destruction, recurrence is the norm. Adjunctive therapy with oral isoprinosine or intralesional α-interferon can be helpful in difficult cases, but the expense of the latter is generally prohibitive. Potent antiretroviral combination is probably the best adjunctive therapy and has led to easier control, if not resolution, of viral warts in many cases (personal communication, Nashat Hanna).

Mollusca contagiosa

These lesions are the result of a pox virus infection and occur sporadically in HIV-negative individuals as isolated papular lesions with a pearly solid central core which can be shelled out. In the HIV-infected patient mollusca often increase rapidly in number and size, especially on the face where shaving spreads the infection. Sheets of mollusca can develop which are difficult to treat. Therapy is similar to that for viral warts. Physical destruction is more effective than chemical agents and should be performed regularly until complete healing is achieved as without this, recurrence is common. Curretage of the central part of the larger lesions is an effective therapy and should not be baulked at, as regeneration of normal skin is relatively rapid. If left untreated, this infection is a major cosmetic problem which can cause great distress to the patient. In the early stages patients are often unaware of the spread of the infection and the importance of early and persistent treatment. As in the previous section, highly active antiretroviral combination therapy often leads to improvement in previously intractable cases and there is a recent report of benefit with cidofovir.[23]

Herpes simplex (HSV)

Recurrent herpetic lesions, both oral and genital, are common in HIV-infected individuals. The incidence of infection with HSV-1 and HSV-2 is highest in patients infected with HIV via the sexual route and is especially prevalent in gay men. Many subjects will self-medicate with acyclovir, either using cream or a short course of tablets, whilst many of those with recurrent disease are on long-term prophylaxis with twice-daily oral aciclovir in varying dosages. Acyclovir is effective as a treatment in almost all cases, although poor absorption of drug may be a problem. A new prodrug of aciclovir, valaciclovir, is marketed and reduces the necessity for five times daily therapy in primary attacks. A related nucleoside analogue prodrug, famciclovir, has also recently been licensed. Aciclovir-resistant strains of HSV are rare and not clinically important at early stages of HIV infection, occurring predominantly when subjects have very low CD4 cell counts.[24] Thymidine kinase mutants are mainly found, followed by DNA polymerase resistance.

Herpes zoster (VZV)

Commonly associated with transient immune suppression during other infections, this condition caused by varicella zoster virus (VZV) is the sequel to primary infection at which time classic chickenpox occurs. In HIV disease herpes zoster may develop at all stages of infection, but in those with more profound immune suppression multidermatomal disease is common. With the advent of HIV disease the finding of herpes zoster should alert the physician to the possible underlying diagnosis.

If infection involves the ophthalmic branch of the trigeminal nerve then urgent referral to an ophthalmologist is vital to preempt ocular complications. Treatment with high-dose acyclovir, or preferably valaciclovir or famciclovir, is effective in most cases, although longer than normal courses of therapy, or higher dosage regimens may be required. Recurrences may be a problem, although secondary prophylaxis is not mandatory. Postherpetic neuralgia does occur but appears to be often less problematic than in the immunocompetent.

Kaposi's sarcoma

See Chapter 17.

ORAL CONDITIONS

The mouth is a site of many HIV-related conditions. Oral hygiene is extremely important since although IgA function is relatively well preserved, a high incidence of gum disease has been reported in HIV-infected subjects. Regular review by dental services is advisable and prompt and diligent attention to any oral pathology is vital. Patients should be encouraged to attend their dentist a minimum of six-monthly and the hygienist three-monthly and to confide their HIV status as this is helpful for the early diagnosis of HIV-related events such as Kaposi's sarcoma.

Gingivitis

Gum recession, bleeding, dental abscesses and the more severe cancrum oris are all sequelae of poor dental hygiene in the setting of HIV-related immune dysfunction. These are unpleasant conditions for the patient and may be important in terms of oral transmission of HIV. Treatment revolves around dental cleansing, debridement or laying open of abscess cavities in conjunction with adequate antibiotic therapy. Prophylaxis may be achieved by regular cleaning and the use of antiseptic mouthwashes.

Aphthous ulceration

Perioral herpes simplex is much commoner than intraoral lesions due to this virus but persistent mouth ulcers should be sampled for microbiology tests to exclude the diagnosis. Other oral pathology, e.g. candidosis, should also be treated, since ulcers will often resolve in the setting of a mouth free of pathogens. However, a significant proportion of HIV-infected subjects will suffer from recurrent bouts of aphthous ulceration, especially of the hard palate and gingival margins.

Treatment of mild ulceration is usually achieved with topical steroids in pellet formulation or with anaesthetic mouthwashes. More severe cases can be healed with short courses of oral corticosteroids or thalidomide.[25] The latter, which is mainly used for leprosy, is also of value in the treatment of oral and genital ulceration in Behçet's disease. It is available on a named patient basis and appears to act as an immunomodulator. Thalidomide has antitumour necrosis factor (TNF) activity and may be useful as adjunctive therapy in HIV opportunistic infections such as tuberculosis or microsporidiosis. It is also under investigation as an inhibitor of HIV-related B-cell activation.

Oral candidiasis

Candida albicans and, less commonly, *C. glabrata* cause oral candidiasis in HIV disease with frequency and severity of infection related to underlying immunosuppression. Although candida may be a problem during seroconversion and also during episodes of antibiotic therapy, when it occurs *de novo* in HIV disease it signifies progression from the asymptomatic phase. White, curdy, removable plaques on the mucosa are the hallmark of candidal infection (pseudomembranous candidiasis) but erythematous forms with red sore patches in the mouth are also seen.

Standard therapy is with topical antifungals, such as amphotericin B lozenges and nystan pastilles, or oral solution. For more severe episodes systemic azole therapy is preferable and is highly effective in the majority of cases with early asymptomatic disease.

FPu-conazole is currently our first-line choice. It is usually effective in a single dose regime (150–400 mg), the higher dose being sufficient for coexistent oesophageal candidiasis. It is well absorbed at all pH levels (compared to itraconazole and ketoconazole which are acid labile and may be poorly absorbed in those subjects with a degree of achlorhydria).

Fluconazole also has less enzyme-inducing ability and, therefore, a lower tendency to drug interactions, particularly with protease inhibitors and non-nucleoside reverse transcriptase inhibitors (see later). Concern has arisen over resistant strains of candida, coupled with clinical resistance, often developing after extended courses of oral azoles and this supports the use of pulse therapy, rather than continuous prophylaxis. In the past this has been difficult to achieve in late-stage HIV disease, although severe persistent candida seems less problematic with the success of current antiretroviral regimes. Should fluconazole resistance occur, success may be achieved with itraconazole, particularly oral solutions, which may have both a topical as well as a systemic effect. Novel azoles and other antifungals are under development.

Oral hairy leucoplakia (OHL)

Epstein–Barr virus infection of the oral mucosa, specifically the margins of the tongue, can result in this condition, which is associated with HIV progression. It presents as white adherent patches with a ribbed appearance and although usually asymptomatic it may cause soreness or pain on eating. Treatment with high-dose oral acyclovir (or prodrugs) for 3–4 weeks is an effective therapy, as is topical applications of aciclovir cream or retinoic acid gel.

SURROGATE MARKERS OF HIV DISEASE

In the first 10 years of antiretroviral drug development there was a need to test treatment tactics, strategies and available treatment markers against clinical endpoints. It has now been convincingly demonstrated that antiretroviral combination therapy prolongs life and extends the disease-free period.[26–30] Many of the benefits of these therapies are predicted by short-term changes in two laboratory markers: CD4 cell count and plasma HIV RNA load.[31,32] Improvements in our understanding of the pathogenesis of HIV disease and viral dynamics[33] have provided both patients and physicians with a greater understanding of the tools used to monitor HIV and have enabled the establishment of new therapy goals. To provide maximal improvement in patient outcome (both length and quality of life), we must arrest viral replication and improve immune function by enabling reexpansion of lost T-lymphocyte populations. The availability of more potent antiretroviral combinations and of routine viral load measurement has enabled the establishment of the level of quantification of a sensitive viral load assay (200–500 copies/ml) as the optimum response to treatment. This response is typically associated with substantial delays in the selection of drug-resistant viral mutants and hence with more prolonged therapeutic responses than observed with less complete suppression of viral replication.

An ideal surrogate marker must be biologically plausible, easily and reproducibly measurable and changes in that marker both on and prior to therapy must accurately predict outcome.[34,35] Whilst no single marker of HIV activity represents a perfect surrogate, combined use of CD4 and viral load measurement both on and off treatment provides a quality predictive indicator. Viral load after seroconversion[36–38] is a powerful predictor of both future disease progression and death,[34] as well as of non-progression.[37,38] In the circumstances of a CD4 cell count above 500/mm^3, the CD4 cell count appears to provide no additional predictive value. A high viral load, greater than 30–50 000 copies/ml, therefore implies increased risk of disease progression or death and suggests the need for consideration of treatment intervention. CD4 represents an excellent staging marker for HIV, indicating whether a patient is at immediate risk of an event: either an opportunistic infection (OI)[39,40] or death.[40–42] It is therefore of considerable value in decisions to commence prophylaxis for some OIs such as *Pneumocystis carinii* pneumonia (PCP).[43] Indeed, the widespread use of prophylaxis for a range of OIs has helped make clinical trials which use clinical events (other than death alone) difficult to interpret. The problems with using clinical events as trial endpoints have been recently reviewed.[44] Clinical events, however, remain a good marker of quality of life. Waiting for a patient to deteriorate clinically before starting treatment or changing a failing therapy is out of step with other areas of medical practice. Given that viral load and CD4 can predict risk of clinical events, it appears reasonable to intervene before disease progression.

CD4 has been used for the last 10 years as the guide to commencing antiretroviral therapy, based on the understanding that as consistent declines in CD4 are associated with disease progression or death, we should intervene to prevent/delay the onset of disease events. Interpretation of CD4 should now be made in the context of viral load measurement but the immune count remains a valuable tool in guiding the initiation of therapy. However, as an on-therapy surrogate marker short-term changes in CD4 are, at best, limited as an individual marker, predicting 50% or less of a treatment effect.[32,45,46] This is not surprising given that antiretroviral therapy may both influence trafficking and function of CD4 and other immunologic cells,[47,48] information not captured by enumeration from a blood sample. Measurement of CD4 alone, therefore, is not adequate for monitoring therapy response.

Plasma viral RNA load represents the best single on-therapy marker currently available. This is biologically plausible, given that it is active HIV replication which drives CD4 depletion and ultimately disease progression. Furthermore, the viral load measured in a plasma sample is largely derived from lymphoid replication where 99% of daily viral turnover occurs.[49,50] Although not necessarily correlating with viral load in all body compartments,[51] an issue which may be critical to failure on initially effective regimens, changes in plasma viral load with therapy correlate well with changes observed from lymph node samples.[52] As an individual marker, a decrease of 75% in plasma viral load predicts 59% of the treatment effect of zidovudine during the first six months of therapy.[32] Responses of 0.25 log or more with didanosine therapy are associated with improved survival relative to non-responders.[53] A 0.3 log decrease correlates with a 27% reduction in the relative hazard of disease progression[54] and a 0.5 log decrease with a 63% reduction.[32] In the AIDS clinical trials group study number 175 (ACTG 175), which included both patients receiving nucleoside analogue combination and monotherapy, a decrease of 1.0 log at week 8 correlated with a 65% reduction in risk or disease progression and at week 56 with a 90% reduction in risk.[31] Similarly, during combination therapy with zidovudine and lamivudine, each 1.0 log change in viral load on therapy was associated with a three fold (adjusted) reduction in risk of disease progression to AIDS.[55] Measurement of plasma viral RNA load can therefore be used as a marker of treatment effect and is now universally accepted as essential in patient monitoring.[56] Plasma viral RNA load can be seen as analogous to the speed of a vehicle moving towards a destination; only when we know the distance from the destination can we estimate the arrival time. The CD4 cell count is the measure of distance. It is not surprising, therefore, that together they represent an excellent surrogate marker and guide to monitoring patients.

O'Brien et al. found that a 75% decrease in plasma HIV RNA with a 10% increase in CD4 count explained 79% of treatment effect, with no additional benefit from measurement of β_2-microglobulin or treatment assignment.[32] Similarly, Phillips et al. found a 1.0 log change in viral RNA with a twofold difference in CD4 cell count provided the best on-therapy predictive value for risk of progression to AIDS during ZDV/lamivudine. This interaction was not reported in ACTG 175.[31] Both CD4 and plasma HIV RNA load are continuous variables which require interpretation in the clinical context and with appreciation of their limitations and their intra- and interpatient variability. Both are affected by OIs and certainly viral load may rise transiently following vaccination.[57,58] Viral load does not provide full information about viral pathogenicity or resistance to antiretrovirals. Similarly, CD4 cell count does not provide full information regarding immune function. However, whilst they provide physicians with the best means to manage therapy with the currently available antiretroviral agents, it remains uncertain whether their validity is maintained in the presence of immumodulators, such as interleukin-2.

LIFE CYCLE OF HIV AND POSSIBLE TARGETS FOR ANTI-HIV THERAPEUTIC AGENTS

Over the last 10 years much has been learnt about the life cycle of HIV, knowledge of which has led to the consideration of a large number of possible therapeutic agents. Many drugs have been tested *in vitro* and found to interfere with one or more stages of retroviral lifecycle from attachment to the final production of mature virions. From this selection process, inhibitors of the reverse transcriptase enzyme were amongst the earliest drugs to be assessed in clinical trials and one of their number, zidovudine (AZT), was rapidly licensed for clinical use in advanced HIV disease (1986 USA, 1987 UK).

Subsequently, other nucleoside analogues (didanosine [ddI], zalcitabine [ddC], lamivudine [3TC] and stavudine [d4T]) and non-nucleoside reverse transcriptase inhibitors (delavirdine, nevirapine) have been licensed. These agents act prior to integration of the virus into the host genome and thus would not be expected to prevent replication and production of virus of already infected cells. Not surprisingly, therefore, there is considerable interest in agents which act later in the replicative cycle, such as viral proteinase inhibitors. The agents, including indinavir, nelfinavir, ritonavir and saquinavir, inhibit the HIV's aspartic proteinase, thus preventing cleavage of *gag* and *gag-pol* polyproteins into structural and functional proteins and resulting in release of immature non-infectious virions. Inhibition is reversible *in vitro* upon withdrawal of drug from cultures. Selectivity with respect to human aspartic proteinases and other proteinase classes is very high (50 000-fold). Acute cytotoxicity has been identified only at micromolar concentrations, affording a high *in vitro* therapeutic index (>1000-fold).

Other proposed therapeutic targets include the viral integrase enzyme, viral regulatory genes (tat, nef, vpr) and surface receptors involved in viral attachment and entry (CCR5).

ANTIRETROVIRAL THERAPY

Primary infection

The immune response is initially highly effective in controlling HIV replication with rapid reduction in the level of plasmas viraemia. Studies are now evaluating aggressive intervention at this time point to establish whether viral eradication is feasible. Results from an intervention study of zidovudine vs placebo for six months in patients presenting with a seroconversion illness demonstrated at six months that the CD4 count in the treated group was an average of 137 cells higher than those who received placebo, which is a several times greater effect than that seen with zidovudine monotherapy at later stages of HIV disease.[59] During subsequent follow-up over 15 months, progression to symptomatic disease (oral candida, oral hairy leucoplakia) was mainly confined to the placebo group (seven events vs one event). Recently a number of more aggressive combination regimens have been assessed and appear to suppress the viral load in these patients reliably and so far durably (6–12 months or more) to undetectable levels.

Treatment of HIV infection at this very early stage may lead to improved responses[60] although data from different investigators are inconsistent. Some have argued that this may possibly lead to a lower viral 'set point' (after discontinuation of treatment), which would lead to improved long-term prognosis. Others, however, have cautioned that alternatively the initial immune response may be diminished with reduced exposure to virus secondary to very early treatment (personal communication, F. Gotch).

Established infection

Evidence of massive viral replication during all stages of HIV infection strongly supports the view that immunological decline and subsequent clinical progression are driven by HIV.[49,50] Therefore, to prolong quality of life in persons with HIV, therapeutic intervention should achieve substantial, preferably complete and prolonged suppression of viral replication, prevent infection of additional cells and, at least, create an environment in which immune regeneration may occur.

In vitro and *in vivo* data demonstrate that arresting viral replication cannot be achieved in a sustainable manner through single-agent therapy. Variants in the viral swarm which are resistant to antiretrovirals exist prior to initiation of therapy[61-63] and may be rapidly selected for during treatment to become the dominant quasispecies, often coinciding with virological failure.[63,64] Furthermore, clinical, immunological and virological responses observed during combination therapy appear to be consistently superior in both magnitude and duration to those seen with antiretroviral monotherapy. A number of two- and three-drug combinations have been observed to reduce viral replication, as measured by plasma viral load assays, to below the lower level of test detectability in the majority of recipients. These responses are associated with an apparent delay in development of resistance, relative to poorer responders, and a substantial rise in CD4 cell counts.

Despite these exciting responses most physicians accept these optimal responders will still ultimately fail and require therapy modification. Separate compartmental turnover of HIV beyond the plasma/lymphoid compartment has been documented in the genital tract and CNS – compartments which may not be well penetrated by all antiretroviral agents and may represent a potential source of resistant virus. Acquisition of virus resistant to ZDV and other antiretroviral drugs is well documented and increasingly common in urban seroconverter cohorts. Additionally, drug interactions, poor tolerability, intercurrent illnesses and episodic adherence failure (a well-documented problem, particularly with three times daily regimens in a range of disease states) are all likely contributors to the circumstances which will enable viral escape. Furthermore, many recipients of antiretroviral therapy do not make optimal responses, necessitating early treatment modification.

Achieving sustained viral suppression with the currently available antiretroviral agents is likely to involve appropriate selection of components of combination regimens to obtain an optimal antiviral response, with rational sequencing or adjustment of regimens to prolong responses. Rational choice may not only use evidence from randomized clinical studies, but will also use biologically plausible data from sources such as *in vitro* and mathematical models. Appropriately chosen combination regimens may not only provide the possibility of synergistic suppression of viral replication, but may also delay the emergence of resistant virus, select for novel patterns of mutations reflecting a less competent viral quasispecies, provide therapy against established resistant strains and cover a wide range of infected cell lines (for example, monocyte-macrophages and lymphocytes, acutely and chronically infected cells), viral phenotypes (such as syncytium inducing) and body compartments (e.g. CSF and lymph nodes).

In order to ensure the most rational and strategic use of the available agents, decisions regarding therapy should not be made solely on the basis of efficacy, but with consideration for a variety of other factors including:[65]

- safety profiles and potential interactions with concomitant medications;
- clinical history (e.g. a history of peripheral neuropathy or pancreatitis);
- current clinical status;
- potential of a given agent or regimen to limit future therapeutic options or efficacy due to selection of crossresistant or multidrug-resistant virus;
- pharmacokinetic and metabolic interactions (e.g. liver enzyme system inhibition or induction);
- intracellular pharmacokinetic interactions;
- *in vitro* synergy or non-antagonism;
- activity in different cell lines;
- convenience of administration.

Creating a regimen which can be readily adhered to by the patient appears critical in attaining and sustaining the best response from a combination. The following section will address the principles guiding initial choice and potential sequencing of antiretroviral regimens in patients commencing therapy, those experienced with zidovudine (ZDV) and other nucleoside analogue therapy and managing intolerance. The choice of agents used in this discussion is based on availability of clinical or surrogate marker data and a stage of clinical development which suggests likely availability in clinical practice within the next 2–3 years. These include the nucleoside analogues zidovudine (ZDV, AZT, Retrovir®), zalcitabine (ddC, HIVID®), didanosine (ddI, Videx®), stavudine (d4T, Zerit®), lamivudine (3TC, Epivir™) and abacavir (1592U89), the NNRTIs nevirapine (Viramune®), delavirdine (Rescriptor®), Efavirenz (Sustiva® or DMP 266) and the proteinase inhibitors saquinavir (Invirase® hard gel saquinavir, Fortovase® soft gel saquinavir), indinavir (Crixivan®), ritonavir (Norvir®), nelfinavir (Viracept®) and Amprenavir (VX-478, 141W94).

INITIAL THERAPY

Multiple studies have demonstrated the superiority of ZDV monotherapy over placebo on clinical endpoints of disease progression and survival in treatment-naive patients with AIDS and AIDS-related complex (ARC).[66,67] Earlier intervention with ZDV monotherapy, in asymptomatic disease or at CD4 counts above 300/mm^3, may provide a delay in disease progression over 1–2 years when compared to intervention at the onset of symptoms or at lower CD4 counts,[68–71] but does not provide any additional survival or quality of life benefit.[69–76] ZDV monotherapy has been shown to be clinically superior to ddC[74] in previously untreated patients with CD4 cell counts below 300/mm^3. However, the efficacy of ddI may be similar.[73,75,76] Initial activity marker responses larger than typically observed with nucleoside analogue monotherapy have been reported in patients during monotherapy with the proteinase inhibitors indinavir, ritonavir and nelfinavir as well as saquinavir at exposures higher than achieved with current dosing recommendations.[77–86] For example, viral load reductions observed during monotherapy with a proteinase inhibitor are typically 1–1.5 $\log_{10}$ compared with a nucleoside analogue of around 0.5 $\log_{10}$. In the majority of monotherapy recipients, however, virological failure is observed within 16–24 weeks, although patients with large initial responses (>2 $\log_{10}$ reductions in viral load) and low (<200 copies/ml) viral load nadirs may experience a more durable response.

Clinical endpoint data from ACTG 175 and Delta 1 have shown that nucleoside analogue combination therapy with ZDV plus ddC or ddI is a superior first-line choice to ZDV monotherapy with no additional toxicity.[75,76,87] Surrogate marker data from both studies demonstrated a correlation between improved clinical outcome and superior CD4 and viral load responses in the combination therapy arms. Data from the CAESAR study, which included 16% treatment-naive patients, also support this view[88] with recent data from Merck study 028 providing clinical evidence for superiority of ZDV/indinavir (or indinavir alone) over ZDV monotherapy.[89]

Multiple surrogate marker studies and preliminary data from clinical endpoint studies[90,91] assessing newer combinations of nucleoside analogues or nucleoside analogues with a proteinase inhibitor also support the superiority of combination therapy and suggest three drug combinations may have greater antiviral activity than two nucleoside analogues. Combinations of saquinavir with a second proteinase inhibitor, such as ritonavir or nelfinavir, also result in dramatic reductions in viral load. Although the studies do not have identical designs and baseline characteristics, comparison of responses illustrates that similar CD4 and viral load changes are observed in treatment-naive patients with each of the leading two nucleoside combination regimens and that inclusion of a protease inhibitor as the third agent or using a double proteinase inhibitor regimen provides substantial additional antiviral effect.

Combinations of NNRTIs with ZDV have, in general, yielded responses which are less impressive and may be less durable than those seen with nucleoside analogue or ZDV/proteinase inhibitor combinations. This is most likely due to the rapid appearance of viral variants resistant to these compounds. However, triple combination with ZDV/ddI/ nevirapine in treatment-naive patients has demonstrated greater antiviral effect over one year than the ZDV/ddI combination[92] with a high proportion of triple-therapy recipients achieving reductions in viral load to below 200 copies/ml. Most clinicians would now consider triple-agent therapy with two nucleoside analogues and a third agent to represent the current standard of care. Given similar activity across a range of triple combinations, recommendations cannot be made solely on this basis. There is, therefore, clearly a need for strategic consideration of other factors, such as the potential to limit future therapeutic options and the observed benefits of different agents in initial versus subsequent therapy regimens, to guide choice in initial therapy.

Additionally, novel approaches such as inclusion of hydroxyurea appears substantially to improve the virological activity of ddI and ddC *in vitro*[93] and at least ddI *in vivo*,[94,98] probably by affecting intracellular deoxynucleoside triphosphate pools, warrant further investigation as a means of maximizing the potential of nucleoside analogues.

While other triple or quadruple combinations of reverse transcriptase inhibitors, reverse transcriptase and proteinase inhibitors or proteinase inhibitor-based combinations are all worthy of or currently under investigation, it seems unlikely that these will result in significantly greater treatment effects than have already been observed with these classes of combinations.

Antiretroviral-experienced patients

Multiple studies have demonstrated that switching to an alternative agent as monotherapy or addition of a second agent is associated with clinical or surrogate marker benefits compared with continuing ZDV monotherapy. Clinical data suggest that the earlier such changes are initiated, the greater the associated therapeutic benefits. Limited data are available on switching from or adding to combination regimens. Many physicians now believe the best benefits are gained by switching at least two components of a treatment regimen, with some recent data supporting this view.

Switching single agents

In patients with CD4 counts below 500/mm^3, significant clinical progression delay may be gained by switching to ddI following prior treatment with ZDV,[76,88,96] although presence of the ZDV resistance-associated mutation at codon 215 may diminish response to ddI.[97] The value of switching to ddC monotherapy is less clear, although a subset of patients may benefit,[98] and the only comparative study showed at least equivalence with ddI switching to d4T (40 mg twice daily) in patients with at least six months prior ZDV experience and CD4 cell counts between 50 and 500/mm^3 is also superior to continued ZDV, significantly delaying disease progression, death or immunological decline.[99] The benefit of switching therapy appears to be independent of the duration of prior ZDV.[100,101]

Adding a single agent

In general, superior responses are observed with addition of agents to an ongoing monotherapy regimen compared to switching to a second monotherapy. Clinical benefits have been reported for the addition of ddI, ddC and 3TC to established ZDV therapy[75,87,88] and for ritonavir in severely immunodeficient patients experienced with and mostly still receiving a range of nucleoside analogues both as mono and combination therapy.[102] Additionally, clinical benefit has been reported with adding 3TC to established ZDV/ddC or ZDV/ddI therapy.[88]

Data from surrogate marker studies also support the strategy of adding an additional therapy, with 24-week response data suggesting similar benefits with a range of agents.

Switching or adding more than one new agent

Available data suggest that greater benefit may be gained by switching to or addition of multiple agents. In particular, using a proteinase inhibitor as one of the additional agents appears to provide the best response in nucleoside analogue-experienced persons. Clinical endpoint data are available demonstrating the benefits of switching to ddC/saquinavir rather than either drug alone and for adding 3TC plus indinavir compared to adding 3TC alone in ZDV-experienced patients. Activity marker data support the value of a range of combination approaches.[29,89] Data involving addition of a nucleoside analogue with an NNRTI have not demonstrated such substantial marker responses as regimens involving a proteinase inhibitor.[103] The optimal use of NNRTIs may therefore be at the time of first therapy initiation, 'saving' the proteinase inhibitors for later use. Whilst the benefit of switching or adding therapies appears independent of the duration of prior ZDV, the presence of ZDV-resistant virus may make virologic response to addition of either ddC, ddI or delavirdine less likely.[97,104] This suggests that these therapies are best added soon after starting ZDV or, preferably, commenced with ZDV. There is currently no consensus regarding appropriate therapy when ZDV resistance is established. However, stavudine has demonstrated significant clinical benefit in persons previously treated with ZDV. Additionally, multinucleoside analogue-resistant virus has occasionally been reported from patients heavily treated with a range of these agents,[105] suggesting that in some circumstances switching within this class may not be of value. However, proteinase inhibitors, which act on a different enzyme, remain active against nucleoside analogue-resistant virus. Both safety and efficacy data indicate that addition of any therapy is best in patients with higher CD4 cell counts, perhaps greater than 100/mm^3. However, switching to an alternative agent may continue to be of value in patients with CD4 counts less than 50 cells/mm^3.[106]

OTHER ISSUES IN ANTIRETROVIRAL USE

Most large clinical or small surrogate marker studies use relatively heterogeneous patient populations not stratified for multiple factors such as SI/NSI phenotype, viral load or presence of resistant virus at baseline. Additionally, most large studies are analysed by intention-to-treat methods, often despite a substantial proportion of patients either changing therapy or being lost to follow-up, which may lead to under- or overestimation of therapeutic effect. Evidence indicating that surrogate endpoints can be used to predict clinical outcome is increasing, potentially allowing for more rapid evaluation of new agents or regimens and suggesting that treatment decisions may be based upon these markers.[65,107] The optimum use of available antiretrovirals, choice of components of a combination regimen and the sequencing of those regimens should depend not only on data from clinical studies with their intrinsic limitations, but on a number of additional factors as outlined below.

Drug interactions

Patient history and awareness of concomitant medications is obviously important if overlapping toxicities and the potential for pharmacokinetic interactions are to be avoided or interactions harnessed to improve bioavailability of an agent (Fig. 15.3). Toxicities with nucleoside analogues often occur through similar mechanisms for example, peripheral neuropathy with ddI, d4T and ddC appears related to inhibition of human mitochondrial γ-DNA polymerase[108] and exacerbation of ddC-related neuropathy has been described with both ddI[109] and 3TC.[110] Concomitant therapy with these agents should therefore proceed with caution. Additionally, in patients commencing therapy with advanced disease it may be best to discontinue ddC, d4T or ddI after several months, as drug-related toxicities such as peripheral neuropathy and pancreatic dysfunction appear to be related to cumulative dosage and are more common in advanced disease.[111,112]

Compatibility of intracellular pharmacokinetics is also particularly relevant when combining nucleoside analogues. As these agents require activation by intracellular triphosphorylation, combination therapy with, for example, two thymidine-based analogues (such as ZDV and d4T) may be less than ideal as they compete for phosphorylation along the same pathway. A similar interaction has been reported between 3TC and ddC[113] but does not appear clinically important. Changes in phosphorylation of ZDV *in vivo* appear to correlate with clinical activity suggesting that interactions which lead to lower concentrations of the active triphosphate should be avoided.[114]

Some combinations of proteinase inhibitors, as well as potentially providing antiviral synergy and convergent selective pressure, may lead to higher free drug concentrations through saturation or inhibition of the P450 CYP3A4 isoenzyme, the enzyme responsible for metabolism of these compounds.[115] For drugs with limited bioavailability, such as saquinavir, this metabolic interaction may be exploited to increase blood levels. This may lead to increased efficacy, albeit with the possibility of increased toxicity. Interactions

● = CAUTION: Co-administration NOT recommended

▼ = CAUTION: Dose modification required

○ = Caution: drug interaction may require dose modification

★ = Caution: potential additive toxicity

		1	2	3	4	5	6	7	8	9	10	11	12	13	14	15	16	17	18	19	20	21	22	23	24	25	26	27	28	29
Zidovudine	1	1			●	★										★	★				★				★	○	○	★		
Didanosine	2		2	★	★													●				○				○	○	★		
Zalcitabine	3		★	3														●				○				○	○	★		
Stavudine	4	●	★	★	4																					○	○			
Lamivudine	5	★				5																								
Saquinavir	6						6	○	▼	○	●	●	●	●					●	●										
Indinavir	7						○	7	○	○	●	●	○	○					▼	●				○						
Ritavir	8						▼	○	8	○	●	●	○	○	○				●	●				○					●	○
Nelfinavir	9						○	○	○	9	●	●	○	○	○				●	●				○					●	
Astemizole/Terfenadine	10						●	●	●	●	10				●								●	●						
Cisapride	11						●	●	●	●		11			●								●	●						
Carbamazepine	12						●	○	○	○			12	○	○									○					○	○
Phenytoin	13						●	○	○	○			○	13		○			○	○									○	○
Clarithromycin	14								○	○	●	●	○		14				○	○										
Sulphonamides (high/dose)	15	★												○		15					★	★			★	★	★			
Pyrimethamine	16	★															16				★									
Pentamidine (iv)	17		●	●														17			○	●			○	●	●			
Rifabutin	18						●	▼	●	●				○	○				18				○	○					○	○
Rifampicin	19						●	●	●	●				○	○					19			○	●					○	○
Flucytosine	20	★														★	★	○			20	★			★	○	○			
Amphotericin B	21		○	○												★		●			★	21			★	●	●			
Fluconazole	22										●	●		○					○	○			22							
Itraconazole/Ketoconazole	23							○	○	○	●	●	○	○					○	●				23						
Ganciclovir	24	★														★		○			★	★			24	○	○	★		
Foscarnet	25	○	○	○	○											★		●			○	●			○	25	●			
Cidofovir + Probenecid	26	○	○	○	○											★		●			○	●			○	●	26			
Vincristine/Vinblastine	27	★	★	★																					★			27		
Oral contraceptive	28								●	●			○	○					○	○									28	
Corticosteroids	29								●				○	○					○	○										29
Methadone	30								○											○										

Fig. 15.3 Major drug interactions.

between PIs and NNRTIs vary: enzyme inducers such as nevirapine and possibly efavirenz reduce levels of some PIs (as assessed by AUC) whilst enzyme inhibitors such as delavirdine may increase levels.

In vitro synergy

In vitro data demonstrate that many antiretroviral combinations have at least additive and often synergistic activity,[116] exceptions being the antagonism observed between ZDV and d4T in the setting of ZDV resistance[117] and possibly SQV-IDV. Such data may be used to guide selection of optimal combinations, although issues including viral strain, cell line, drug concentrations and timing of drug administration relative to viral exposure should be considered when interpreting *in vitro* data.

Differential activity between cell lineages and phenotypes

Choice of therapy may also be driven by the need to combine agents which are most active in stimulated cells (for example, ZDV or d4T) with those most active in resting cells (such as ddC, ddI and 3TC)[118] or both cell types (proteinase inhibitors); those most active in acutely infected cells (nucleoside analogues, NNRTIs) and those active in both acutely and chronically infected cells (proteinase inhibitors), with compounds within the same activity group being substitutable in a regimen.

The presence of virus with an aggressive biological syncytium-inducing (SI) phenotype, with high *in vitro* replicative capacity and extensive T-cell tropism is associated with accelerated disease progression and unresponsiveness to ZDV therapy.[119–121] Activity of ddI appears to be maintained *in vivo* in the presence of SI variants.[122] Indeed, ddI has been reported to facilitate reversal of SI variants to the NSI phenotype.[122] Saquinavir has also been noted to inhibit syncytium formation *in vitro*,[123] suggesting proteinase inhibitors should be used if SI virus is present. Data on the activity of other antiretrovirals in the presence of SI virus are currently lacking.

Compartment penetration

Effective control of HIV replication will require the penetration of sufficient inhibitory concentrations of antiretrovirals into all body compartments. The CNS in particular may have a distinct virus population[124–126] with drug resistance developing more slowly,[127] an issue which may necessitate continued use of a CNS-penetrating compound in a regimen despite the presence of resistant virus in the plasma. ZDV has the highest CSF:plasma ratio of the available drugs (around 0.6) and appears to have a protective effect against AIDS dementia.[128] ZDV-resistant virus has, however, been isolated from both CSF and brain tissue.[129] Of the other nucleoside analogues, ddC, ddI and d4T all have CSF:plasma ratios of around 0.2 or more whilst 3TC may penetrate less well. Combinations of ZDV/3TC and d4T/3TC have recently been reported to have a similar effect on CSF HIV RNA levels. CNS penetration of proteinase inhibitors is not established but thought generally to be low. High protein binding of several of these compounds may mean that CSF:plasma ratios do not accurately reflect tissue levels.

Resistance and crossresistance

Evidence linking the presence of drug-resistant viral quasispecies to virological and clinical failure is increasing and information on patterns of resistance and crossresistance should therefore be considered when deciding how best to sequence and/or combine agents, with optimum sequences or combinations comprising agents which select non-overlapping resistance patterns and maintain the widest possible base of future treatment.[130] To date, HIV has proven to be a highly mutable virus whose enzymes exhibit remarkable plasticity and concerns exist for the potential of selecting for multidrug-resistant HIV. Again, caution must be used in translating data from interactions observed *in vitro*, even with clinical isolates, to clinical practice. Prevention of resistance appears feasible only when viral replication is fully arrested in all body compartments where antiviral drug levels (hence selective pressures) are achieved. The availability of rapid probes to detect resistance-associated mutations has the potential to both contribute to data-driven decision making and expand our understanding of the clinical importance of resistance.

Resistance data from ACTG 116B/117 revealed a strong correlation between the presence of phenotypic (IC_{50} >1.0 µM) or genotypic (presence of 215 and 41 mutations) ZDV resistance and disease progression.[131,132] Importantly, the increased risk of progression and death with ZDV resistance was independent of the benefits associated with switching to ddI in this trial. Patients with ZDV-resistant virus were at increased risk of disease progression whether they continued on ZDV or switched to ddI, implying that the benefits associated with change of therapy are not directly related to suppression of ZDV-resistant virus. *In vitro* observations of increased cytopathogenicity[133] and increased replicative capacity of ZDV-resistant virus compared to wild-type virus in drug-free stimulated peripheral blood mononuclear cells (PBMCs)[134] may help explain these findings. Quantitative assessment of plasma HIV RNA with or without the 215 mutation has shown that addition of ddI to ongoing ZDV therapy results in a decrease in wild-type RNA but not mutant RNA, despite the mutant virus being sensitive to ddI *in vitro*.[97] Similarly, patients with ZDV-resistant virus are significantly less likely to achieve a virological response to the addition of ddC than those with wild-type virus at baseline.[104] These data are in keeping with a report suggesting that for every 10-fold reduction in viral susceptibility to ZDV, there is a corresponding 2.2- and two-fold decrease in sensitivity to ddI and ddC, respectively.[135] Additionally, both viral and cellular resistance to ZDV may limit the future utility of d4T,[136,137] an agent potentially useful in combination regimens.

It would therefore appear that ZDV resistance has negative consequences for patients, both in terms of disease progression and limitation of subsequent treatment options with nucleoside analogues. This may have important implications regarding the choice of cotherapies with ZDV, timing of modification of antiretroviral therapy and, particularly, when to stop ZDV.

Monotherapy with ddI selects for a mutation at codon 74 in 56% of patients at six months, which is associated with both virological failure[138] and crossresistance to ddC.[139] However, resistance to ddI is generally infrequent when ddI is combined with ZDV in initial regimens. Similarly, the 184V mutation which develops almost universally by 12 weeks *in vivo* during both monotherapy and combination therapy in patients treated with 3TC is associated with reduced susceptibility (up to eight-fold) to both ddC and ddI,[140] raising concerns with this compound regarding limitation of subsequent therapeutic options, an issue which requires clarification. As 3TC appears active in a range of clinical contexts, including persons with advanced disease and substantial prior ZDV experience, it may be prudent to save this compound for later in the therapy sequence. Unfavourable changes in sensitivity to both ZDV and ddI have been reported to arise during monotherapy with d4T.[141] Reduced susceptibility to ddC appears slow to develop, with the most well-characterized mutation at codon 69[142,143] not impacting on viral sensitivity to other nucleoside analogues, although mutations at other sites leading to crossresistance to ddI and/or 3TC have occasionally been reported during ddC therapy. It has therefore been suggested that ddC may be better placed early in a treatment sequence as it is less likely to limit future activity of other agents.[130]

Salvage after failure of a proteinase inhibitor is equally problematic with a lack of clear data to guide rational decision making. The proteinase enzyme is surprisingly flexible with maintenance of good function despite numerous mutations in its structure. A number of mutations selected by both ritonavir and indinavir have been described, both *in vitro* and *in vivo*, that result in crossresistance to each other.[144,145] Indinavir has been reported to select for virus which is crossresistant to saquinavir and amprenavir.[144] Ritonavir-resistant virus appears frequently crossresistant to nelfinavir. The key mutations associated with saquinavir resistance at codon 90 and, less frequently, 48 of the proteinase appear to arise slowly during treatment.[145] Although it has been suggested that this may be due to low selective pressure of saquinavir with the hard gel formulation, which has limited oral bioavailability, low-dosage indinavir applying similar selective pressure is associated with rapid development of resistance and multiple mutations (suggesting the number of mutations required for resistance to this agent in no way represent a 'genetic barrier').[144] Additionally, higher exposure to saquinavir does not result in more rapid selection of resistant virus or other mutations.[86] It may be because saquinavir-resistant virus is functionally compromised, as has been suggested for the double 48 and 90 mutation.[147] After one year's therapy with saquinavir, about 14% of isolates have some reduced sensitivity to the drug and some of these also demonstrate reduced sensitivity to a range of proteinase inhibitors.[148] It is unlikely that the codon 90 mutation would facilitate subsequent indinavir resistance, given this mutation is infrequently observed during indinavir therapy[144] and other indinavir resistance-associated mutations are commonly observed as part of the background polymorphism of the proteinase enzyme.[17,22] These findings suggest that saquinavir may be preferred for first-line therapy.[146] The predominant resistance patterns for two other proteinases, nelfinavir and amprenavir, appear to be different from the three established proteinase inhibitors.[130] While there are not sufficient data yet to know whether or not mutations selected *in vivo* will be crossresistant to the other proteinase inhibitors, it is possible that these two drugs may be effective against virus which is resistant to another proteinase inhibitor. Hence, they could either be positioned early or after failure of the other drugs.

Cross-class resistance at codons 103 and 181 may limit the value of sequencing or combining NNRTIs. Continued clinical effectiveness of some drug regimens may be achieved with well-tolerated agents (e.g. NNRTIs, proteinase inhibitors) by dosage increase to above the inhibitory concentration for resistant virus. However, this approach may result in selection of more highly resistant mutants, within the constraints of replication competency.

Delaying the development of resistance

Viral replication in the presence of the selective pressure of antiretrovirals represents the ideal circumstances for selection of resistant virus. Reduction in viral replication to lowest achievable levels therefore appears the best strategy for delaying resistance. Additionally, some combinations of mutations may represent unacceptably dysfunctional changes for HIV and lead to the delay of resistance appearance to one or more components of a combination or selection of an increasingly compromised virus.

As ZDV-resistant virus is associated with a poorer outcome and limitation of effect of some other reverse transcriptase inhibitors, prolonged ZDV may be best avoided, especially in patients with advanced disease in whom resistance develops most rapidly.[131,132] As data suggest that transmission of ZDV-resistant virus is common, with ZDV resistance-associated mutations being detected in 10–25% of seroconverting patients in urban cohorts,[149,150] at least one of these agents may also be best included in initial therapy regimens.

Arguments are available to support both convergent (single-target) and divergent (multitarget) combination strategies with interest focusing on the possibility of selecting for virus with reduced replication capacity. Combination of ZDV with ddC or ddI appears to delay the appearance of resistance to these compounds but not to ZDV.[130] However, novel mutation patterns may emerge during combination therapy. Mutants resistant to both ZDV plus ddI and to this combination plus the NNRTI nevirapine have been selected *in vitro*, with both mutants exhibiting apparently normal growth patterns.[151] Isolates with unique patterns of amino acid substitutions at codons 62, 75, 77, 116 and 151 have occasionally been isolated from patients receiving prolonged combination therapy with ZDV plus ddI or, less frequently, ddC which are resistant to both drugs[105,152,153] and confer crossresistance to d4T and 3TC. The reason these novel mutations are not seen during monotherapy probably relates to their failure to compete for dominance with those mutants which do emerge.

Resistance to 3TC does not appear to be delayed by combination use with ZDV.[154] Addition of 3TC to ongoing ZDV therapy has been reported to lead to a resensitization of ZDV-resistant virus to this nucleoside analogue *in vitro* and *in vivo*,[154] although dual resistance to both ZDV and 3TC has been reported *in vitro*[155,156] and *in vivo*. Data from Glaxo study NUCA 3002 at week 12 indicate that the presence of the 3TC 184V mutation is not uniformly associated with reversal of phenotypic ZDV resistance. In this study, phenotypic resistance to 3TC was detected in 82% of patients on combination therapy at this time, with resensitization of previously ZDV-resistant isolates noted in only 4/10 of these and dual ZDV/3TC phenotypic resistance in 5/10.[157] Prior use of ZDV/ddC or ZDV/ddI does not appear to affect the benefit of 3TC compared to patients only pretreated with ZDV.[30]

Development of reduced susceptibility to saquinavir and indinavir appears to be delayed by combination therapy with nucleoside analogues compared to that seen during monotherapy with the proteinase inhibitor. Combinations of NNRTIs with ZDV may also delay the appearance of high-level resistance to these compounds, but not ZDV. However, dual- and triple-resistant virus has been described.[151]

Finally, resistance to all or most therapies appears delayed when substantial ($>2 \log_{10}$) reductions in viral load are achieved although inability to isolate virus in these patients prevents the testing for viral resistance. Duration of treatment response may therefore be predicted with at least some compounds by the virologic nadir.[130]

OTHER APPROACHES TO THERAPY, INCLUDING VACCINES AND IMMUNOMODULATORS

The failure of current antiretroviral approaches to prevent eventual progression of HIV disease has stimulated considerable interest in the use of immunomodulatory therapies, either alone or adjunctive to antiviral treatment. There have been a number of approaches and it is only possible to summarize a few of them here.

Enhancement of HIV-specific immune responses

Vaccines
Traditionally, vaccines have been used to prevent acquisition of infections but several vaccine candidates have been used in the attempt to stimulate immunity and thus prevent or slow down progression. These include antiidiotype vaccines against the gp120 binding site on CD4 or various prophylactic vaccines, particularly recombinant vaccines against membrane proteins such as gp120 and gp160. Results to date are inconclusive.

Transfer of humoral immunity to HIV-infected individuals

Neutralizing antibodies
Use of polyclonal antisera pooled from HIV-positive patients with high titre neutralizing antibody has been advocated, although to date no statistically significant

benefits in terms of survival or immunological or virological markers have been shown, although a French study showed a decrease in OIs (mainly cerebral toxoplasmosis).[158] There remains a concern about safety using these blood products, although plasma is inactivated with β-propriolactone (UK) in all centres plus pasteurization (France) or filtering (USA).

In children with HIV, standard intravenous immunoglobulin preparations have been shown to decrease the frequency of bacterial infection to which they are particularly prone, but there appears to be no benefit compared to those already on co-trimoxazole for PCP prophylaxis.[159]

Monoclonal antibodies against specific neutralizing epitopes such as those directed at C_3 loop antigens, CD4 or gp 41 are being explored. However, drawbacks include antihuman immune responses and the need to use a combination of these antibodies to decrease the frequency of resistance and increase the neutralizing potential.

Cellular transfer of HIV-specific cell populations is also being explored.

Cytokines

Inhibition of cytokines, particularly tumour necrosis factor α (TNFα) has been stimulated by observations that high levels of TNFα may be found in HIV infection and that TNFα upregulates HIV expression *in vitro*. Particularly high levels are found in late-stage disease patients with active pulmonary tuberculosis and studies with pentoxifylline and thalidomide (which both suppress TNFα) commenced. Effects on clinical benefit and on viral replication are being evaluated.[159,160]

Various cytokines such as IL-2 and IL-12 and interferons are being investigated for their immune enhancement. However, there is concern here in that preliminary studies suggested viral replication might be increased although this was more likely to occur at low (< 200–250) CD4 counts. In an attempt to circumvent this drawback investigations using combinations of cytokines plus antiretrovirals are being pursued.[161] The belief that immune activation generally may be detrimental has led to studies using thalidomide and cyclosporins to see if these will delay progression, particularly in those who have markers of immune activation such as raised $β_2$-microglobulin, neopterin and TNFα.

CONCLUSION

The following broad treatment principles have recently been proposed by the British HIV Association.

- Treatment should be offered prior to substantial immunodeficiency (CD4 <350).
- Initial treatment should include combinations of at least three drugs.
- Switches in therapy for failure should involve substitution or addition of at least two new agents.
- Viral load and CD4 measurement are essential for optimum patient management.
- Reducing viral load to below the detection level of a sensitive assay represents the optimal treatment response. Failure to achieve this response in persons starting therapy or loss of this virological control in persons established on therapy should prompt consideration of therapy modification. This response appears most reliably achieved with combinations of two nucleoside analogues plus a third agent (a protease inhibitor, a NNRTI or possibly a third nucleoside analogue) or two protease inhibitors.

REFERENCES

1. Barre-Sinoussi, F., Chermann, J.C., Rey, F. *et al.* (1983) Isolation of a T-lymphotropic retrovirus from a patient at risk for acquired immune deficiency syndrome. *Science*, **220**, 868–871.
2. Pedersen, C., Lindhardt, B.O., Lokke, B. *et al.* (1989) Clinical course of primary HIV infection: consequences for subsequent course of infection. *Br. Med. J.*, **229**, 154–157.
3. Nielsen, C., Pedersen, C., Lundgren, J. and Gerstoft, J. (1993) Biological properties of HIV isolates in primary HIV infection: consequences for the subsequent course of infection. *AIDS*, **7**, 1035–1040.
4. Centers for Disease Control (1986) Classification system for human T-cell lymphotropic virus type II/lymphadenopathy-associated virus infections. *MMWR*, **35**, 334–339.
5. Klein, R.S., Harris, C.A., Small, C.B. *et al.* (1984) Oral candidiasis in high risk patients as the initial manifestation of the acquired immune deficiency syndrome. *N. Engl. J. Med.*, **311**, 354–358.
6. Greenspan, D., Greenspan, J.S., Hearst, N.G. *et al.* (1987) Relation of oral hairy leucoplakia to infection with the human immunodeficiency virus and the risk of developing AIDS. *J. Infect. Dis.*, **155**, 475–481.
7. Haynes, W.S., Ascher, M.S., Wink, W. *et al.* (1993) Use of T-lymphocyte subset analysis in the case definition of AIDS. *J. Acq. Immun. Defic. Syndr.*, **6**, 287–294.

8. Peddecord, K.M., Benenson, A.S., Hofherr, L.K. et al. (1993) Variability of reporting and lack of adherence to consensus guidelines in human T-lymphocyte immunophenotyping reports: results of a case series. *J. Acq. Immun. Defic. Syndr.*, **6**, 823–830.
9. Giorgi, J.V., Liu, Z., Hultin, L.E. et al. (1993) Elevated levels of CD38+CD8+ cells in HIV infection adds to the prognostic value of low CD4+ T cell levels: results of 6 years follow-up. *J. Acq. Immun. Defic. Syndr.*, **6**, 904–912.
10. Graham, N.M., Park, C.P., Praitadosi, S. et al. (1995) Prognostic value of combined response markers among human immunodeficiency virus – infected persons: possible aid in the decision to change zidovudine monotherapy. *Clin. Infect. Dis.*, **20**, 352–362.
11. Cooper, D.A., Maclean, P., Finlayson, R. et al. (1985) Acute AIDS retrovirus infection. Definition of a clinical illness associated with seroconversion. *Lancet*, **i**, 537–540.
12. Kinloch de Loes, S., de Saussure, P., Saurat, J.H. et al. (1993) Symptomatic primary infection due to human immunodeficiency virus type 1: review of 31 cases. *Clin. Infect. Dis.*, **17**, 59–65.
13. Baumberger, C., Kinloch, S., Yerly, S. et al. (1994) High levels of circulating RNA in patients with symptomatic primary HIV-1 infection. *AIDS*, **7**, S59–S64.
14. Boag, F., Dean, R., Hawkins, D.A. et al. (1992) Abnormalities of liver function during HIV seroconversion illness. *Int. J. STD and AIDS*, **2**, 46–48.
15. Keet, I.P.M., Krijnen, P., Koot, M. et al. (1993) Predictors of rapid progression to AIDS in HIV-1 seroconverters. *AIDS*, **7**, 51–57.
16. Connor, E.M., Spelling, R.S., Gelber, R. et al. (1994) Reduction of maternal–infant transmission of human immunodeficiency virus Type I with zidovudine treatment. *N. Engl. J. Med.*, **331**, 1173–1180.
17. Webster, A. (1991) Cytomegalovirus as a possible co-factor in HIV disease progression. *J. Acq. Immun. Defic. Synd.*, **4**, (suppl 1) S47–52.
18. Youle, M.S. et al. (1994) Effects of high-dose oral acycloyir on herpes virus disease and survival in patients with advanced HIV disease; a double blind placebo controlled study. *AIDS*, **8(5)**, 641–649.
19. Laga, M., Drallo, M.O. and Buvel, A. (1994) Inter-relationships of sexually transmitted diseases and HIV: where are we now? *AIDS*, **8**, (suppl): S119–S124.
20. Yarchoan, R., Venzon, D.J., Pluda, J.M. et al. (1991) CD4 count and the risk for death in patients infected with HIV receiving antiretroviral therapy. *Ann. Intern. Med.*, **115(3)**, 184–189.
21. Moore, P.S. and Chang, Y. (1995) Detection of herpes virus-like DNA sequences in Kaposi's sarcoma in patients with and without HIV infection. *N. Engl. J. Med.*, **332**, 1181–1185.
22. Koopmans, P.P. et al. (1995) Pathogenesis of hypersensitive reactions to drugs in patients with HIV infection: allergic or toxic? *AIDS*, **9**, 217–222.
23. Pavia, A.T., Meadows, K.P., Tyring, S.K. and Rallis, T.M. (1998) Treatment of recalcitrant molluscum contagiosum with cidofovir. 5th Conference on Retroviruses and Opportunistic Infections, Chicago, abstract 504.
24. Youle, M.S., Hawkins, D.A., Collins, P. et al. (1988) Acyclovir resistant herpes in AIDS treated with foscarnet *Lancet*, **ii**, 341–342.
25. Youle, M., Clarbour, J., Farthing, C. et al. (1989) Treatment of resistant aphthous ulceration with thalidomide in HIV seropositive patients. *Br. Med. J.*, **298**, 432.
26. Saravolatz, L.D., Winslow, D.L., Collins, G. et al. (1996) Zidovudine alone or in combination with didanosine or zalcitabine in HIV-infected patients with the acquired immune deficiency syndrome or fewer than 200 CD4 cells per cubic millimeter. *N. Engl. J. Med.*, **335**, 1099–1106.
27. Hammer, S.M., Katzenstein, D.A., Hughes, M.D. et al. (1996) A trial comparing nucleoside monotherapy with combination therapy in HIV-infected adults with CD4 cell counts from 200–500 per cubic millimeter. *N. Engl. J. Med.*, **335**, 1081–1090.
28. DELTA Co-ordinating Committee (1996) DELTA: a randomised double-blind controlled trial comparing combinations of zidovudine plus didanosine or zalcitabine or zidovudine alone in HIV infected individuals. *Lancet*, **348**, 283–291.
29. Lalezari, J., Haubrich, R., Burger, H.U. et al. (1996) Improved survival and decreased disease progression of HIV in patients treated with saquinavir plus HIVID. IXth International Conference on AIDS, Vancouver, July, abstract LB.B.6033.
30. Katlama, C. on behalf of the CAESAR Co-ordinating Committee (1996) Clinical and survival benefit of 3TC in combination with zidovudine-containing regimens in HIV-1 infection: interim results of the CAESAR study. 3rd International Congress on Drug Therapy in HIV Infection, Birmingham, November, abstract SS2.1.

31. Katzenstein, D.A., Hammer, S.M., Hughes, M.D. et al. (1996) The relation of virologic and immunologic markers to clinical outcomes after nucleoside therapy in HIV-infected adults with 200 to 500 cells per cubic millimeter. *N. Engl. J. Med.*, **335**, 1091–1098.
32. O'Brien, W.A., Hartigan, P., Martin, D. et al. (1996) Changes in plasma HIV-1 RNA and CD4+ lymphocyte counts and the risk of progression to AIDS. *N. Engl. J. Med.*, **334**, 426–443.
33. Fauci, A.S., Pantaleo, G., Stanley, S. and Weissman, D. (1996) Immunopathogenic mechanisms of HIV infection. *Ann. Intern. Med.*, **124**, 654–663.
34. Prentice, R.L. (1989) Surrogate endpoints in clinical trials. *Stat. Med.*, **8**, 431–440.
35. Hughes, M.D., De Gruttola, V. and Welles S.L. (1995) Evaluating surrogate markers. *J. Acq. Immune Defic. Syndr. Hum. Retrovirus*, **10** (suppl 2): S1–S8.
36. Mellors, J.W., Rinaldo, C.R., Gupta, P. et al. (1995) Prognosis in HIV-1 infection predicted by the quantity of virus in the plasma. *Science*, **272**, 1167–1169.
37. Cao, Y., Qin, L., Zhang, L., Safrit, J. and Ho D.D. (1995) Virologic and immunologic characterization of long-term survivors of human immunodeficiency type I infection. *N. Engl. J. Med.*, **332**, 201–208.
38. Pantaleo, G., Menzo, S., Vaccarezza, M. et al. (1995) Studies in subject with long-term non-progressive human immunodeficiency virus infection. *N. Engl. J. Med.*, **332**, 209–216.
39. Crowe, S.M., Carlin, J.B., Stewart, K.I., Lucas, C.R. and Hoy, J.F. (1991) Predictive value of CD4 lymphocyte numbers for the development of opportunistic infections and malignancies in HIV-infected persons. *J. AIDS*, **4**, 770–776.
40. Fernandez-Cruz, E., Desco, M., Garcia Montes, M. et al. (1990) Immunological and serological markers predictive of progression to AIDS in a cohort of HIV-infected drug users. *AIDS*, **4**, 987–994.
41. Philips, A.N., Elford, J., Sabin, C. et al. (1992) Immunodeficiency and the risk of death in HIV infection. *J. Am. Med. Assoc.*, **268**, 2662–2666.
42. Hanson, D.L., Horsburgh, C.R., Fann, S.A., Havlik, J.A. and Thompson, S.E. (1993) Survival prognosis of HIV infected patients. *J. Acq. Immune. Defic. Syndr.*, **6**, 624–629.
43. Masur, H., Ognibene, F.P., Yarchoan, R. et al. (1989) CD4 counts as predictors of opportunistic pneumonias in human immunodeficiency virus (HIV) infection. *Ann. Intern. Med.*, **111**, 223–231.
44. Neaton, J.D., Wentworth, D.N., Rhame, F., Hogan, C., Abrams, D.I., Deyton, L for the Terry Beirn Community Programs for Clinical Research on AIDS (CPCRA) (1994) Methods of studying interventions: considerations in choice of a clinical endpoint for AIDS trials. *Stat. Med.*, **13**, 2107–2125.
45. Choi, S., Lagakos, S.W., Schooley, R.T. and Volberding, P.A. (1993) CD4+ lymphocytes are an incomplete surrogate marker for clinical disease progression in persons with asymptomatic HIV infection taking zidovudine. *Ann. Intern. Med.*, **118**, 674–680.
46. Lin, D., Fischl, M.A. and Schoenfeld, D. (1993) Evaluating the role of CD4-lymphocyte counts as surrogate endpoints in human immunodeficiency virus clinical trials. *Stat. Med.*, **12**, 835–842.
47. Levy, J.A., Ramachandran, B., Barker, E., Guthrie, J. and Elbeik, T. (1996) Plasma viral load. CD4+ lymphocyte counts and HIV-1 production by cells. *Science*, **271**, 670–671.
48. Kelleher, A.D., Carr, A., Zaunders, J. and Cooper, D.A. (1996) Alterations in the immune response of human immunodeficiency virus (HIV)-infected subjects treated with an HIV-specific protease inhibitor, ritonavir. *J. Infect. Dis.*, **173**, 321–329.
49. Ho, D.D., Neumann, A.U., Perelson, A.S. et al. (1995) Rapid turnover of plasma virions and CD4 lymphocytes in HIV-1 infection. *Nature*, **373**, 123–126.
50. Wei, X., Ghosh, S.K., Taylor, M.E. et al. (1995) Viral dynamics in human immunodeficiency virus type 1 infection. *Nature*, **373**, 117–122.
51. Liuzzi, G., Chiriani, A., Clementi, M. et al. (1996) Analysis of HIV-1 load in blood, semen, and saliva: evidence for different viral compartments in a cross-sectional and longitudinal study. *AIDS*, **10**, F51–F56.
52. Lafeuillade, A., Djediouane, A., Poggi, C. et al. (1996) Kinetics of viral clearance in plasma, peripheral blood mononuclear cells and lymph nodes. *AIDS*, **10**, 801–802.
53. Yerly, S., Kaiser, L., Mermillod, B. et al. (1995) Response of HIV RNA to didanosine as a predictive marker of survival. *AIDS*, **9**, 159–163.
54. Coombs, R.W., Welles, S.L., Hooper, C. et al. (1996) Association of plasma human immunodeficiency virus type 1 RNA level with risk of clinical progression in patients with advanced infection. *J. Infect. Dis.*, **174**, 704–712.
55. Philips, A.N., Eron, J.J., Bartlett, J.A. et al. (1996) HIV-1 RNA levels and the development of clinical disease. *AIDS*, **10**, 859–865.

56. Saag, M.S., Holodniy, M., Kuritzkes, D.R. et al. (1996) HIV viral load markers in clinical practice. *Nature Med.*, **2**, 625–629.
57. Staprans, S.I., Hamilton, B.L., Follansbee, S.E. et al. Activation of virus replication after vaccination of HIV-infected individuals. *J. Exp. Med.*, **182**, 1727–1737.
58. Donavon, R.M., Bush, C.E., Markowitz, N.P., Baxa, D.M. and Saravolatz, L.D. (1996) Changes in virus load markers during AIDS-associated opportunistic diseases in HIV-infected persons. *J. Infect. Dis.*, **174**, 401–403.
59. Kinloch de Loes, S., Hirschel, B.J., Koen, B. et al. (1995) A controlled trial of zidovudine in primary human immunodeficiency virus infection. *N. Engl. J. Med.*, **333**, 408–413.
60. Rosenberg, E.S., Billingsley, J.M., Caliendo, A.M. et al. (1997) Vigorous HIV-1-specific CD4+ T cell responses associated with control of viremia. *Science*, **278**, 1447–1450.
61. Nàjera, I., Richman, D.D., Olivares, I. et al. (1994) Natural occurrence of drug resistance mutations in the reverse transcriptase of human immunodeficiency virus type 1 isolates. *AIDS Res. Hum. Retroviruses*, **10**, 1479–1488.
62. Nàjera, I., Holguin, A., Quinones-Mateu, E. et al. (1995) *Pol* gene quasispecies of human immunodeficiency virus: mutations associated with drug resistance in virus from patients undergoing no drug therapy. *J. Virol.*, **69**, 23–31.
63. Frost, S.D.W. and McLean, A.R. (1994) Quasispecies dynamics and the emergence of drug resistance during zidovudine therapy of HIV infection. *AIDS*, **8**, 323–332.
64. Kellam, P., Boucher, C.A.B., Tijnagal, J.M.G.H. et al. (1994) Zidovudine treatment results in the selection of human immunodeficiency virus type 1 variants whose genotypes confer increasing levels of drug resistance. *J. Gen. Virol.*, **75**, 341–351.
65. Gazzard, B.G., Moyle, G.J., Weber, J. et al. (1997) British HIV Association guidelines for antiretroviral treatment of HIV seropositive individuals. *Lancet*, **349**, 1086–1092.
66. Fischl, M.A., Richman, D.D., Hansen, N. et al. (1990) The safety and efficacy of zidovudine (AZT) in the treatment of subjects with mildly symptomatic human immunodeficiency virus type 1 (HIV) infection: a double-blind, placebo-controlled trial. *Ann. Intern. Med.*, **112**, 727–737.
67. Volberding, P.A., Lagakos, S.W., Koch, M.A. et al. (1990) Zidovudine in asymptomatic human immunodeficiency virus infection – a controlled trial in persons with fewer than 500 CD4-positive cells per cubic millimeter. *N. Engl. J. Med.*, **322**, 941–949.
68. Hamilton, J.D., Hartigan, P.M., Simberkoff, M.S. et al. (1992) A controlled trial of early versus late treatment with zidovudine in symptomatic human immunodeficiency virus infection. *N. Engl. J. Med.*, **326**, 437–443.
69. Cooper, D.A., Gatell, J.M., Kroon, S. et al. (1993) Zidovudine in persons with asymptomatic HIV infection and CD4+ cell counts greater than 400 per cubic millimeter. *N. Engl. J. Med.*, **329**, 297–303.
70. Concorde Coordinating Committee (1994) Concorde: MRC/ANRS randomized double-blind controlled trial of immediate and deferred zidovudine in symptom-free HIV infection. *Lancet*, **343**, 871–881.
71. Lenderking, W.R., Gelber, R.D., Cotton, D.J. et al. (1994) Evaluation of the quality of life associated with zidovudine treatment in asymptomatic human immunodeficiency virus infection. *N. Engl. J. Med.*, **330**, 738–743.
72. Volberding, P.A., Lagakos, S.W., Grimes, J.M. et al. (1994) The duration of zidovudine benefit in persons with asymptomatic HIV infection. *J. Am. Med. Assoc.*, **272**, 437–442.
73. Noticeboard (1993) Didanosine and ACTG 116A. *Lancet*, **341**, 109.
74. Follansbee, S., Drew, L., Olsen, R. et al. (1993) The efficacy of zalcitabine (ddC, HIVID) versus zidovudine (ZDV) as monotherapy in ZDV naive patients with advanced HIV disease: a randomized, double-blind comparative trial (ACTG 114; N3300) [abstract no PO-B26–2113]. IXth International Conference on AIDS, Berlin.
75. Hammer, S., Katzenstein, D., Hughes, M. et al. (1996) A trial comparing nucleoside monotherapy with combination therapy in HIV-infected adults with CD4 cell counts from 200 to 500 per cubic millimeter. *N. Engl. J. Med.*, **335**, 1081–1090.
76. Katzenstein, D.A., Hammer, S.M., Hughes, M.D. et al. (1996) The relation of virologic and immunologic markers to clinical outcomes after nucleoside therapy in HIV-infected adults with 200 to 500 CD4 cells per cubic millimeter. *N. Engl. J. Med.*, **335**, 1091–1098.
77. Massari, F., Staszewski, S., Berry, P. et al. (1995) A double-blind, randomized trial of indinavir (MK-639) alone or with zidovudine vs zidovudine alone in zidovudine naive patients [abstract no. LB-6]. 35th Interscience Conference on Antimicrobial Agents and Chemotherapy, San Francisco.

78. Massari, F., Conant, M. and Mellors, J. (1996) A phase II open-label, randomized study of the triple combination of indinavir, zidovudine, and didanosine versus indinavir alone and zidovudine/didanosine in antiretroviral naive patients [abstract no. 90]. 3rd Conference on Retroviruses and Opportunistic Infections, January.
79. Danner, S.A., Carr, A., Leonard, J.M. et al. (1995) Safety, pharmacokinetics and antiviral activity of ritonavir, an inhibitor of HIV-1 protease [abstract no. 75]. 4th International Workshop on HIV Drug Resistance, Sardinia.
80. Danner, S.A., Carr, A., Leonard, J.M. et al. (1995) A short-term study of the safety, pharmacokinetics, and efficacy of ritonavir, an inhibitor of HIV-1 protease. N. Engl. J. Med., **333**, 1528–1533.
81. Markowitz, M., Saag, M., Powderly, W.G. et al. (1995) A preliminary study of ritonavir, an inhibitor of HIV-1 infection. N. Engl. J. Med., **333**, 1534–1539.
82. Molla, A., Korneyeva, M. and Gao, Q. (1996) Ordered accumulation of mutations in HIV protease confers resistance to ritonavir. Nature Med., **2**, 760–766.
83. Norbeck, D., Hsu, A., Granneman, R. et al. (1995) Virologic and immunologic response to ritonavir (ABT-538), an inhibitor of HIV protease [abstract no. 70]. 4th International Workshop on HIV Drug Resistance, Sardinia.
84. Markowitz, M., Conant, M., Hurley, A. et al. (1995) Phase I/II dose range-finding study of the HIV proteinase inhibitor Ag 1343 [abstract no. LB-4]. 35th Interscience Conference on Antimicrobial Agents and Chemotherapy, San Francisco.
85. Moyle, G., Youle, M., Chapman, S. et al. (1995) A phase II dose-escalation study of the agouron proteinase inhibitor Ag 1343 [abstract no. LB-3]. 35th Interscience Conference on Antimicrobial Agents and Chemotherapy, San Francisco.
86. Schapiro, J.M., Winters, M.A., Kozal, M.J. et al. (1995) Saquinavir monotherapy trial: prolonged suppression of viral load and resistance mutations with higher dosage [abstract no. LB5]. 35th Interscience Conference on Antimicrobial Agents and Chemotherapy, San Francisco.
87. Delta Coordinating Committee (1996) Delta: a randomised double-blind controlled trial comparing combinations of zidovudine plus didanosine or zalcitabine with zidovudine alone in HIV-infected individuals. Lancet, **348**, 283–291.
88. Katlama, C. (1996) Clinical and survival benefit of 3TC in combination with zidovudine-containing regimens in HIV-infection: interim results of the CAESAR study [abstract no. SS2.1]. AIDS, **10**(suppl 2), S9.
89. Merck Study 028: Executive summary.
90. The Pisces (SV14604) Writing Committee, Stellbrink, H.I., Hawkins, D.A., Clumeck, N. et al. (1998) A randomized, multicentre phase II study of saquinavir plus zidovudine plus zalcitabine in previously untreated or minimally pre-treated HIV-infected patients. Submitted to AIDS for publication
91. Hammer, S.M., Squires, K.E., Hughes, M.D. et al. (1997) A controlled trial of two nucleosides analogues plus indinavir in persons with human immunodeficiency virus infection and CD4 cell counts of 200/mm^3 or less. AIDS clinical trials group 320 study team. N. Engl. J. Med., **337**, 725–733.
92. Conway, B., Montaner, J.S.G., Cooper, D. et al. (1996) Randomised double blind one year study of the immunological and virological effects of nevirapine, didanosine and zidovudine combinations among antiretroviral naïve, AIDS-free patients with CD4 200–600 [abstract no. OP7.1]. AIDS, **10**(suppl 2), S15.
93. Gao, W.Y., Johns, D.G., Mitsuya, H. (1994) Anti-HIV-1 activity of hydroxyurea in combination with 2′, 3′-dideoxynucleosides. Mol. Pharmacol., **46**, 767–772.
94. Biron, F., Lucht, F., Peyramond, D. et al. (1995) Anti-HIV activity of the combination of didanosine and hydroxyurea in HIV-1-infected individuals. J. Acq. Immune Defic. Syndr., **10**, 36–40.
95. Montaner, J.S.G., Zala, C., Raboud, J.M. et al. (1996) A pilot study of hydroxyurea (HO-urea) as adjuvant therapy among patients with advanced HIV disease receiving didanosine (ddI) therapy [abstract no. 406]. 3rd Conference on Retroviruses and Opportunistic Infections, Washington.
96. Kahn, J.O., Lagakos, S.W., Richman, D.D. et al. (1992) A controlled trial comparing continued zidovudine with didanosine in human immunodeficiency virus infection. N. Engl. J. Med., **327**, 581–587.
97. Holodniy, M., Katzentein, D., Mole, L. et al. (1996) Human immunodeficiency virus reverse transcriptase codon 215 mutations diminish virologic response to didanosine-zidovudine therapy in subjects with non-syncytium-inducing viral phenotype. J. Infect. Dis., **174**, 854–857.
98. Fischl, M.A., Stanley, K., Collier, A.C. et al. (1995) Combination and monotherapy with zidovudine and zalcitabine in patients with advanced HIV disease. Ann. Intern. Med., **122**, 24–32.
99. Gottlieb, M., Peterson, D., Adler, M. et al. (1995) Comparison of safety and efficacy of two doses

of stavudine (Zerit, d4T) in a large simple trial in the US parallel track program [abstract no. I171]. 35th Interscience Conference on Antimicrobial Agents and Chemotherapy, San Francisco.
100. Thompson, M., Creagh, T., Morris, A. et al. (1993) Impact of antiretroviral therapy on survival and CD4 decline in Georgia [abstract no. PO-B26–2110]. IXth International Conference on AIDS, Berlin.
101. Torres, R.A., Barr, M.R., McIntyre, K.I. et al. (1995) A comparison of zidovudine, didanosine, zalcitabine and no antiretroviral therapy in patients with advanced HIV disease. Int. J. STD and AIDS, 6, 19–26.
102. Cameron, B., Heath-Chiozzi, M., Kravcik S. et al. (1996) Prolongation of life and prevention of AIDS in advanced HIV immunodeficiency with ritonavir [abstract no. LB6a]. 3rd Conference on Retroviruses and Opportunistic Infections, Washington.
103. D'Aquila, R.T., Hughes, M.D., Johnson, V.A. et al. (1996) Nevirapine, zidovudine, and didanosine compared with zidovudine and didanosine in patients with HIV-1 infection. A randomized, double-blind, placebo-controlled trial. Ann. Intern. Med., 124, 1019–1030.
104. Sylvester, S., Caliendo, A., An, D. et al. (1995) HIV-1 resistance mutations and plasma RNA during ZDV+ddC combination therapy. J. Acq. Immune Defic. Syndr. Hum. Retroviruses, 10 (suppl 3), 23.
105. Shirasaka, T., Kavlick, M.F., Veno, T. et al. (1995) Emergence of human immunodeficiency virus type 1 variants with resistance to multiple dideoxynucleosides in patients receiving therapy with dideoxynucleosides. Proc. Natl Acad. Sci. USA, 92, 2398–2402.
106. Kazempour, K., Kammerman, L.A. and Farr, S.S. (1995) Survival effects of ZDV, ddI, and ddC in patients with CD4≤50 cells/mm^3. J. Acq. Immune Defic. Syndr., 10 (suppl 2), S97–106.
107. Gazzard, B.G. and Moyle, G.J. (1995) Individualisation of HIV therapy: the clinician's perspective. Br. J. Clin. Prac., 49, 145–147.
108. Keilbaugh, S.A., Prusoff, W.H., Simpson, M.V. (1991) The PC12 cell as a model for studies of the mechanism of induction of peripheral neuropathy by anti-HIV 1-dideoxynucleoside analogs. Biochem. Pharmacol., 42, R5–8.
109. LeLacher, S.F. and Simon, G.I. (1991) Exacerbation of dideoxycytidine-induced neuropathy with dideoxyinosine. J. AIDS, 4, 538–539.
110. Cupler, E.J. and Dalakas, M.C. (1995) Exacerbation of peripheral neuropathy by lamivudine. Lancet, 345, 460–461.
111. Moyle, G.J., Nelson, M.R., Hawkins, D.A. et al. (1993) The use and toxicity of didanosine (ddI) in HIV antibody positive individuals intolerant to zidovudine (AZT). Quart. J. Med., 86, 155–163.
112. Abrams, D.I., Goldman, A.I., Launer, C. et al. (1994) A comparative trial of didanosine or zalcitabine after treatment with zidovudine in patients with human immunodeficiency virus infection. N. Engl. J. Med. 330, 657–662.
113. Back, D.J., Haggard, P.G., Veal, G.J. et al. (1995) Intracellular phosphorylation interactions between nucleoside analogues [abstract no. 41]. 5th European Conference on Clinical Aspects and Treatment of HIV Infection, Copenhagen.
114. Stretcher, B.N., Pesce, A.J., Frame, P.T. et al. (1994) Correlates of zidovudine phosphorylation with markers of HIV disease: progression and drug-toxicity. AIDS, 8, 763–769.
115. Cohen, C., Sun, E., Camkeron, D. et al. (1996) Ritonavir-saquinavir combination treatment in HIV-infected patients [abstract no. LB7b]. 36th Interscience Conference on Antimicrobial Agents and Chemotherapy, New Orleans.
116. Johnson, V.A. (1994) Combination therapy: more effective control of HIV type 1? AIDS Res. Hum. Retroviruses, 8, 907–912.
117. Merrill, D.P., Moonis, M., Chou, T-C. et al. (1996) Lamivudine or stavudine in two- and three-drug combinations against human immunodeficiency virus type 1 replication in vitro. J. Infect. Dis., 173, 355–364.
118. Gao, W.Y., Agbaria, R., Driscoll, J.S. et al. (1994) Divergent anti-human immunodeficiency virus activity and anabolic phosphorylation of 2'3'-dideoxynucleoside analogs in resting and activated human cells. J. Biol. Chem., 269, 12633–12638.
119. Koot, M., Keet, I.P.M., Vos, A.H.V. et al. (1993) Prognostic value of HIV-1 syncytium-inducing phenotype for rate of CD4+ cell depletion and progression to AIDS. Ann. Intern. Med., 118, 681–688.
120. Richman, D.D. and Bozette, S.A. (1994) The impact of syncytium-inducing phenotype of human immunodeficiency virus on disease progression. J. Infect. Dis., 169, 968–974.
121. Schellekens, P.T.A., Koot, M., Roos, M.T.L. et al. (1995) Immunologic and virologic markers determining progression to AIDS. J. Acq. Immune Defic. Syndr., 10(suppl 2), S62–66.
122. Delforge, M-L., Liesnard, C., Debaisieux, L. et al. (1995) In vivo inhibition of syncytium-inducing variants of HIV in patients treated with didanosine. AIDS, 9, 89–90.

123. Moyle, G. (1996) Saquinavir: a review of its development, pharmacological properties and clinical use. *Exp. Opin. Invest. Drugs*, **5**, 155–167.
124. Epstein, L.G., Kuiken, C., Blumberg, B.M. *et al.* (1991) HIV-1 V3 domain variation in brain and spleen of children with AIDS: tissue-specific evolution within host-determined quasispecies. *Virology*, **180**, 583–590.
125. Haggerty, S. and Stevenson, M. (1991) Predominance of distinct viral genotypes in brain and lymph node compartments of HIV-1-infected individuals. *Viral Immunol.*, **4**, 123–131.
126. Ball, J.K., Holmes, E.C., Whitwell, H. *et al.* (1994) Genomic variation of human immunodeficiency virus type-1 (HIV-1): molecular analysis of HIV-1 in sequential blood samples and various organs obtained at autopsy. *J. Gen. Virol.*, **75**, 867–879.
127. Wildemann, B., Haas, J., Ehrhart, K. *et al.* (1993) In vivo comparison of zidovudine resistance mutations in blood and CSF of HIV-1 infected patients both simultaneously in plasma but also between different body compartments such as blood and CSF. *Neurology*, **43**, 2659–2663.
128. Portegies, P. (1994) AIDS dementia complex: a review. *J. Acq. Immune Defic. Syndr.*, **7**(suppl 2), S38–49.
129. Di Stephano, M., Norkrans, G., Chiodi, F. *et al.* (1993) Zidovudine-resistant variants of HIV-1 in brain. *Lancet*, **342**, 865.
130. Moyle, G.J. (1996) Use of viral resistance patterns to antiretroviral drugs in optimising selection of drug combinations and sequences. *Drugs*, **52**, 168–185.
131. D'Aquila, R.T., Johnson, V.A., Welles, S.L. *et al.* (1995) Zidovudine resistance and HIV-1 disease progression during antiretroviral therapy. *Ann. Intern. Med.*, **122**, 401–408.
132. Japour, A.J., Welles, S., D'Aquila, R.T. *et al.* (1995) Prevalence and clinical significance of zidovudine resistance mutations in human immunodeficiency virus isolated from patients after long-term zidovudine treatment. *J. Infect. Dis.*, **171**, 1172–1179.
133. Tremblay, M., Rooke, R. and Wainberg, M.A. (1992) Zidovudine-resistant and -sensitive HIV-1 isolates from patients on drug therapy: *in vitro* studies evaluating level of replication-competent viruses and cytopathogenicity. *AIDS*, **6**, 1445–1449.
134. Caliendo, A., Savara, A. and An, D. (1995) Zidovudine-resistance mutations increase replication in drug-free PBMC stimulated after infection [abstract]. *J. Acq. Immune Defic. Syndr. Hum. Retrovirol.*, **10**(suppl 3), 2–3.
135. Mayers, D.L., Japour, A.J., Arduino, J-M. *et al.* (1994) Dideoxynucleoside resistance emerges with prolonged zidovudine therapy. *Antimicrob. Agents Chemother.*, **38**, 307–314.
136. Rooke, R., Parniak, M.A., Tremblay, M. *et al.* (1991) Biological comparisons of wild-type and zidovudine-resistant isolates of human immunodeficiency virus type 1 from the same subjects: susceptibility and resistance to other drugs. *Antimicrob. Agents Chemother.* **35**, 988–991.
137. Dianzani, F., Antonelli, G., Turriziani, O. *et al.* (1994) Zidovudine induces the expression of cellular resistance affecting its antiviral activity. *AIDS Res. Hum. Retroviruses*, **10**, 1471–1478.
138. Kozal, M.J., Kroodsma, K., Winters, M.A. *et al.* (1994) Didanosine resistance in HIV-infected patients switched from zidovudine to didanosine monotherapy. *Ann. Intern. Med.*, **121**, 263–268.
139. Gu, Z., Gao, Q., Parniak, M.A. *et al.* (1992) Novel mutation in the human immunodeficiency virus type 1 reverse transcriptase gene encodes cross resistance to 2′,3′-dideoxyinosine and 2′,3′-dideoxycytidine. *J. Virol.*, **66**, 7128–7135.
140. Gao, Q., Gu, Z.X., Parniak, M.A. *et al.* (1993) The same mutation that encodes low-level human immunodeficiency virus type 1 resistance to 2′,3′-dideoxyinosine and 2′,3′-dideoxycytidine confers high-level resistance to the (−) enantiomer of 2′,3′-dideoxy-3′-thiacytidine. *Antimicrob. Agents Chemother.*, **37**, 1390–1392.
141. Lin, H.J., Myers, L.E. *et al.* (1994) Multicenter evaluation of quantification methods for plasma human immunodeficiency virus type 1 RNA. *J. Infect. Dis.*, **170**, 553–562.
142. Fitzgibbon, J.E., Howell, R.M., Haberzettl, C.A. *et al.* (1992) Human immunodeficiency virus type 1 pol gene mutations which caused decreased susceptibility to 2′,3′-dideoxycytidine. *Antimicrob. Agents Chemother.*, **36**, 153–157.
143. Gulick, R., Mellors, J., Havlir, D. *et al.* (1996) Potent and sustained antiretroviral activity of indinavir (IDV), zidovudine (ZDV) and lamivudine (3TC) [abstract no. B931]. XIth International Conference on AIDS, Vancouver.
144. Condra, J.H., Schleif, W.A., Blahy, O.M. *et al.* (1995) In vivo emergence of HIV-1 variants resistant to multiple protease inhibitors. *Nature*, **374**, 569–571.
145. Moyle, G. and Gazzard, B.G. (1996) Current knowledge and future prospects for the use of HIV proteinase inhibitors. *Drugs*, **5**, 155–167.
146. Roberts, N.A. (1995) Drug-resistance patterns of saquinavir and other proteinase inhibitors. *AIDS*, **9**(suppl 2), S27–32.

147. Wilson, S.I., Phylip, L.H. and Mills, J.S. (1995) Interactions of substrates and inhibitors with escape mutants of HIV-1 proteinase [abstract]. *J. Acq. Immune Defic. Syndr. Hum. Retrovirol.* **10**(suppl 3), 32.
148. Jacobsen, H., Haengoi, M., Ott, M. *et al.* (1995) Reduced sensitivity to saquinavir: an update on genotyping from phase I/II trials [abstract no. 68]. 4th International Workshop on HIV Drug Resistance, Sardinia.
149. Perrin, L., Yerly, S., Rakik, A. *et al.* (1995) Transmission of 215 mutants in primary HIV infection and analysis after 6 months of ZDV [abstract no. 21]. 4th International Workshop an HIV Drug Resistance, Sardinia.
150. Harzic, M., Patey, O., Ferchal, F. *et al.* (1995) AZT resistance mutations in virus from patients with recent HIV infection [abstract no. 26]. 4th International Workshop on HIV Drug Resistance, Sardinia.
151. Larder, B.A., Kellam, P., Kemp, S.D. (1993) Convergent combination therapy can select viable multidrug resistant HIV-1 in vitro. *Nature*, **365**, 451–453.
152. Shafer, R.W., Kozal, M.J., Winters, M.A. *et al.* (1994) Combination therapy with zidovudine and didanosine selects for drug resistant human immunodeficiency virus type 1 strains with unique patterns of pol gene mutations. *J. Infect. Dis.* **169**, 722–729.
153. Shafer, R.W., Iversen, A.K.N., Winters, M.A. *et al.* (1995) Drug resistance and heterogeneous long-term virologic responses of human immunodeficiency virus type 1-infected subjects to zidovudine and didanosine combination therapy. *J. Infect. Dis.*, **172**, 70–78.
154. Larder, B.A., Kemp, S.D. and Harrigan, P. (1995) Potential mechanism for sustained antiretroviral efficacy of AZT-3TC combination therapy. *Science*, **269**, 696–699.
155. Goulden, M.G., Cammack, N., Hopewell, P.L. *et al.* (1996) Selection in vitro of an HIV-1 variant resistant to both lamivudine (3TC) and zidovudine. *AIDS*, **10**, 101–102.
156. Nijhuis, M., de Jong, D., van Leeuwen, R. *et al.* (1996) The rise in HIV-1 RNA load during 3TC/AZT combination therapy is associated with the selection of viruses resistant to both 3TC and AZT [abstract no. 7]. 3rd Conference on Retroviruses and Opportunistic Infections, Washington.
157. Johnson, V.A., Quinn, J.B., Benoit, S.L. *et al.* (1996) Drug resistance and viral load in NUCA 3002: lamivudine (high or low dose)/zidovudine (ZDV) combination therapy versus ZDV/dideoxycytidine combination therapy in ZDV-experienced (greater than or equal to 24 weeks) patients (CD4 cells 100–300/mm3) [abstract no. 113]. 3rd Conference on Retroviruses and Opportunistic Infections, Washington.
158. Vittecoq, D., Cherret, S., Moranjonbert, L. *et al.* (1995) Passive immunotherapy in AIDS – a double blind randomized study based on transfusion of plasma rich anti-human immunodeficiency virus I antibodies vs. transfusions of seronegative plasma. *Proc. Natl. Acad. Sci. USA*, **92**, 1195–1199.
159. Spector, S.A., Gelber, R.D., McGrath, N. *et al.* (1994) A controlled trial of intravenous immune globulin for the prevention of serious bacterial infections in children receiving zidovudine for advanced human immunodeficiency virus infection. Paediatric AIDS Clinical Trials Group. *N. Engl. J. Med.*, **331**(18), 1181–1187.
160. Marriott, J.B., Cookson, S., Carlin, E. *et al.* (1997) A double-blind placebo-controlled phase II trial of thalidomide in asymptomatic HIV-positive patients: clinical tolerance and effect on activation markers and cytokines. *AIDS Res. Hum. Retroviruses*, **13**(18), 1625–1631.
161. Kovacs, J.A., Vogel, S., Albert, J.M. *et al.* (1996) Controlled trial of interleukin-2 infusions in patients infected with the human immunodeficiency virus. *N. Engl. J. Med.*, **335**(18), 1350–1356.

16 The management of opportunistic infections in AIDS

A. Pozniak

As HIV infection causes progressive deterioration of the host immune system, the patient becomes increasingly likely to develop opportunistic infections. These range from minor trivial skin infections to severe life-threatening conditions involving major organs such as the lungs or brain.

The management of these conditions rests on the firm principles of medical investigation to provide a definitive diagnosis and the provision of specific therapy for individual infectious agents. However, as is often the case in medicine, treatments are started before results are available to the clinician which is why clinical experience, careful history taking and examination are vital components of management of patients with HIV infection and AIDS.

PNEUMOCYSTIS CARINII PNEUMONIA (PCP)

Pneumocystis carinii pneumonia remains the most commonly reported AIDS-defining disease in the UK.[1] Both prophylaxis against PCP and dual nucleoside antiretroviral therapy have decreased the incidence of PCP.[2] HAART (highly active antiretroviral therapy) may have a more dramatic effect in preventing patients from becoming at risk of PCP. Whether the risk will be reduced when the CD4 counts of HAART patients rise above 200 mm^3 is unknown. Some preliminary data suggest that it does.

Common presenting symptoms[3]

Duration of symptoms varies widely from days to several weeks. Symptoms include:

- fever
- non-productive cough
- progressive shortness of breath
- chest tightness intensified with inspiration or coughing.
- fatigue and weight loss.

N.B. A fulminant illness lasting less than a week can occur. Patients who have been receiving prophylaxis with aerosolized pentamidine may have unusual presentations or have extrapulmonary involvement, e.g. eyes, intestine, thyroid, CNS, etc.[4,5]

Examination

- Adds little to the diagnosis of PCP.
- Examination of lungs is usually normal.
- Fine crackles or wheezes are occasionally heard, usually with severe episodes.
- Pleurisy is rare.

Investigations

Chest radiograph/CT
The chest radiograph is a useful diagnostic screening test.

- It is abnormal in 90% of patients with PCP (5–14% can have a normal radiograph).[6]

- The commonest radiographic appearance is a diffuse increase in interstitial then alveolar markings.
- Bilateral perihilar ground-glass opacities are obvious in severe cases.
- Extensive consolidation or atypical features[7] which include cysts, cavities or pneumothorax can occur, especially if patients have been on aerosolized pentamidine (AP) prophylaxis.
- Pleural effusions and lymphadenopathy are rare.
- CT may show diffuse ground-glass shadowing.

Exercise oximetry

Three per cent fall in oxygen saturation after 10 min exercise or a resting value of less than 90% saturation suggests PCP[8] (100% sensitive, 77% specific).

Blood gases

Ninety-two percent of patients with PCP have either an abnormal alveolar-arterial gradient (A-a) at rest (>15 mmHg)[9] on air or a resting PaO_2 <80 mm Hg. *NB. Many pulmonary conditions can affect this gradient.* A room air A-a gradient >35 mmHg is associated with a poor outcome.

Pulmonary function tests

- The reduction in TLCO (single-breath carbon monoxide transfer factor) is the most sensitive of all standard pulmonary function tests. It lacks specificity, perhaps due to direct lung damage produced by HIV itself.[10]
- A sudden fall in sequential measurements of TLCO nearly always represents pulmonary disease.
- A value of 70% or less below baseline has been suggested as a cut-off point for disease.

Radioisotope scanning[11]

- ^{99M}TcDTPA clearance has been used as a measure of pulmonary epithelial permeability.
- Clearance is increased in:
  - patients with pneumocystis pneumonia;
  - smokers (but in pneumocystis a biphasic clearance curve is seen).
- The technique is time consuming, costly and requires a nuclear medicine department.
- Gallium-67 citrate scans have a low specificity (50%).
- Human polyclonal immunoglobulin (HIG) labelled with indium-II may help in differentiating pulmonary infection from pulmonary KS.[12]

Routine haematology/biochemistry

- Liver function tests, urea and electrolytes are not normally deranged by PCP.[13]
- ESR is elevated non-specifically.
- The WBC count is rarely elevated.

Lactate dehydrogenase (LDH)[14]

- This has been used as a marker of pulmonary inflammation but lacks specificity.
- LDH concentration rises with the onset of symptoms and falls slowly with therapy.
- Elevation is due largely to fraction five which predominates in pulmonary tissue.
- LDH can be raised in up to 20% of patients with other pulmonary conditions.

Induced sputum[15,16]

- The patient inhales 20–30 ml hypertonic (3%) saline generated by ultrasonic nebulizer (Ultraneb or Devilbiss) over 10–20 min.
- Irritation of airways plus osmotic effect drains fluid into the interstitium.
- Deep coughing is encouraged and sputa collected and stained with Giemsa, toluidine blue-0, silver and immunofluorescence.
- Problems with severe coughing and nausea can occur.
- Patients on AP have decreased positive diagnostic rate.
- Variability in diagnostic rate is dependent on dedicated personnel and enthusiasm from staff, patients and laboratory.[17]

It is essential that transmission of airborne pathogens (e.g. MTB) is not increased by this procedure. Negative pressure rooms, respiratory cubicles or hoods with appropriate ventilation are needed.

Bronchoalveolar lavage (BAL) and transbronchial biopsy (TBB)

- There is no major benefit in performing TBB as well as BAL in suspected PCP.[15] TBB is contraindicated in respiratory failure, when the patient is being ventilated or in intractable coagulopathy, whereas BAL is safe in these patients.
- TBB has a 9% incidence of pneumothorax (no effect on this rate even if fluoroscopic screening is used).
- BAL is a sensitive test for PCP (>90%) but is operator dependent and requires an experienced and efficient laboratory service.

Open lung biopsy

This is reserved for those in whom BAL/TBB is non-diagnostic and/or empirical treatment is not successful.

Stains for PCP

- Cyst wall
    Gomori's methenamine silver
    - sensitivity/specificity 95%
    - up to 10% false negatives

Toluidine blue-0
Calcofluor white
- Non-cyst wall stains
    Wright Giemsa
    Diff-Quick
    False negative 10–15% cases
- Immunochemical
    Monoclonal antibodies
    – 90% sensitive
    – false positives can occur.

Polymerase chain reaction (PCR)[18]

The use of the PCR to diagnose active pneumocystis in serum, sputum or saliva has been developed. As with all highly sensitive tests, attention must be paid to assay conditions and controlling contamination by host tissues and other microbes to ensure that findings are specific for *P. carinii*. The mere finding of *P. carinii* by PCR does not necessarily indicate the presence of active disease.

Presumptive diagnosis

Even with a lack of confirmatory laboratory data, a presumptive diagnosis can be made on clinical, radiological and oximetry findings without resort to bronchoscopy.[19]

If empirical therapy is given beware that:

- bacterial pneumonia can respond to co-trimoxazole, especially pneumococcus and *Haemophilus influenzae*;
- non-specific pneumonitis can resolve spontaneously;
- TB, Kaposi's sarcoma and CMV can resemble PCP;
- patients with these will not respond to, and may deteriorate on, empirical therapy.

Treatment

Manipulation of the inflammatory response[20–23]

Four controlled studies have shown that the administration of corticosteroids within 72 h of starting specific anti-pneumocystis therapy can reduce or prevent the initial decline in oxygenation and improve prognosis by halving the risk of pulmonary failure and reducing the risk of death by a third.

Indications

Corticosteroids should be given when PaO_2 on air ≤9.3 kPa (70 mmHg) or an a-A O_2 gradient on air ≥4.7 kPa (35 mmHg).

Dosage

Prednisolone 40 mg bd for five days, then 40 mg od for five days, then 20 mg od for 11 days[20] or intravenous methylprednisolone 1 g od for three days, 0.5 g for three days then oral prednisolone 40 mg to zero over 10 days (give ranitidine from days 1–10). Steroids must be completed before discontinuing definitive PCP therapy, otherwise recrudescence of disease may occur.

Supportive care

Continuous positive airway pressure (CPAP)[24] by face mask improves oxygenation in patients with desaturation refractory to standard masks. Problems with CPAP are:

- pulmonary barotrauma;
- gastric aspiration;
- facial pressure necrosis.

ICU Care[25]

There has been an increase in survival from 15% to 38–55% in patients admitted to ICU with PCP.[26] Poor prognosis is associated with:

- low albumin;
- pH <7.35;
- need for positive end expiratory pressure (PEEP) >10 cm H_2O after 96 h in ICU.

Drug treatment[27–31]

Three comparative trials[29–31] using co-trimoxazole and intravenous pentamidine have shown no difference in efficacy but mortality appeared lower and improvements were seen more quickly with co-trimoxazole.

Side effects

Most adverse reactions develop between seven and 14 days of therapy. Patients who developed significant toxicity due to one drug rarely did so to an alternative drug. The addition of folinic acid does not prevent haematological toxicity.

Monitoring recovery

Eighty-five percent of patients with PCP will respond. Most will respond to therapy within the first 5–7 days and failure to improve by 7–10 days is unusual. Transient worsening of signs and symptoms within the first 2–3 days of therapy is common, probably due to compartmental fluid shifts. Defervescence of fever, improved blood gas exchange and reduction of dyspnoea occur on about the seventh day. The chest radiograph improves slowly and may take more than two or three weeks to become normal.

Non-responders to treatment

There is no routine microbiological or biochemical method readily available to determine whether a patient is not responding to treatment because the organism is resistant to the agent chosen or because there is an immunological dysfunction or because of another factor. Resistance to co-trimoxazole has been reported but is rare. Patients generally improve after 4–8 days of therapy and it makes little sense to switch agents before that time. If there is no initial improvement, approaches that rely on parenteral rather than oral routes of administration are generally preferred. Switching from co-trimoxazole to parenteral pentamidine or switching to clindamycin and primaquine or to atovaquone in mild to moderate PCP are all useful strategies. When switching from co-trimoxazole to IV pentamidine, the co-trimoxazole is usually continued for an overlap period of 48–72 h because IV pentamidine takes 2–4 days to reach therapeutic levels. Combining pentamidine and co-trimoxazole offers no advantage and causes more unwanted effects. Trimetrexate or eflornithine can be used as salvage treatment when deterioration occurs despite first-line therapies.

Other treatments in mild to moderate PCP

Clindamycin, primaquine,[32,33] dapsone-trimethoprim and atovaquone[34] have all been used and found to be effective in mild to moderate PCP.

Prophylaxis against PCP[35–38]

The incidence of PCP in people known to have been HIV positive for nine months or more is decreasing in the UK. The majority of cases occur in those patients who have an HIV diagnosis at or within three months of presentation. These data suggest that the use of primary prophylaxis has a beneficial effect.

Primary prophylaxis
Susceptibility to infection with PCP increases as immunosuppression progresses, so primary prophylaxis should be offered when a patient has:

- a CD4 count which falls below 200 cells/ml (0.2×10^9/l);
- oropharyngeal thrush (unrelated to antibiotic or corticosteroid therapy);
- persistent unexplained fever (>37.7°C) for two weeks or more.

Other patients might also be offered prophylaxis including:

- those who have had an AIDS-defining illness regardless of CD4 count;
- those who are aged ≥45 years and whose CD4 count falls below 250 cells/ml;
- those with rapidly declining CD4 count when the count reaches below 250 cells/ml.

Secondary prophylaxis
This should be offered to those who have already had PCP.

Drugs for prophylaxis
Oral low-dose co-trimoxazole (960 mg daily or 960 mg three times a week) has proved to be superior to nebulized pentamidine as an agent in both primary and secondary prophylaxis. Most physicians use dapsone or nebulized pentamidine if patients are intolerant of co-trimoxazole.[39] Atovaquone can also be used.[40]

With co-trimoxazole in prophylactic doses, rash occurs in up to 20%. There is no adverse interaction with AZT or other nucleotides

When PCP develops in patients on PCP prophylaxis, most clinicians institute therapy with a different agent from the one used for prophylaxis. There are no data to validate the advantage of this approach.

Specific drug treatments for PCP

Co-trimoxazole
Trimethoprim and sulphamethoxazole (1:5 parts) Septrin

Preparations
Tablet 480 mg, 960 mg
Ampoule 480 mg/5ml

Dosage and administration
120 mg/kg/day for 21 days in two or more divided doses. The dose is usually given IV but after fever and symptoms have settled, the patient can be changed to oral therapy at the same dosage. Many patients tolerate the IV route better than the tablets.

What to monitor
Full blood count – twice a week
U&Es – twice a week.
Liver function tests – twice a week

Side effects/toxicity
Nausea and vomiting
- Use antiemetics IV or PR regularly.
- Increase the duration of the infusion.
- Decrease the dose to 75% of the initial dosage.
- Change to IV route if the patient is taking tablets.
- Change to second-line therapy.

Rash (up to 44%)
The action taken will depend on the severity of the rash and the duration of therapy completed, i.e. in the third week, unless the rash is very severe, treatment should be continued and topical steroids and oral antihistamines used for symptomatic relief. If the rash is

very severe (e.g. Stevens–Johnson syndrome [SJS] has been reported with co-trimoxazole):

- stop co-trimoxazole and start alternative therapy;
- give systemic corticosteroids (no evidence they are useful in SJS).

Skin rashes may be prevented by concurrent steroid therapy and appear when the steroids are stopped.

Impaired renal function
Co-trimoxazole is predominantly renally excreted so the dose should be adjusted in patients with renal impairment.

Haematological toxicity
Neutropenia and thrombocytopenia can occur with co-trimoxazole therapy.

Elevated liver enzymes
Hepatitis, cholestatic jaundice and liver failure have been reported as being associated with co-trimoxazole therapy in non-AIDS patients. Elevated liver enzymes on either prophylactic or treatment doses of co-trimoxazole should be monitored closely. If liver enzymes are more than five times the upper limit of normal, the patient should be changed to second-line therapy.

Drug levels
Reducing the dose of co-trimoxazole by 25% to keep trimethoprim concentration at around 5–8 mg/ml minimizes toxicity without reducing efficacy.

Drug interactions
Absolute contraindications for treatment doses of co-trimoxazole are didanosine (ddI) and zalcitabine (ddC), with which there is an increased risk of pancreatitis. ddI or ddC can be restarted after a 1–2-week washout period following completion of high-dose co-trimoxazole therapy.

With treatment doses of sulphadiazine, there is an increased risk of renal precipitation of sulphadiazine and haematological toxicity.

Prophylactic regimes of sulphadiazine and co-trimoxazole can be prescribed concurrently.
Use with caution:

- warfarin: increased anticoagulant activity
- phenytoin: increased serum *phenytoin* levels
- myelosuppressive agents, e.g. ganciclovir, zidovudine (AZT): increased toxicity
- dihydrofolate reductase inhibitors, e.g. pyrimethamine: increased haematological toxicity

Folinic acid 15–30 mg po daily can be given to prevent toxicity.

Intravenous pentamidine

Preparations
Vial   300 mg

Dosage and administration
4 mg/kg/day for 21 days as a single IV infusion. Each dose should be added to 250 ml of 0.9% sodium chloride or 5% dextrose and given over one hour with the patient supine.

What to monitor
Blood pressure – regularly during infusion
Full blood count – twice a week
U&Es – twice a week
Serum calcium and magnesium – once a week
Fasting blood glucose – daily
Liver function tests – once a week

Blood glucose testing should be continued at regular intervals following completion of therapy, especially if hypoglycaemia occurred during treatment, as hyperglycaemia and diabetes mellitus have been reported up to several months after completion of therapy.

Serum amylase and ECG monitoring should be undertaken where clinically indicated.

Side effects/toxicity
- Rapidly deteriorating renal function
  This can initially manifest itself as an increase in serum urea and/or potassium. Serum creatinine often mirrors this increase after 2–4 days. Change to alternative therapy.
- Slowly deteriorating renal function
  Continue with full dose for seven days, then reduce to 75% of full dose for the remainder of the course. If renal function continues to deteriorate, change to third-line therapy.

Hypotension
- Increase duration of infusion, e.g. give over 2 h.
- Give a plasma volume expander prior to the infusion, e.g. 500 ml gelofusine.
- Change to alternative therapy.

Hypocalcaemia
Give calcium supplementation.

Pronounced and repeated disturbances in blood glucose
Give glucose or insulin as appropriate. It has been reported that severe prolonged hypoglycaemia is caused by destruction of the B-cells in the pancreas, predisposing the patient to later diabetes mellitus. If this occurs, the patient should be changed to alternative therapy.

Drug interactions
Reduced excretion of didanosine (ddI) zalcitabine (ddC) may occur if renal function is compromised by pentamidine. Additive toxicity may predispose to pancreatitis. Stop ddI and ddC during pentamidine treatment and allow a two-week washout period before restarting.

With nephrotoxic drugs, e.g. foscarnet, amphotericin B, cumulative renal toxicity may occur.

Use with caution:

- aminoglycoside antibiotics; e.g. gentamicin: monitor serum levels.
- High-dose acyclovir; may precipitate in the kidneys
- High-dose sulphadiazine: may precipitate in the kidneys and cause additive pancreatitis.

Clindamycin and primaquine[41,42]
Preparations
Clindamycin:
Capsule   150 mg
Ampoule   300 mg/2 ml
Primaquine:
Tablet   7.5 mg (*unlicensed product*)

Dosage and administration
Clindamycin: 600–1200 mg six-hourly po/IV for 21 days
Primaquine: 30 mg once daily po for 21 days

- Used in mild to moderate PCP
- 85% improved
- Rash 50%
- Leucopenia, nausea 21%

Clindamycin injection should be further diluted prior to administration with either 0.9% sodium chloride (normal saline) or 5% dextrose.

What to monitor
G6PD – prior to treatment
Liver function tests – twice a week
Full blood count – twice a week
U&Es – twice a week

A stool chart should be kept throughout the treatment period.

Side effects/toxicity
Clindamycin

Diarrhoea
- Discontinue treatment immediately.
- Culture for *Clostridium difficile*.
- Start vancomycin 125 mg six-hourly po for 7–10 days.
- Isolate patient.

Elevation of liver function tests
Transient elevations of liver function tests may occur after starting therapy. If these are sustained or jaundice develops, discontinue treatment.

Rash
The action taken will depend on the severity of the rash and the duration of therapy completed, i.e. in the third week, unless the rash is very severe, treatment should be continued and topical steroids and antihistamines used for symptomatic relief. If the rash is very severe:

- discontinue treatment;
- give systemic corticosteroids.

A skin rash may be masked by concurrent steroid therapy.

Primaquine

Haemolytic anaemia
This can occur in patients with G6PD deficiency, in whom primaquine is contraindicated. Treatment should be stopped.

Methaemoglobinaemia
This can occur occasionally.

- Treatment should be stopped
- If severe, methylene blue can be given 1–2 mg/kg IV over five minutes. Repeated doses may be given.

Neutropenia
If neutrophils $<0.7 \times 10^9/l$; give folinic acid 15 mg daily po and monitor carefully. If neutrophils $<0.5 \times 10^9/l$:

- treatment should be stopped;
- may need growth factors GM-CSF, G-CSF.

Drug interactions
No significant drug interactions occur with either clindamycin or primaquine. Primaquine should be used with caution with other drugs causing haematological disturbances.

Dapsone and Trimethoprim[43]
Preparations
Dapsone:
Tablet   50 mg
Trimethoprim:
Tablet   100 mg
Ampoule 100 mg/5 ml

Dosage and administration
Dapsone: 100 mg daily po for 21 days

Trimethoprim: 20 mg/kg/day po/IV in two divided doses for 21 days

- Used in mild/moderate PCP
- 90% improve
- One-third develop toxicity/adverse reactions – rash (10%), nausea (7%), neutropenia (3%), methaemoglobinaemia (3%).

Trimethoprim can be given as a slow IV bolus but as the volume of the dose is likely to be in excess of 25 ml, an infusion may be preferred.

What to monitor
Full blood count – twice a week
U&Es – twice a week
Liver function tests – twice a week

Side effects/toxicity
Trimethoprim

Renal impairment
If CrCl <25 ml/min, give normal dose for three days then halve the dose for the remainder of the course.
If CrCl <15 ml/min, give half the normal dose.

Haematological toxicity
If neutrophils $0.5–0.7 \times 10^9/l$ and platelets $50–70 \times 10^9/l$:

- reduce to 75% of initial dose;
- give folinic acid 15 mg daily po.

If neutrophils $<0.5 \times 10^9/l$ and platelets $<50 \times 10^9/l$:

- stop treatment;
- give folinic acid 15 mg daily po;
- may need growth factors GM-CSF and G-CSF.

Dapsone

Anaemia
Decrease the dosage, e.g. to 75 mg daily for treatment or 50 mg thrice weekly for prophylaxis. Transfuse as necessary.

Nausea and vomiting
Give regular antiemetics.

Rash
The action taken will depend on the severity of the rash and the duration of therapy completed, i.e. in the third week, unless the rash is very severe, treatment should be continued and topical steroids and antihistamines used for symptomatic relief. If the rash is very severe:

- discontinue treatment;
- give systemic corticosteroids.

A skin rash may be masked by concurrent steroid therapy.

Elevation of liver function tests
If a rise in liver function tests is sustained or jaundice develops, stop treatment.

Drug interactions
Trimethoprim
Use with caution:

- phenytoin: increased phenytoin levels
- warfarin: increased anticoagulant activity
- didanosine (ddI): decreased oral absorption of trimethoprim due to the buffer in ddI. When given concurrently, leave a two-hour gap between doses
- rifampicin: may increase the elimination of trimethoprim.

Dapsone
Use with caution:

- rifampicin: decreased plasma concentration of dapsone
- probenecid: reduced excretion of dapsone
- drugs that increase gastric pH, e.g. H_2 antagonists, didanosine (ddI): dapsone requires an acidic environment for absorption. When given concurrently leave a two-hour gap between doses.

Atovaquone[44]

Atovaquone is licensed for the oral treatment of mild to moderate PCP in patients who are intolerant of co-trimoxazole.

A dose of 750 mg eight-hourly has been used in patients unable to tolerate other regimens. A daily prophylactic dose of 1500 mg has been used.[45]

Preparations
Tablet 250 mg

Dosage and administration
750 mg eight-hourly po for 21 days. Doses should be taken with food to maximize absorption. Atovaquone suspension appears to be superior.

What to monitor
Full blood count – twice a week
U&Es – twice a week
Liver function tests – twice a week

Side effects/toxicity
Rash
This may be related to the plasma levels obtained. The action taken will depend on the severity of the rash and the duration of therapy completed, i.e. in the third week, unless the rash is very severe, treatment should be continued and topical steroids and antihistamines used for symptomatic relief.

If the rash is very severe:

- discontinue treatment;
- give systemic corticosteroids.

A skin rash may be masked by concurrent steroid therapy.

Neutropenia
This occurs more commonly in patients receiving the higher dose regime.

- If neutrophils $<0.5 \times 10^9/l$, stop atovaquone.
- May need growth factors G-CSF, GM-CSF.

Elevation of liver function tests
Mild, transient rises in liver enzymes have been reported. If these rise to five times the upper limit of normal, atovaquone should be stopped.

Renal impairment
The use of atovaquone in patients with severe renal impairment has not been studied to date. However, the lack of significant renal elimination would suggest that excretion is unlikely to be affected.

Drug interactions
Atovaquone is highly bound to plasma protein (>99.9%). Therefore, caution should be used when administering it with other highly plasma protein-bound drugs with a narrow therapeutic index (e.g. phenytoin, warfarin), as competition for binding sites may occur.

Primary and secondary prophylaxis against PCP

Co-trimoxazole
960 mg daily po on Monday, Wednesday and Friday[46]
or
960 mg or 480 mg daily po seven days a week (daily dosing may aid compliance)

What to monitor
Full blood count – every 1–3 months
U&Es – every 1–3 months
Liver function tests – every 1–3 months
If rash occurs and is mild, desensitization can be attempted in hospital or day care (Table 16.1).

Table 16.1 Desensitization regime for co-trimoxazole. Rapid protocol – day care or inpatient

Hour	Dose TMP/SMX
0	0.004/0.02 mg
1	0.04/0.2 mg
2	0.4/2 mg
3	4/20 mg
4	40/200 mg
5	160/800 mg

Use oral suspension 40 mg/200 mg per 5 ml. Take 100 ml fluid after each dose

Nebulized pentamidine
Preparations
Nebule   300 mg/5 ml

Nebulized pentamidine can be used for primary and secondary prophylaxis against PCP in patients who are intolerant to co-trimoxazole: *it should not be used for the treatment of PCP*. Nebulized pentamidine should not be given to patients:

- with a history of chronic obstructive airways disease (COAD) or asthma;
- with potential undiagnosed TB;
- with disseminated PCP;
- with a high risk of pneumothorax.

Dosage and administration
300 mg every 2–4 weeks. Pentamidine nebules should be administered using a suitable nebulizer, e.g. a Respigard II, modified Acorn system 22 or an equivalent device with either a compressor or piped oxygen at a flow rate of 6–10 l/min. A filter should be attached to the exhaust line to reduce atmospheric pollution. Patients should nebulize in a well-ventilated, smoke-free environment and no one else should be present. In hospitals or health-care settings nebulization should only take place in facilities with adequate ventilation. The period of nebulization should be continuous.

There is a much lower incidence of systemic side effects with nebulized pentamidine. However, it should be used with caution in patients who have renal or hepatic dysfunction, hypertension or hypotension, hyperglycaemia or hypoglycaemia, leucopenia, thrombocytopenia or anaemia.

Local reactions involving the upper respiratory tract (URT) can occur, ranging in severity from cough, shortness of breath and wheezing to bronchospasm; these are more common in patients who smoke or have a history of asthma. Pneumothorax has also been reported. If URT symptoms do occur, inhaled or nebulized salbutamol can be administered prior to pentamidine.

Pentamidine IV
4 mg/kg once a month IV. This method should only be used for PCP prophylaxis if oral agents and nebulized pentamidine are contraindicated – there is little published evidence of its efficacy.

What to monitor
BP – prior to dosing
Full blood count – once a month
U&Es – once a month
Ca^{2+}, Mg^{2+} – once a month
Blood sugar – once a month
Liver function tests – once a month

Dapsone +/– Pyrimethamine[47,48]
Preparations
Dapsone:
Tablet 100 mg

+/– Pyrimethamine:
Tablet 25 mg

Administration
Dapsone can be used alone or in combination with pyrimethamine. Different dosage regimes have been recommended, including:

- dapsone 100 mg + pyrimethamine 25 mg three times each week;
- dapsone 50 mg daily.

N.B. Folinic acid 15 mg po daily should be given to patients taking pyrimethamine.

What to monitor
Full blood count – every 1–3 months
U&Es – every 1–3 months
Liver function tests – every 1–3 months

TOXOPLASMOSIS[49,50]

Toxoplasma gondii is a protozoan that infects all mammals and some birds and is usually acquired by eating foods contaminated with the oocyst or meat from animals previously infected.

The prevalence of infection is dependent on geographical area. In the United States between 10% and 60% of adults are seropositive for toxoplasma, in France 80% and in the UK 40%. Reactivation of this infection (usually in a severely immunocompromised host) causes toxoplasmosis.

The major forms of toxoplasmosis are:

- encephalitis with focal intracerebral lesions;
- pneumonitis and choroidoretinitis.

Toxoplasma encephalitis (TE)[51]

Prior to HAART it had been calculated that approximately 27–40% of AIDS patients who are seropositive for toxoplasma will ultimately develop TE. Toxoplasma can infect any cell in the brain and therefore the clinical syndrome of TE is non-specific and may include both focal and non-focal signs and symptoms of central nervous system dysfunction.

Clinical presentation

The clinical presentation ranges from an insidious process evolving over weeks to a fulminant acute

Table 16.2 Grading of severity of *P. carinii* pneumonia. This is important in making decisions about outpatient vs inpatient treatment, low-dependency vs high-dependency care, admission to ICU and the use of corticosteroids

	Mild	Moderate	Severe
Symptoms and signs	Dyspnoea on exertion (brisk walking), with or without cough and sweats	Dyspnoea on minimal exertion, occasional dyspnoea at rest, fever ± sweats	Dyspnoea at rest, tachypnoea at rest, persistent cough
Blood gas tensions (room air)	PaO_2 normal, SaO_2 falling on exercise	PaO_2 8.1–11.0 kPa	PaO_2 < 8.0 kPa
Chest X-ray	Normal or minor perihalar shadowing	Diffuse interstitial shadowing	Extensive shadowing with or without diffuse alveolar shadowing sparing costophrenic angles and apices

PaO_2 Partial pressure of oxygen.
SaO_2 Arterial oxygen saturation

[Algorithm flowchart — Fig. 16.1]

IgG +ve / −ve
- −ve → Diagnosis of TE unlikely
- +ve → CT/MRI
  - Single lesion → Do MRI
    - If no MRI → Treatment for TE
    - Single lesion → Consider other causes → Consider brain biopsy
  - Multiple lesions → Treatment for TE
  - No lesion → Work-up for other causes; if more found, repeat MRI at 48–72 h
- Treatment for TE:
  - Improved by 7 days → Presumptive diagnosis of TE treatment 3–6/52 then maintenance
  - No improvement clinically 7–14 days or deteriorated by 3 days

Pregnant women – spiromycin oral 100 mg tds
Adjunctive treatment – corticosteroids given to:
  • reduce cerebral oedema
  • reduce mass effect
but no difference in response rates if given or not*
  • may be useful if level of consciousness deteriorates

Prognosis
  • fatal if untreated
  • response 68–95%
  • 40% have long-term neurological sequelae
  • median survival 225–490 days

*Colin, J., McMeeking, A., Cohen, W., Jacons, J. and Holman R.S. (1989) Evaluation of the policy of empiric treatment of suspected toxoplasma encephalitis in patients with AIDS. Am. J. Med. **86** 521-527.

Fig. 16.1 Algorithm for suspected TE.

confusional state. Treatment is usually begun on the basis of a presumptive diagnosis dependent on:

- the characteristic findings on CT or MRI scan;
- a positive serological test. In one study this combination had a predictive value as high as 80%.

In 95% of patients with TE symptoms include:

- subacute onset;
- headache;
- focal neurological signs;
- convulsions.

Focal signs
Commonest are:

- hemiparesis;
- abnormalities of speech;
- cranial nerve lesions.

Rarely:

- focal dystonia, parkinsonism, tremor, hemichorea, panhypopituitarism, diabetes insipidus, SIADH;
- neuropsychiatric symptoms may be dominant;
- spinal cord disease has been described.

Differential diagnosis

- CNS lymphoma
- Progressive multifocal leucoencephalopathy (PML)
- Infection of CNS, such as with CMV, HSV, VZV or with fungi, bacteria or mycobacteria.

Investigations

Computed tomography (CT)[52]
Patients with toxoplasma encephalitis usually have focal or multifocal abnormalities demonstrable on CT scan. These focal areas of encephalitis may be multiple, are hypodense and are usually ring enhancing after contrast. Double-dose contrast may show them more easily. They tend to occur at the corticomedullary junction and frequently involve the basal ganglia. They resolve on treatment in 2–3 weeks. The earliest sign of response is reduction in size of lesions. Complete resolution with treatment can be seen on a CT scan between three weeks and six months. Patients who clinically respond to therapy will usually have radiographic evidence of improvement within three weeks of initiation of therapy.

Magnetic resonance imaging (MRI)[53]
- High signal abnormalities are seen on T2-weighted images.
- Deep lesions are 1–3 cm in diameter; superficial 1 cm.
- The centre of the lesion has complex high and low signal intensities.
- Small lesions resolve in 3–5 weeks.
- Gadolinium enhancement may show focal TE lesions in patients with a normal CT.
- For patients with single lesions on CT, MRI should be performed as it is more sensitive in detecting multiple lesions.

Differential diagnosis on CT/MRI
Diffuse TE may give normal CT/MRI. The major CT/MRI differential diagnosis is lymphoma. Brain abscess,

secondary metastases, tuberculous and pyogenic/fungal abscess can look similar, as can PML sometimes.

Serology[54]
- Results of serological tests may vary between laboratories.
- Toxoplasma latex agglutination (IgG) test may be negative but less than 3% of patients have a negative Sabin–Feldman (IgG) dye test at the time of their toxoplasma encephalitis.
- Rise in IgG with active infection is uncommon.
- Absence of antibodies in patients with suspected TE suggests another aetiology.
- IgM antibodies are rarely found in TE.

Histology from brain biopsy
Demonstration of toxoplasma tachyzoite is proof of active infection.

CSF findings
- Non-specific
- Mild mononuclear pleocytosis
- Mild elevation of CSF protein
- Intrathecal production of toxoplasma IgG is associated with TE

PCR
Active toxoplasmosis can be diagnosed by using the polymerase chain reaction for the B1 gene but its role needs further elucidation.

Culture
Infected tissue inoculated into mice takes 2–3 weeks for a diagnosis.

Treatment

Pyrimethamine and sulphadiazine are the mainstays of treatment. They sequentially block folic acid metabolism and thereby act synergistically against *T. gondii*.

Side effects/toxicity

Toxicity is high, up to 40% of the patients. Drug rash is the most frequent dose-limiting problem and haematological toxicity is usually due to cytopenias and may occur at any point during therapy. Folinic acid is a useful adjunct to prevent bone marrow toxicity.

Other regimes

Other regimes used include clindamycin and pyrimethamine, co-trimoxazole, dapsone, pyrimethamine, spiromycin, trimetrexate, 5-fluorouracil. Novel agents such as atovaquone can be used for the treatment of toxoplasma encephalitis. The macrolides clarithromycin and azithromycin have also been used, alone or in combination with pyrimethamine. Case studies show conflicting data.

Adjunctive therapy

Corticosteroids have been used when papilloedema is present. Their routine use is unproven. Anticonvulsants may be required for control of fits.

Response to treatment

Fifty percent of patients who ultimately respond to therapy have initial neurological improvement by day three; 86% have clinically improved by day seven. In those who do not respond or who have other eventual diagnoses such as lymphoma, new clinical abnormalities usually appear between days five and 12 of therapy.

Brain biopsy

For those patients who do not improve within 10–14 days of appropriate treatment, brain biopsy should be seriously considered, with or without a change in therapy.

Specific drug treatment for toxoplasmosis

Sulphadiazine, pyrimethamine and folinic acid
Preparations
Sulphadiazine:
Tablet   500 mg
Ampoule 1 g/4 ml

Pryimethamine:
Tablet   25 mg

Folinic acid:
Tablet   15 mg

Dosage and administration
Sulphadiazine: 1.5–2 g six-hourly po/IV
Pyrimethamine: 200 mg po on day one, 75 mg daily po thereafter
Folinic acid: 15 mg daily po

Treatment doses should be continued until clinical signs and symptoms have improved, usually 4–6 weeks.

Sulphadiazine injection should be diluted further prior to administration. Each 2 g dose should be added to 100 ml of 0.9% sodium chloride (normal saline) and given over one hour.

Folinic acid is given prophylactically to reduce the potential for myleosuppression: *do not give folic acid*.

What to monitor
Full blood count – twice a week
U&Es – twice a week
Liver function tests – twice a week

Side effects/toxicity
Sulphadiazine

Nausea and vomiting
- Give regular antiemetics.
- Give oral doses after food.
- Decrease the dose where possible.
- Change to alternative treatment.

Rash
The action taken will depend on the severity of the rash, i.e. if the rash is mild, treatment can be continued and topical steroids and antihistamines used for symptomatic relief. If the rash is very severe:

- stop sulphadiazine and change to alternative treatment;
- Give systemic corticosteroids.

A skin rash may be masked by concurrent steroid therapy.

Crystalluria
Crystalluria is more common with sulphadiazine than with co-trimoxazole, due to its lower solubility. Adequate hydration will prevent this occurring; a urine output of at least 1200 ml/day should be maintained during treatment. If a patient shows signs of renal deterioration or 'gravel' appears in the urine, sulphadiazine crystalluria should be suspected.

- Stop sulphadiazine.
- Increase the solubility of sulphadiazine by alkalating the urine (pH >7).
- Give sodium bicarbonate 1.26% 500 ml six-hourly IV or 3 g six-hourly po.

If the patient can maintain an adequate fluid input, sulphadiazine can be reintroduced. However, it may be preferable to change to alternative therapy.

Decreased renal function
Severe renal impairment, i.e. CrCl <10 ml/min:

- stop sulphadiazine.

Moderate renal impairment, i.e. CrCl 20–50 ml/min:

- reduce sulphadiazine dosage;
- change to oral therapy if the patient is receiving sulphadiazine intravenously.

Sulphadiazine and pyrimethamine

Haematological toxicity
If neutrophils $<0.7 \times 10^9/l$:

- increase dose of folinic acid to 30 mg daily po;
- stop (where possible) concurrent myelosuppressive drugs, e.g. zidovudine.

If neutrophils $<0.5 \times 10^9/l$:

- discontinue both agents and start clindamycin, reintroducing pyrimethamine at a lower dose;
- may need growth factors, G-CSF and GM-CSF.

Pyrimethamine may cause a macrocytic anaemia resembling folic acid deficiency. Check serum folate levels and increase folinic acid if found to be low.

Elevated liver enzymes
If liver enzymes are more than five times the upper limit of normal, stop sulphadiazine. N.B. Pyrimethamine has been reported to cause hepatotoxicity when used in combination with sulfadoxine, i.e. Fansidar.

Drug interactions
Absolute contraindications:

- didanosine (ddI) and zalcitabine (ddC): additive toxicity with sulphadiazine and increased risk of pancreatitis
- PCP treatment doses of co-trimoxazole: increased risk of urinary precipitation with sulphadiazine and additive haematological toxicity

Use with caution:

- myelosuppressive agents, e.g. ganciclovir, zidovudine (AZT): additive toxicty with both agents
- warfarin: increased anticoagulant activity with sulphadiazine
- phenytoin: increased phenytoin levels with sulphadiazine

There is evidence that sulphadiazine and pyrimethamine are effective agents for PCP prophylaxis. Patients taking other regimens should be on PCP prophylaxis and this should be continued during both the treatment and maintenance phases of toxoplasmosis therapy.

Clindamycin, pyrimethamine and folinic acid[55,56]
Preparations
Clindamycin:
Capsule   150 mg
Ampoule   300 mg/2ml

Pyrimethamine:
Tablet   25 mg

Folinic acid:
Tablet   15 mg

Dosage and administration
Clindamycin: 600–1200 mg six-hourly po/IV
Pyrimethamine: 200 mg po on day one, 75 mg daily po thereafter
Folinic acid: 15 mg daily po

Treatment should be continued until clinical signs and symptoms have improved, usually 4–6 weeks.

What to monitor
Full blood count – twice a week
U&Es – twice a week
Liver function tests – twice a week

A stool chart should be kept during treatment.

Side effects/toxicity (*see* PCP, pp. 214–218)

Atovaquone
Atovaquone is not currently licensed in the UK for the treatment of cerebral toxoplasmosis. It is available on a named patient/compassionate use basis for patients who fulfil the following criteria.

- AIDS diagnosis.
- Able to take oral medication with food.
- Have active cerebral toxoplasmosis.
- Intolerant or unresponsive to standard therapy (i.e. pyrimethamine/sulphadiazine and pyrimethamine/clindamycin).

Preparations
Tablet   250 mg

Dosage and administration
750 mg six-hourly po. Doses should be taken with food to maximize absorption.

What to monitor
Full blood count – twice a week
U&Es – twice a week
Liver function tests – twice a week

Side effects/toxicity (*see* PCP, pp. 214–218)
Experimental/investigational treatments
Folate inhibitors
- Trimethoprim 20 mg/kg/day
- Trimetrexate
- Piritrexin
- Dapsone 100 mg od

plus

- Pyrimethamine 25 mg od
- Fansidar IM

Macrolides/azalides
- Clarithromycin 1 g bd plus pyrimethamine
- Roxythromycin
- Doxycycline 100 mg tds IV

Others
Arprinocid (purine analogue), quinghausu, 5-FU and pentamidine.

Secondary prophylaxis

Sulphadiazine, pyrimethamine and folinic acid
Sulphadiazine: 1 g 8–12-hourly po
Pyrimethamine: 25 mg daily po
Folinic acid: 15 mg daily po
Secondary prophylaxis should be continued for life: patients can relapse on maintenance therapy.

What to monitor
Full blood count – every 4–8 weeks
U&Es – every 4–8 weeks
Liver function tests – every 4–8 weeks

Clindamycin, pyrimethamine and folinic acid
Clindamycin: 300 mg six-hourly po
Pyrimethamine: 25 mg daily po
Folinic acid: 15 mg daily po

Secondary prophylaxis should be continued for life: patients can relapse on maintenance therapy.

What to monitor
Full blood count – every 4–8 weeks
U&Es – every 4–8 weeks
Liver function tests – every 4–8 weeks

Atovaquone
No dosage recommendation is available for secondary prophylaxis with atovaquone but a treatment dose of 250 mg 6 hourly has been used in patients unable to tolerate other regimes. Atovaquone has been used in combination with pyrimethamine and folinic acid for prophylaxis.

Prophylaxis against toxoplasmosis

Primary prophylaxis
Data from Alpha (MRC-INSERM Trial) suggest that patients with CD4 <100 who also have positive IgG toxoplasma serology should receive primary prophylaxis as there was a 25.4% incidence of TE in this group. The one-year survival after TE was 29.6% (RR of death 1.8). Co-trimoxazole in the same doses used for PCP prophylaxis appears to be effective against TE as well.[57]

Alternatives such as dapsone/pyrimethamine can be used but monotherapy with these or the macrolide atovaquone cannot be recommended.

Prevention of new infection with *T. gondii* in HIV-positive patients

Screen at-risk patients by serology for IgG antibodies. Prevent new infection by:

- washing fruit and vegetables;
- washing hands, utensils and surfaces after contact with raw meat;
- cooking meat to well done (smoked or cured in brine is non-infectious);
- avoiding dried meat;
- avoiding cat faeces – wear gloves when handling cat litter and gardening.

CANDIDA

Oral and oropharyngeal candidiasis

Oral candidiasis is the most common fungal infection in HIV-infected patients. In AIDS patients prior to new antiviral therapies, its prevalence was about 50%.[56] The presence of oral candida is predictive of progressive immune suppression and development of AIDS.[58]

A diagnosis of pseudomembranous candidiasis can be confirmed by the presence of typical white plaques which are easily removed to reveal a reddened or bleeding mucosa. Erythematous and atrophic oral candidiasis can be subtle and difficult to diagnose. These forms may require demonstration of fungal organisms by staining material scraped from patches on the tongue, buccal mucosa or palate. Angular cheilitis may occur.

Culture is less helpful since candida species are part of the normal oral flora. Empiric antifungal therapy is often appropriate.

True systemic candidiasis and fungaemia are unusual (1% incidence in AIDS patients).

Treatment[59,60]

Even in severely immunosuppressed patients, there is some evidence to suggest that intermittent courses of topical therapy that are not systemically absorbed are as effective as the use of oral azoles. In one study, 10 mg clotrimazole troches sucked five times a day produced a similar clinical response to 100 mg orally of fluconazole at the end of a 14-day treatment period.[59] Although mycological response and culture were superior with fluconazole, there was no difference in clinical outcome. The relapse rate for the two groups was the same at long-term follow-up (one month).

The disadvantages of fluconazole are the cost and the need to take a systemic drug.

Oral solutions of fluconazole and itraconazole have both a local and systemic effect. Itraconazole is more effective against a wide range of candida species.

Other treatment possibilities include:

- oral ketoconazole;
- local amphotericin, nystatin or miconazole gel (especially useful in angular cheilitis);
- occasionally intravenous amphotericin B has to be used in clinically problematic cases.

Secondary prophylaxis

Prior to HAART, current candidiasis was an inevitable problem in severely immunocompromised HIV-infected patients. Relapse after acute infection occurred in about 50% at one month.

For those requiring secondary prophylaxis, the minimum drug and dose necessary to prevent disease recurrence once acute control of candida or oropharyngeal candidiasis is achieved are currently unknown.

Vaginal candidiasis

Recurrent and persistent vaginal candida is a problem in HIV-infected women and tends to occur before CD4 decline. Episodes can be treated topically or systemically but there are no controlled trials to define optimum therapy. The CPCRA 010 trial of fluconazole 200 mg weekly versus placebo to prevent candidiasis did see some benefit in the treatment arm in women who had more than two episodes per year of vaginal candida.[61]

Oesophageal candida

This occurred in approximately 10% of AIDS patients. In patients with advanced HIV disease and oral thrush, if dysphagia or odynophagia is present, oesophagitis can be presumptively diagnosed. Empiric antifungal therapy without endoscopic confirmation is appropriate.

Endoscopy should be considered in patients with:

- oesophageal symptoms but no oral candida;
- persistent oesophageal symptoms despite empiric therapy.

The diagnosis will be established by demonstrating the presence of yeast by fungal staining of oesophageal plaques seen at endoscopy.

Treatment[62-65]

Orally administered ketoconazole, itraconazole and fluconazole are effective therapies. Itraconazole oral solution may be more effective.

After a treatment course, the incidence of relapse by one month has been as high as 50%, due to either recurrence or reinfection. Therefore indefinite suppressive antifungal therapy has been used in some centres. However, intermittent dosing depending on symptoms has been advocated to prevent the emergence of drug resistance. Daily or even weekly therapy (using fluconazole 150 mg once a week) has been suggested.

Some feel that the high cost of suppressive therapy and the potential for inducing resistance discourage routine suppressive therapy except in patients who experience frequent recurrences and those with fluconazole-resistant infections.

Disseminated candida

This is rare and occurs in AIDS patients receiving parental nutrition or in those with indwelling catheters and should be treated with amphotericin B.

Drug resistance

In vitro susceptibility of antifungal drugs to fungal organisms is of questionable value because of the limited correlation of the results with clinical response[66] and lack of standardized tests. Susceptibility tests should be done only in reference laboratories.

Treatment failures are increasing among HIV-positive patients using intermittent or continuous fluconazole. Factors associated with fluconazole failure include more than three episodes of thrush in the previous six months, a history of oesophageal candida, recent prior oral thrush and previous opportunistic infection, especially MAC or toxoplasmosis. Resistance both *in vitro* and clinically has emerged.[67]

Topical therapy for oral candida

Nystatin pastilles/amphotericin lozenges
Dissolve one in the mouth four times or more each day after food. The dose can be titrated against response: up to eight pastilles/lozenges a day may be required.

Nystatin suspension
1–5 ml rinsed around and held in the mouth four times each day after food.

Miconazole oral gel
Place 5–10 ml in the mouth four times each day after food. Miconazole gel can be used under dentures: place on the fitting surface before insertion.

Systemic therapy for oral/oesophageal candida

Systemic therapy should be used for patients in whom topical treatment has been ineffective and those with oesophageal candidiasis. Start therapy with the lowest possible dose and give for seven days, then review. Initially patients can be maintained on intermittent short courses of a systemic drug (i.e. 2–7 days). As patients become more immunosuppressed, prophylactic therapy may be required. The dose and frequency of this should be titrated against symptoms. Patients with oesophageal candida may need a higher dose.

Ketoconazole
Starting dose: 200 mg po daily after food. Can be increased to 400 mg daily.

Itraconazole capsules
Starting dose: 200 mg po daily after food. Can be increased to 400 mg daily.

Itraconazole solution
This is a clear, slightly viscous liquid containing 10 mg of itraconazole and 400 mg hydroxypropyl-β-cyclodextrin (as solubilizer) per 1 ml. The supply of the solution is restricted to a maximum of three months continuous therapy or six months total exposure due to concern over the long-term effects of the solubilizer. In AIDS patients who have responded to itraconazole solution and who cannot be maintained on itraconazole capsules, this period can be extended.

Fluconazole
Starting dose: 50 mg po daily. Can be increased to 400 mg daily.

Fluconazole solution
This has recently been made available. Starting dose: 100 mg po twice daily. Can be increased to a maximum of 5 mg/kg/day.

Amphotericin infusion
Amphotericin in the treatment and prophylaxis of superficial candida is used in the same treatment regime as that for cryptococcal meningitis. Treatment is continued until the candida has been cleared, after which the patient can be maintained on a dose of 1 mg/kg for a minimum number of days per week, e.g. 2–3.

What to monitor

Systemic agents
Full blood count – every two weeks
U&Es – every two weeks
Liver function tests – every two weeks

Itraconazole
Serum cholesterol baseline – every three months

Patients on amphotericin infusions will require more intensive monitoring.

Side effects/toxicity

Raised liver enzymes
If there is a significant rise in liver enzymes, i.e. greater than five times the upper limit of normal, or clinical signs of hepatotoxicity during treatment with systemic azole drugs, therapy should be stopped. Ketoconazole has been associated with fatal hepatotoxicity.

Decreased renal function
Fluconazole is excreted predominantly in the urine as unchanged drug. The usual dose should be given on days one and two. From day three the dosage should be modified according to the patient's creatinine clearance.

Amphotericin[68]
Nephrotoxicity
Amphotericin is nephrotoxic: it can cause a rise in blood urea, renal tubular acidosis and hypokalaemia. Irreversible degeneration of the renal tubules can occur following cumulative doses greater than 5 g.

If CrCl <10 ml/min increase the dosage interval from 24 to 48 h.

If renal function deteriorates during amphotericin treatment, i.e. the serum creatinine rises above 260 mmol/l:

- use alternate-day dosing;
- reduce the dose;
- consider changing to liposomal amphotericin;
- discontinue treatment.

There has been some work to suggest that the renal toxicity of amphotericin can be reduced by using intralipid 10% as the infusion fluid. Each dose should be added to 100 ml. However, the efficacy of this combination has not been proven in any large clinical studies and its use to date is largely anecdotal.

Fever, chills and malaise
Give 100 mg hydrocortisone IV and 10 mg chlorpheniramine IV prior to infusion. Some centres add 50 mg pethidine to the amphotericin infusion, however, there are no data concerning the stability of this.

Electrolyte disturbances
Supplement as appropriate with IV and oral preparations. If there is difficulty in replacing potassium or magnesium, the use of potassium-sparing diuretics, e.g. amiloride and/or spironolactone, may ameliorate the situation.

Drug interactions

Ketoconazole
Absolute contraindications:

- rifampicin: decreased GI absorption of rifampicin, increased metabolism of ketoconazole
- terfenadine: increased serum levels of terfenadine, increased potential for cardiac toxicity.

Use with caution:

- drugs that increase gastric pH, e.g. antacids, H_2 antagonists, ddI: decreased absorption of ketoconazole. Give ketoconazole two hours before any of the above
- warfarin: increased anticoagulant effect
- isoniazid: reduced plasma levels of ketoconazole
- phenytoin: metabolism of both agents may be altered. Plasma phenytoin levels may be increased and ketoconazole levels decreased

Fluconazole
Use with caution:

- rifampicin: decreased bioavailability (25%) and decreased half-life (20%) of fluconazole when given concomitantly with rifampicin. Larger doses of fluconazole will be required, e.g. increase by 1/3–1/2
- warfarin: increased anticoagulant activity
- phenytoin: increased phenytoin levels

Itraconazole
Absolute contraindications:

- Rifampicin: decreased absorption of rifampicin. increased metabolism of itraconazole
- terfenadine: increased serum levels of terfenadine, increased potential for cardiac toxicity

Use with caution:

- drugs that increase gastric pH, e.g. antacids, H_2 antagonists, ddI: decreased absorption of itraconazole. Give itraconazole two hours before any of the above
- warfarin: increased anticoagulant effect
- phenytoin: metabolism of both agents may be altered. Plasma phenytoin levels may be increased and itraconazole levels decreased.

Amphotericin
Absolute contraindications: nephrotoxic drugs, e.g. pentamidine, treatment doses of foscarnet.

Use with caution:

- aminoglycosides: monitor serum levels of aminoglycoside
- corticosteroids: may increase potassium loss
- flucytosine: increased toxicity of flucytosine with concomitant amphotericin

CRYPTOCCOCAL MENINGITIS[69,70]

In most areas of the world cryptococcal meningitis occurs in approximately 7% of patients with AIDS. In some parts of Africa the incidence is 30%. Meningitis is the most common manifestation of cryptococcosis, occurring in up to 90% of cases. Concurrent pulmonary involvement occurs in up to one-third and is occasionally severe.

Unusual sites of dissemination include the prostate, urinary tract, retina, skin or bone.

Diagnosis

- Commonest complaints are fever, headache and malaise (>60% of patients).
- Nausea, vomiting (50%) or mental state changes (15–30%).
- Focal deficits (10%).
- Stiff neck and photophobia. (20% only).
- Fundi may show multiple exudates ± haemorrhage.
- Skin lesions, sometimes resembling molluscum, can occur.
- Mouth, lung, bones, joints, liver, kidneys and adrenals can all be involved.

Peripheral blood findings

- WBC usually <4000/mm^3.
- Serum cryptococcal antigen (CRAG) positive.
- Blood cultures can be positive in up to 50% of cases.

CSF findings

- Lumbar puncture can confirm the diagnosis of cryptococcal meningitis based on detection of CRAG in the CSF (usually lower than blood).
- India ink (negrosin) stain is positive in more than 70%.
- Isolation of *Cryptococcus neoformans* from spinal fluid occurs in 90–100% of cases.
- CSF, glucose and protein are normal or only slightly abnormal in >50% of cases.

Prognosis

Historically, in HIV-negative populations, the success rate with treatment is 75–85%. Patients with cryptococcal meningitis who are obtunded or who have high initial CSF pressure, decreased CSF glucose, low CSF leucocyte count or high cryptococcal antigen titres tend to have a poor prognosis.

Elevated CSF pressure at baseline reduces the probability of CSF clearance and increases the risk of relapse and death.

Treatment[71]

Induction phase – first two weeks:

- amphotericin B – 0.7 mg/kg
- flucytosine – 100 mg/kg in four divided doses

CSF pressure – with or without pressure symptoms:

- asymptomatic and CSF pressure >320 mmHg.
- symptomatic and CSF pressure >180 mmHg

then:

1. daily removal of 25–30 ml of CSF with a drain ±
2. acetazolamide treatment
3. intraventricular shunting/lumbar drain

Continuation phase:

- fluconazole – 800 mg a day for two days then 400 mg a day.

Repeat lumbar punctures should be done routinely at two and 10 weeks to detect persistence/recurrence.

Other drugs

Liposomal amphotericin has been used but studies were uncontrolled. For prophylaxis, itraconazole may be a suitable alternative for patients unable to take fluconazole.

Specific drug treatment

Amphotericin B
Preparations
Vial 50 mg

Dosage and administration
0.3 mg–1 mg/kg/day IV as a single daily dose.

Most trials in patients with AIDS-associated cryptococcal meningitis have used doses of up to 0.7 mg/kg/day: however, the maximum recommended dosage of amphotericin B in seriously ill patients with systemic fungal infection is 1.5 mg/kg/day.

Each dose should be administered as an IV infusion in 5% dextrose (maximum concentration 1 mg/10 ml) and given over six hours. Faster rates of infusion have been used.

A test dose of 1 mg/100 ml given over one hour should be given first.

If the patient experiences no adverse effect within one hour, the treatment infusion can be started.

The duration of amphotericin treatment is dependent on patient factors. Generally, however, patients can be changed to oral fluconazole after 14 days.

What to monitor
U&Es (including magnesium) – twice a week
Liver function tests – twice a week
Full blood count – twice a week

Side effects/toxicity
Renal impairment
Amphotericin is nephrotoxic: it can cause a rise in blood urea, renal tubular acidosis and hypokalaemia. Irreversible degeneration of the renal tubules can occur following cumulative doses greater than 5 g.

If CrCl <10 ml/min increase the dosage interval from 24 to 48 h.

If renal function deteriorates during amphotericin treatment, i.e. the serum creatinine rises above 260 mmol/l:

- use alternate-day dosing;
- reduce the dose;
- consider using intralipid 10% as infusion fluid or changing to liposomal or colloidal amphotericin;
- discontinue treatment.

Fever, chills and malaise
Give 100 mg hydrocortisone IV and 10 mg chlorpheniramine IV prior to infusion.

Some centres add 50 mg pethidine to the amphotericin infusion; however there are no data concerning the stability of this.

Electrolyte disturbances
Supplement as appropriate with IV and oral preparations. If there is difficulty in replacing potassium or magnesium, the use of potassium-sparing diuretics, e.g. amiloride and/or spironolactone, may ameliorate the situation.

Drug interactions
Absolute contraindications are nephrotoxic drugs, e.g. pentamidine, treatment doses of foscarnet.
Use with caution:

- aminoglycosides: monitor serum levels
- corticosteroids: may increase potassium loss
- flucytosine: increased toxicity of flucytosine with concomitant amphotericin

Liposomal amphotericin B
This is a relatively new formulation of amphotericin which is less nephrotoxic than amphotericin B. Much larger doses can be used, e.g. up to 5 mg/kg/day. However, the cost is high, i.e. 1×50 mg vial = £117 compared with £3.70 for 1×50 mg vial of amphotericin B (1998 prices). At present, there is insufficient evidence to recommend its first-line use for the treatment of cryptococcal meningitis associated with AIDS.

Flucytosine
Preparations
IV infusion 10 mg/ml 250 ml infusion bottle. Give dose over 20–40 min.

What to monitor
Blood levels – predose levels should be 25–50 mg/l (200–400 mmol/l)
Full blood count – twice a week
U & Es – twice a week
Liver function tests – twice a week

Side effects/toxicity
- Nausea, vomiting, diarrhoea, rashes
- Confusion, headaches
- Hepatitis
- Thrombocytopenia
- Leucopenia
- Aplastic anaemia
- Renal failure

Reduce total dose by giving 12-hourly when CrCl <90 ml/min and 24-hourly when CrCl <50 ml/min.

Raised liver enzymes
If there is a significant rise in liver enzymes, i.e. greater than five times the upper limit of normal, during fluconazole therapy, it should be stopped and other agents considered.

Drug interactions
Use with caution:

- rifampicin: decreased bioavailability (25%) and decreased half-life (20%) of fluconazole when given concomitantly with rifampicin. Larger doses of fluconazole will be required, e.g. increase by 1/3–1/2
- warfarin: increased anticoagulant activity
- phenytoin: increased phenytoin levels

Maintenance

Fluconazole and itraconazole
Fluconazole 800 mg od for two days, then 400 mg po daily *or*

Itraconazole 600 mg od for three days, then 200 mg bd

What to monitor
Full blood count – once a month
U&Es – once a month
Liver function tests – once a month

Side effects/toxicity
Decreased renal function
Fluconazole is excreted predominantly in the urine as unchanged drug. The usual dose should be given on days one and two: from day three the dosage should be modified according to the patient's creatinine clearance:

CrCl (ml/min)	Dosage interval/daily dose
>40	24 h
21–40	48 h or half normal daily dose
10–20	72 h or one-third normal daily dose

Amphotericin
Amphotericin can be used for secondary prophylaxis in patients who are intolerant to or relapse on fluconazole. A dose of 1 mg/kg given once or twice weekly has been used.

What to monitor
Full blood count – once a week
U & Es, magnesium – once a week
Liver function tests – once a week

HISTOPLASMOSIS[72–75]

Histoplasmosis is a serious opportunistic infection in patients with AIDS. It is uncommon in the UK but in the USA occurs in 2–5% of patients with AIDS from endemic areas, mainly Central and South Central USA. It often represents the first manifestation of severe immune suppression. Primary infection is usually pulmonary, followed by dissemination.

Symptoms and signs

The infection usually presents with fever and weight loss of 1–2 months' duration. Respiratory complaints can occur in up to half the patients and hepatosplenomegaly occurs in one-third. Ten to 20% of patients present with a meningitis or a syndrome resembling septicaemia, while 5–10% of patients have skin or GI manifestations including diarrhoea, intestinal obstruction, perforation, bleeding or peritonitis.

Diagnosis

Chest X-rays show diffuse infiltrates, often a reticular nodular pattern in 60% of cases.[73] *Histoplasma capsulatum* has been isolated from blood, bone marrow, respiratory secretions (especially bronchoalveolar lavage or transbronchial biopsy) or localized lesions in over 85% of cases. Isolation of the organism may take up to four weeks. Detection of a glycoprotein antigen in body fluids is providing an important diagnostic method. It takes only one day for diagnosis of histoplasmosis and the test can be used in monitoring success of treatment.[72]

Antigen is present in:

- the urine (95%);
- blood (86%);
- BAL fluid and CSF (70%).

Staining tissue sections or peripheral blood smears also permits rapid diagnosis.

Treatment[72,74,76]

An 'induction phase' with amphotericin B (at least 50 mg od until a total cumulative dose of 1–2 g is reached) induces remission in 80% of patients, usually within the first week of therapy.[72] Moderate to severe cases, i.e. those with hypertension, hypoxia, (PaO_2 less than 60 mmHg), meningitis or those with severely abnormal laboratory results such as neutropenia <750 cells/ml, platelets <50 000 cells/ml, LFTs five times normal should receive amphotericin B 50 mg daily for two weeks followed by treatment with itraconazole (*not* ketoconazole). For mild histoplasmosis itraconazole 200 mg can be used twice daily,[74] while for those intolerant of itraconazole, fluconazole 400–800 mg od can be used.

Itraconazole levels can be checked 2–4 h postdose during the second week of treatment (should be at least 2mg/ml).

Treatment failures occur in patients who are severely ill or who have CNS histoplasmosis.

Secondary prophylaxis

Relapse is common. Secondary prophylaxis should be with itraconazole 200 mg twice daily or amphotericin B weekly or biweekly.[72]

COCCIDIOCOMYCOSIS[77,78]

This occurs in endemic areas of the USA, especially Texas, Arizona and Southern California, usually after the rainy season. Patients at risk are those with:

- CD4 <200;
- skin test positive for coccidiocomycosis.

Symptoms and signs

Most patients present with fever, anorexia and shortness of breath and the chest X-ray shows diffuse pulmonary reticulonodular infiltrate. Some patients present with meningitis and others have disseminated disease.

Diagnosis

Diagnosis is made by fungal stain and culture of sputum/bronchoalveolar lavage or histology. Blood cultures are infrequently positive while serological tests are often negative in patients with AIDS.

Treatment

Amphotericin B (with a total cumulative dose of about 2 g) has been recommended for patients with diffuse interstitial infiltrates on X-ray. Intrathecal amphotericin has been used in some patients. Fluconazole 400–800 mg daily is an alternative for those with mild illnesses.

Secondary prophylaxis

Chronic maintenance is required, usually with fluconazole 200–400 mg daily or amphotericin B administered weekly.

ASPERGILLOSIS[80,81]

Invasive aspergillosis occurs in less than 1% of patients with AIDS. The CD4 count is usually <50 cells/ml. Pulmonary disease is commonest, occurring in 48–66% of cases; chest radiographs show diffused reticulonodular infiltrates, patchy infiltrates, lobar infiltrates, cavities and occasional pleural effusions. Sinus infection occurs in half and disseminated disease in one-third of patients.

Diagnosis

The diagnosis is made by:

- seeing fungal hyphae by fungal stains, or
- cultures of respiratory secretions or tissues.

N.B. Isolation from respiratory secretions does not establish the diagnosis as aspergillus species may merely colonize the airways.

Treatment

This is problematic. Amphotericin B is used as first-line therapy (1–1.5 mg/kg) and itraconazole in doses of up to 600 mg daily has helped some patients. Rifampicin and flucytosine have not demonstrated any survival benefit when added to amphotericin.

Secondary prophylaxis

Lifelong maintenance is required, using either weekly amphotericin or itraconazole.

BLASTOMYCOSIS[82]

This is rare in the UK. Patients usually present with localized pulmonary, CNS or disseminated disease and a septicaemic illness may occur.

Diagnosis

Diagnosis is based on the demonstration of organisms or fungal stain by culture of BAL fluid or lung tissue. An enzyme immunoassay is being developed.

Treatment

Amphotericin B is usually effective and itraconazole has been used successfully.

Secondary prophylaxis

Chronic maintenance is required with itraconazole 200–400 mg daily or ketoconazole.

MYCOBACTERIAL DISEASES

MYCOBACTERIUM AVIUM COMPLEX (MAC)[84-86]

Prior to HAART, MAC caused disseminated disease in as many as 15–40% of patients with HIV infection. Disseminated MAC characteristically occurs in patients with very advanced HIV disease and peripheral blood CD4 T-lymphocyte counts below 100 cells/μl.

M. avium complex is ubiquitous in the environment and can be recovered from water, soil, dairy products and a wide variety of animals including chickens, pigs, dogs, cats and insects. Isolates from local environmental sources often differ substantially from local human isolates with regard to serotype and plasmid profile. The contribution that environmental sources make to human disease is not clear. Potable water or aerosols may be one source of infection. The respiratory and GI tract have been proposed as portals of entry for MAC but both of these tracts are common sites of colonization. Although localized disease can occur, the commonest clinical situation is dissemination.

Presenting features

The commonest are fever, weight loss, night sweats, diarrhoea, abdominal pain, anaemia or abnormalities of liver function, especially alkaline phosphatase. Patients may have intraabdominal lymphadenopathy and hepatosplenomegaly.

Prognosis[85,87]

No prospective studies precisely define the effects of untreated disseminated *M. avium* complex disease on prognosis. Cohort case control studies suggest that this disease is associated with reduced survival, four months compared with 11 months among patients without dissemination. A prospective cohort study of more than 1000 patients with advanced HIV disease showed that patients in whom *M. avium* complex developed had significantly increased risk of death even when other predictors of mortality were controlled.[88]

Screening

The predictive value of screening for MAC by culturing stools or sputum is poor and it cannot be recommended as a routine.

Diagnosis

A single positive blood culture is considered diagnostic of disseminated MAC disease. However, there are some data to suggest that up to 10% of patients may have transient bacteraemia without persistent disease. Positive culture from other normally sterile sites, bone-marrow and liver, probably has the same treatment decision implication as positive blood cultures.

Ten ml of blood can be collected in a suitable tube and centrifuged and then innoculated onto Middlebrook 7H10 or 7H11 agar or BACTEC 12B broth medium or both. Unprocessed blood can be inoculated directly into BACTEC 13A or other suitable broth. MAC is commonly detected with a BACTEC system in 7–14 days and on agar plates from 14–21 days. DNA probes can be used to identify the species of any subsequent growth.

Table 16.3 Recommendations for treatment of disseminated MAC in HIV-infected patients

Option 1	Option 2

Initial regimen:
    Clarithromycin 500 mg bid
    Ethambutol 15 mg/kg daily
	Initial regimen:
	    Clarithromycin 500 mg bid
	    Ethambutol 15 mg/kg daily
	    Rifabutin 300–450 mg daily

Salvage regimen:
    Rifabutin 300–450 mg daily
    Amikacin 10 mg/kg daily
	Salvage regimen:
	    Amikacin 10 mg/kg daily
	    Ciprofloxacin 750 mg bid

Treatment[89,90]

Studies have documented that drug therapy can reduce or eliminate MAC bacteraemia over several weeks or months (Table 16.3). It is unknown what effect treatment has. Many patients are coming off treatment when their HIV viral load is controlled by antiviral therapy for extended periods.

Drug susceptibility test

There is no published evidence that any currently available methods of susceptibility testing using drugs other than the macrolides are of value in guiding or modifying therapeutic regimens in treatment of disease. Primary macrolide resistance has been reported.

Single-agent therapy[89]

Clarithromycin, azithromycin, ethambutol, clofazimine and rifampicin have all been used in monotherapy. Although an initial clinical and microbiological improvement is made with monotherapy, most patients who continue monotherapy will have a recurrence of their symptoms and will also relapse microbiologically. Studies so far suggest that monotherapy with these agents does not eradicate *M. avium* complex and that drug resistance can develop.

Multiple-agent therapy[90,91]

With the use of two-drug therapy, the possibility that one isolate will be resistant is theoretically less likely. The initial trials using four-drug therapy showed clinical benefit and reductions in bacteraemia. One recommendation for treatment is to use a macrolide such as clarithromycin or azithromycin plus ethambutol as a second drug. Another starting regimen is to use the above and add rifabutin as a third agent.

The dose of clarithromycin should not exceed 500 mg bd.

Isoniazid and pyrazinamide have no role in therapy of MAC and clofazamine should not be used. If patients were on rifabutin prophylaxis and then developed MAC, rifabutin can be continued as part of a treatment regimen.

Other drugs used in salvage regimens include amikacin and ciprofloxacin.

Monitoring outcome

Clinical manifestations should be monitored and biological response with blood cultures every four weeks during initial therapy. A clinical improvement may take 4–6 weeks and blood cultures may take 4–12 weeks to become sterile.

Primary prophylaxis[90,92–94]

The United States Public Health Task Force has recommended that patients with HIV infection and less than 50 CD4 T-lymphocytes/ml should receive azithromycin prophylaxis 1 g a day or clarithromycin 500 mg bd for life. Rifabutin 300 mg od can be used as a second-line agent.

Evaluation before prophylaxis

If a decision is made to commence prophylaxis, patients should be assessed to ensure that they do not have active disease due to MAC, MTB or any other mycobacterial disease (Table 16.4).

After clinical evaluation, investigations should include:

- chest X-ray;
- tuberculin skin testing;
- blood cultures for mycobacteria.

Antiviral drugs

If patients are on protease inhibitors bear the following in mind.

- Azithromycin – no significant interaction
- Clarithromycin – levels may increase on PIs. (NB: Do not combine with DMP-266 efavirenz)
- Rifabutin – reduce rifabutin dose by 50% and give nelfinavir or indinavir

Specific drug treatment

Clarithromycin, ethambutol and rifabutin

Preparations
Clarithromycin:
Tablet   250 mg

Ethambutol:
Tablet   100 mg, 400 mg

Rifabutin:
Capsule 150 mg

Table 16.4 Distinguishing MAC from MTB in HIV-infected patients

MAC	MTB
More than 86% have preexisting AIDS	Approximately 70% have not had an AIDS-defining illness
More than 96% have CD4 counts <100	86% have a CD4 count <200, but infection can occur at any CD4 count
Pneumonitis is unusual (4–10%)	Pulmonary involvement in more than 70%
Abnormal CXRs are uncommon	CXRs are abnormal in 83%
Hilar lymphadenopathy (10–15%), cavitary disease (less than 5%) and pleural effusions (10–18%) are rare	Hilar lymphadenopathy, cavitary disease and pleural effusions are common (25%)
Sputum smears are positive in only 10–18%	Sputum smears are positive in 71–83%
Bacteraemia occurs in 86–98%	Bacteraemia occurs in 2–12%
Blood cultures are positive in 1–6 weeks (median 14 days)	Blood cultures may take 6–8 weeks

Dosage and administration
Clarithromycin: 1–2 g daily po in divided doses (depending on patient tolerance)
Ethambutol: 15 mg/kg/day po as a single dose
Rifabutin: 300–450 mg daily po as a single dose.

What to monitor

	First month of treatment	Remainder of treatment
Full blood count	Once a week	Once a month
U&Es	Once a week	Once a month
Liver function tests	Once a week	Once a month
Ophthalmic examination	Baseline for ethambutol	

Side effects/toxicity
Nausea and vomiting
Clarithromycin can cause nausea, vomiting and diarrhoea. If this occurs the tablets should be given in smaller doses after meals, e.g. 500 mg six-hourly. If the patient is still unable to tolerate this, the dose should be

reduced. Antidiarrhoeal agents can be given symptomatically.

Decreased renal function
Rifabutin
If CrCl <30 ml/min, reduce dose by 50%.
Clarithromycin
If CrCl < 30 ml/min, reduce dose by 50%.
Ethambutol
The dosage of ethambutol may need to be reduced in patients with decreased renal function as determined by blood levels.

Raised liver function tests
Rifabutin
Use with caution in patients with severe hepatic impairment. It can cause pseudojaundice (yellow skin and sclera but normal bilirubin).
Clarithromycin
Use with caution in patients with severe hepatic impairment.

Visual impairment
Ethambutol can cause optic neuritis which appears to be related to the dose and length of treatment. It is generally reversible. Ethambutol should be stopped.
Rifabutin can cause uveitis. Recently, a treatment dose of 600 mg a day rifabutin has been shown to cause uveitis and pseudojaundice (yellowing of the skin but not the sclera and a normal bilirubin).
It has been recommended that the treatment dose of rifabutin be dropped to 450 mg a day **maximum** because of the high rate of recurrence of these problems, especially in patients on concomitant clarithromycin and fluconazole.
If uveitis occurs, the drug should be stopped and the patient referred to an ophthalmologist. Local steroids almost always lead to full recovery.

Hearing loss
The high doses of clarithromycin used may cause loss of hearing. Audiometry is recommended for patients before starting treatment.

Drug interactions
Rifabutin
This is reported to induce hepatic microsomal enzymes to a lesser extent than rifampicin. However, patients taking drugs that are metabolized by the enzymes of the cytochrome P450 system (*see Rifampicin*, p.228) should be monitored for decreased effect.

Clarithromycin
Use with caution:

- carbamazepine: reduced rate of excretion of carbamazepine
- theophylline: increased theophylline levels
- warfarin: increased anticoagulant activity
- digoxin: effect of digoxin may be potentiated

Ethambutol
No significant interactions occur with ethambutol.

Ciprofloxacin and amikacin
Preparations
Ciprofloxacin:
Tablet   250 mg

Amikacin:
Vial   500 mg/2 ml

Dosage and administration
Ciprofloxacin: 500–750 mg twice daily po
Amikacin: 15 mg/kg/day IV/IM in two divided doses

Amikacin can be given by the IV or IM routes. If given IV, it can be administered either as a slow bolus or as an infusion in 100 ml of sodium chloride 0.9% or glucose 5% over 30 min.

What to monitor

	First month of treatment	Remainder of treatment
Full blood count	Once a week	Once a month
U&Es	Once a week	Once a month
Liver function tests	Once a week	Once a month

Amikacin: serum level monitoring
Blood samples should be drawn immediately prior to the third dose (trough level) and one hour after the dose (peak level). If an infusion is used, blood should be taken one hour after the start of the infusion. Further monitoring is recommended twice weekly, but this will depend on the clinical condition of the patient.

- Usual trough level < 10 µg/ml
- Usual peak level <30 µg/ml

Side effects/toxicity
Decreased renal function
Ciprofloxacin
Dosage adjustments are only required for patients with severe renal impairment. If CrCl < 20 ml/min, halve dosage and maintain fluid input.
Amikacin
Reduce dose and/or increase dosage interval to prevent accumulation. The manufacturer's recommendation is

to multiply the patient's serum creatinine (mg/100 ml) by 9 to give a figure (in hours) for the interval between doses, e.g. serum creatinine = 4.0 mg/100 ml × 9 = 36 h. Serum level monitoring is essential in patients with renal impairment.

Skin reactions
Ciprofloxacin can cause rash, pruritus, urticaria and photosensitivity. If any of these occur, treatment should be stopped.

Gastrointestinal disturbances
Nausea, diarrhoea, vomiting, dyspepsia, abdominal pain, anorexia, flatulence and dysphagia can occur with ciprofloxacin. Taking tablets after food may decrease the incidence of these.

Hepatic impairment
Transient rises in liver enzymes or serum bilirubin (particularly in patients with previous liver damage), hepatitis, jaundice and major liver disorders including hepatic necrosis have been reported with ciprofloxacin. Stop treatment if liver enzymes rise to three times the upper limit of normal.

Ototoxicity
Amikacin can cause ototoxicity in overdose.

Drug interactions
Ciprofloxacin
Use with caution:

- magnesium, aluminium and iron salts: ciprofloxacin should not be administered within 4 h of medications containing magnesium, aluminium (e.g. antacids, ddI – due to the buffer) or iron as interference with the absorption of ciprofloxacin may occur
- theophylline: increased plasma levels of theophylline may occur
- warfarin: increased anticoagulant effect

Amikacin
Use with caution:

- nephrotoxic drugs, e.g. pentamidine, amphotericin, foscarnet: additive nephrotoxicity may occur
- rapid-acting diuretic agents, particularly when given by the IV route, e.g. frusemide, bumetanide

Drugs used in prophylaxis

Azithromycin
Preparations
Capsule   250 mg
Oral suspension   200 mg/5 ml

Table 16.5 Prevention of disseminated MAC in HIV-infected patients

1. Begin prophylaxis for patients with CD4 count <50/mm^3. Rule out active tuberculosis by clinical exam and appropriate laboratory studies.
2. Obtain mycobacterial blood culture to rule out subclinical bacteraemia.
3. Preventive regimens:
    First line   Azithromycin 1000–1200 mg once per week
    OR
    Clarithromycin 500 mg twice per day
    Second line   Rifabutin 300 mg once per day

Dosage
1 g weekly

Side effects
- Nausea and vomiting, abdominal pain, diarrhoea
- Rashes
- Reversible hearing loss
- Mild neutropenia

MYCOBACTERIUM KANSASII[95]

A slow-growing mycobacteria that can cause dissemination in patients with AIDS; 0.2% of patients at time of AIDS diagnosis had *M. kansasii* infection.

Clinical features

Same as for *M. avium*.

Diagnosis

Same as for *M. avium*.

Treatment

In vitro susceptibility testing and trials in HIV-negative patients suggest rifampicin and ethambutol should be used as first-line agents.[96] This two-drug regimen is used successfully in pulmonary disease in immunocompetent hosts. *M. Kansasii* is usually isoniazid resistant.

Additional drugs such as clarithromycin or azithromycin (as with *M. avium*) may be needed and prolonged therapy may be required in AIDS patients, but response may be poor.

TUBERCULOSIS[97]

The number of cases of tuberculosis is increasing worldwide and the reason for this increase is largely related to the HIV epidemic. TB is important as it is contagious, treatable and preventable.

Clinical presentation

It is widely accepted that the clinical manifestations of TB in HIV-infected people are largely dependent upon the underlying degree of immunosuppression. Patients who are less immune suppressed with higher CD4 counts are likely to have classic symptoms, radiology and laboratory findings consistent with 'reactivation' pulmonary TB.

As immunosuppression progresses pulmonary disease remains common but is more frequently atypical and often accompanied by extrapulmonary disease, most often involving lymph nodes, liver, bloodstream, bone marrow and meninges. In HIV-infected persons tuberculosis appears to have a more rapid and aggressive clinical course.

Primary TB that progresses rapidly is more likely to occur in profoundly immune-suppressed HIV-infected people who acquire new infection with TB.

Chest radiography

Reported radiographic abnormalities include a higher incidence of lymphadenopathy, pleural effusions, diffuse interstitial or miliary shadows, alveolar consolidation without cavitation and pericardial disease.

Tuberculin skin tests

Usually 0.1 ml 1:1000 tuberculin infradermally is used (10 tuberculin unit). Most would accept 10 mm^3 induration (not erythema) as signifying active infection. The US Public Health Service recommends 5 mm induration after 5TU Mantoux.

The prevalence of tuberculin reactivity is related to cutaneous anergy which is inversely proportional to the absolute CD4 count. Skin test non-responsiveness is common in HIV patients with CD4 <400 cells/μl. Patients with CD4 counts >600 cells/μl are no more likely to be anergic than controls. Non-responsiveness to tuberculin can be partly differentiated from true anergy by multiple skin testing using antigens such as candida, mumps, etc. However, anergy testing has limited usefulness and is not routinely recommended.

Sputum smears

The presence of acid-fast bacilli in sputum is more likely in patients with extensive or diffuse pulmonary infiltrates or those with cavities. HIV-positive patients with TB tend to have less cavitation and are therefore less likely to be sputum smear positive.

The yield of acid-fast positive smears may be improved by a method which concentrates specimens by centrifugation.

An RNA probe has been developed to differentiate MTB in sputum from other mycobacterial species.

Culture

Standard techniques use solid (Lowenstein–Jensen) or liquid Middlebrook media. Identification and drug susceptibility can take 6–8 weeks. Some culture systems (e.g. BACTEC) have enhanced rapid detection of MTB and shorten the time required for drug susceptibility testing.

When coupled with rapid species identification by nucleic acid probes, identification and susceptibility testing can be reduced to 10–21 days.

Molecular diagnosis

The polymerase chain reaction (PCR) based on the amplification of a unique repetitive insertion sequence of the MTB genome (IS6110) has been developed. Both false-positive results and detection of persistent mycobacterial DNA in individuals with prior, treated or inactive TB can hinder interpretation.

Another PCR method uses restriction enzyme analysis of PCR products of the gene coding for a 65 kD mycobacterial protein. This method does not require hybridization to species-specific probes or involve the use of radioactivity.

RNA amplification methods are commercially available in order to distinguish between live and dead organisms.

Single-strand conformation polymorphism and gene sequencing can be used to detect drug resistance.

A Luciferase gene expressed by reporter phages which only infect viable live organisms makes the mycobacteria 'glow' and easy to detect. This technique is under evaluation for drug resistance.

Treatment[99–101]

It is vitally important that patients with tuberculosis should be co-managed with colleagues with experience in TB. Response to standard regimens is usually good and patients should be managed as indicated in the Guidelines for the Treatment of Tuberculosis in the

United Kingdom published by the Joint Tuberculosis Committee, has been updated in 1998. That is, unless drug resistance is suspected, rifampicin, isoniazid and pyrazinamide and ethambutol should be given for the initial two months and rifampicin and isoniazid should then be continued for a further four months for all disease, with the exception of meningeal and miliary disease where treatment might continue for up to one year.

Secondary prophylaxis

On the basis of current evidence, the BTS have suggested that lifelong secondary prophylaxis with isoniazid should not be routinely given. Patients should be followed carefully to detect relapse.

Drug resistance[101,102]

Treatment of patients with drug-resistant TB should only be carried out by those experienced in the management of such cases.

Initial drug resistance in previously untreated white UK-born patients in England and Wales is uncommon and only resistance to isoniazid is usually of clinical importance. HIV-positive patients have an increased risk of drug resistance Higher levels of isoniazid resistance occur in ethnic minority groups in 4–6%. Resistance rates to other first-line drugs, rifampicin, pyrazinamide and ethambutol, are much lower still. Resistance to streptomycin is more common but this is not important in the United Kingdom where streptomycin is not a first-line drug.

To ensure the overall drug resistance rate remains low in the UK, recognized treatment guidelines should be followed.

Isoniazid resistance

If isoniazid resistance is known before treatment is started, a fully supervised regimen of streptomycin, pyrazinamide, rifampicin and ethambutol for two months followed by rifampicin and ethambutol for seven months has been shown to give good results.

If definite pretreatment resistance to isoniazid is reported after the start of standard four-drug chemotherapy, isoniazid may be stopped but ethambutol should be given with rifampicin for a minimum of 12 months.

MULTIRESISTANT TUBERCULOSIS[101,102] (RESISTANCE TO AT LEAST ISONIAZID PLUS RIFAMPICIN)

Compared with HIV-infected people with drug-susceptible TB, those patients with multidrug-resistant disease are more likely:

- to develop combined pulmonary and extrapulmonary disease;
- to have atypical chest radiographic abnormalities, disseminated disease and higher mortality.

In the USA there have been recent epidemics of multidrug-resistant pulmonary tuberculosis. Most of the epidemic strains were resistant to isoniazid and rifampicin but the patterns of resistance varied among the outbreaks. A large proportion of the isolates were also resistant to streptomycin and ethambutol and some were resistant to third-line agents. Median survival in one study was 1.5 months for patients with an AIDS diagnosis and multidrug-resistant TB whereas it was 14.3 months for patients with an AIDS diagnosis and drug-susceptible TB. Despite aggressive multidrug treatment, 72–89% of more than 200 patients were dead in 4–19 weeks with 38–70% of the deaths caused by tuberculosis.

The limited data show a dismal outcome for immunodeficient patients with advanced MDR tuberculous infection in whom the diagnosis is delayed. However, early diagnosis, clinically or microbiologically, and appropriate therapy dramatically improve prognosis. Those with an improved prognosis have CD4 >25 cells and have at least two weeks treatment with two drugs to which their organism is susceptible.

Treatment plan

When initiating treatment for HIV-infected people exposed in settings where outbreaks or transmission of multidrug-resistant tuberculosis have occurred or when retreating an HIV-infected patient with a prior history of antituberculous therapy, five or six drugs should be given. The choice of drugs should be guided by susceptibility patterns from outbreak or exposure strains. In the absence of this information, isoniazid, rifampicin, pyrazinamide, ethambutol and streptomycin should be combined with second-line drugs to ensure that at least three and some would recommend four drugs to which the strain may be susceptible, which the patient has not previously received, are taken.

Treatment for multidrug-resistant TB should be continued for 18–24 months and for at least 12 months beyond culture conversion to negative. Directly observed therapy regimens with drugs given daily for 2–4 weeks then twice weekly have markedly improved compliance.[103,104] Such treatment must be closely monitored as full compliance is essential to prevent the emergence of further drug resistance.

Unlike the situation in the USA, multiple drug resistance is uncommon in Britain.

Drug reactions/interactions

Compliance can be a problem in some groups such as intravenous drug users. Supervised chemotherapy (directly observed therapy [DOT]) is required for MDR-TB.

Reactions to antituberculous drugs are more common in HIV-positive individuals. In the United Kingdom, haematological and hepatic reactions to rifampicin, isoniazid and pyrazinamide are commoner and occasionally anaphylactoid reactions to rifampicin can occur.

Rifampicin and isoniazid can interact with ketaconazole and fluconazole, reducing serum concentrations of fluconazole and making antifungal treatment ineffective.

If ketaconazole and rifampicin are taken at the same time, rifampicin absorption is inhibited and failure of the TB treatment can result.

Patients on protease inhibitors have to stop these if on rifampicin or change the rifampicin to rifabutin 150 mg od/300 mg od and use indinavir or nelfinavir.

Prevention strategies

BCG vaccination

This is contraindicated in known HIV-positive individuals.

Although it is possible that previous BCG vaccination may protect HIV-positive individuals against subsequently developing TB, there are currently no data supporting this. BCG vaccination is recommended for infants and young children of parents who are at increased risk of tuberculosis and can be given to such children provided that neither the mother nor the child is known or suspected to be HIV positive. There is a small but definite risk of generalized BCG infection and a theoretical risk that BCG can increase the progression of HIV disease by causing immune activation.

Primary chemoprophylaxis[105]

The IUTLD have suggested that all tuberculin-positive, HIV-positive patients be given chemoprophylaxis. Isoniazid is usually given for 6–12 months. The adult dose is 300 mg od.

In a study of HIV-infected PPD-positive injecting drug users in New York City, no active case of tuberculosis occurred during 63.4 person-years of follow-up for those who received 12 months of isoniazid.

Pape et al.[105] showed that isoniazid significantly reduced the risk of active TB amongst HIV-infected PPD-positive Haitian adults. Isoniazid was given in a dose of 300 mg a day for adults and 10 mg/kg per day, not to exceed 300 mg, for children.

Isoniazid plus rifampicin and pyrazinamide plus rifampicin have been suggested for three-month chemoprophylactic regimens.

Difficulties with chemoprophylaxis are:

- tuberculin tests can be difficult to interpret accurately if prior BCG has been given (?use 2 mm cut-off in HIV-positive patients);
- compliance;
- toxicity;
- cost;
- may not be of long-term benefit.

If chemoprophylaxis is not given then the BTS have recommended three-monthly clinical and chest radiographic monitoring of HIV-positive patients.

Control of crossinfection[106,107]

In order to prevent nosocomial spread of HIV in healthcare centres in the UK, units caring for immunesuppressed patients should carefully review their policy on airborne pathogen spread.

Although most tuberculosis can be treated at home, if hospitalization is required patients with sputum-positive TB should be segregated in a single room with appropriate ventilation and infectious disease control measures as outlined in the BTS policy and DOH policies. More rigorous measures have been suggested, e.g. UV lights, particulate filters, etc., especially in North America.

Notification and contact tracing

It is important that all patients diagnosed as having tuberculosis of any site, whether or not it is bacteriologically confirmed, be notified to the proper officer of the local authority according to statutory requirements. This allows contact tracing and may reveal the source of infection.

Specific drug treatment for patients in whom drug resistance is unlikely

Rifampicin plus isoniazid plus pyrazinamide plus ethambutol. Ethambutol might be excluded only if rates of background isoniazid resistance are less than 4%.

Preparations

Rifampicin:
Tablet   150 mg, 300 mg

Isoniazid:
Tablet   50 mg, 100 mg

Rimactazid:
Tablet   150 mg, 300 mg

Pyrazinamide:
Tablet 500 mg

Ethambutol:
Tablet 100 mg, 400 mg

Rimactazid 150 = rifampicin 150 mg and isoniazid 100 mg
Rimactazid 300 = rifampicin 300 mg and isoniazid 150 mg
Rifater = rifampicin 120 mg, isoniazid 50 mg, pyrazinamide 300 mg

Dosage and administration
Rifampicin: 600 mg daily po if patient >50 kg, 450 mg daily po if patient <50 kg
Isoniazid: 300 mg daily po + pyridoxine 10 mg daily po
Pyrazinamide: 2 g daily po if patient >50 kg, 1.5 kg daily po if patient <50 kg
Ethambutol: 15 mg/kg

All drugs should be taken as a *single dose in the morning before breakfast* to aid compliance. If the tablets make the patient feel nauseated then they can be taken in divided doses, e.g. twice a day. If the patient still experiences nausea the tablets should be taken after food, despite the decrease in bioavailability.

Therapy should be taken for the initial two months. Rifampicin and isoniazid should then be continued for a total of six months. Following completion of treatment, long-term prophylaxis with isoniazid is not necessary.

Check the sensitivities and amend treatment if necessary. MDR–TB is uncommon in the UK. A high level of suspicion should be assumed for patients who have travelled to areas where drug resistance has been reported, e.g. New York City, sub-Saharan Africa, Asia.

What to monitor

	First month of treatment	Remainder of treatment
Full blood count	Once a week	Once a month
U&Es	Once a week	Once a month
Liver function tests	Once a week	Once a month
Ophthalmic examination	Baseline for ethambutol	

Side effects/toxicity
Elevated liver enzymes
Rifampicin, isoniazid and pyrazinamide can all cause an elevation in liver enzymes. Mild, transient and symptomless increases in serum hepatic enzyme concentrations are usual during the early weeks of treatment whatever the drug regime. Treatment should not be stopped unless clinically evident hepatitis occurs or liver enzymes reach five times the upper limit of normal. If the hepatitis has been caused by the MTB drug regime, liver function usually returns rapidly to normal once administration of the regime has been stopped.

If it becomes necessary to stop treatment the drugs should be reintroduced individually at lower doses. *This should be done by those with experience and following a strict protocol.*

Rifampicin is a potent inducer of hepatic microsomal enzymes and may increase the metabolism of concomitantly administered drugs. The initial effects of rifampicin may be detected within the first two days of therapy but it generally takes a week or more until the effects of maximal enzyme induction are manifested.

Hyperuricaemia
Pyrazinoic acid (the main metabolite of pyrazinamide) inhibits the renal tubular secretion of uric acid, causing a rise in serum uric acid concentration. This is usually symptomless but a high serum uric acid concentration can precipitate acute gout in patients with the disease.

Rash
The symptoms of a minor rash can be controlled with an antihistamine. If a more severe skin reaction occurs e.g. exfoliative dermatitis, all drugs should be stopped until the reaction has subsided. The patient can then be carefully rechallenged according to the regime above.

Decreased renal function
Rifampicin
In the presence of impaired renal function, the elimination half-life of rifampicin becomes prolonged only at doses exceeding 600 mg. Provided that hepatic excretory function is normal, the dosage in patients with impaired renal function does not need to be reduced below 600 mg daily.

Isoniazid
Isoniazid is acetylated and hydrolysed in the liver. Acetylation is the most important metabolic step, the rate being dependent on whether or not the patient is a rapid or slow acetylator. In slow acetylators with severely impaired renal function, some accumulation of isoniazid may occur, necessitating a reduction in dosage.

Ethambutol
The dosage of ethambutol may need to be reduced in patients with decreased renal function as determined by blood levels.

Visual impairment

Ethambutol can cause optic neuritis which appears to be related to the dose and length of treatment. The patient complains of loss of colour (green) vision as one of the earliest signs. It is generally reversible. It is unusual on doses of 15 mg/kg per day but if it occurs, ethambutol should be stopped immediately.

Absolute contraindications

1. Ketoconazole and itraconazole. Rifampicin can induce the metabolism of *ketoconazole* and *itraconazole*, leading to decreased serum levels and hence decreased efficacy. Ketoconazole can inhibit the absorption of *rifampicin*.

    Fluconazole should be prescribed for patients taking rifampicin who require systemic antifungal therapy (see below).
2. Protease inhibitors.
3. Non-nucleosides.

Use with caution:

- Fluconazole: decreased bioavailability (25%) and decreased half-life (20%) of *fluconazole* when given concomitantly with rifampicin. Larger doses of fluconazole may be required
- warfarin: decreased anticoagulant effect
- phenytoin: decreased serum levels of phenytoin
- dapsone: decreased plasma concentration of dapsone
- corticosteroids: reduced effect of corticosteroids
- Antacids, anticholinergics and opiates: absorption of rifampicin reduced

Other drugs that interact with rifampicin include cyclosporin A, oral antidiabetic agents, sex hormones, oral contraceptives, digoxin, quinidine, methadone, theophylline, chloramphenicol, azathioprine, β-blockers, verapamil and cimetidine. The activity of these drugs may be impaired and dosage adjustment may be necessary.

Isoniazid

Isoniazid is a hepatic enzyme inhibitor. Plasma levels of phenytoin, carbamazepine and sodium valproate may be increased. Doses may need to be reduced.

Pyrazinamide

No significant interactions occur with pyrazinamide.

Ethambutol

No significant interactions occur with ethambutol.

VIRAL DISEASES

CMV RETINITIS[108,109]

CMV retinitis is a common and severe manifestation of immunosuppression historically affecting 20–30% of patients with AIDS. It is usually a late complication and occurs when the CD4 count is <100 cells/ml, the average being 30 cells/ml.

In two recent observational studies the risk of developing active CMV disease was noted to be 21.4% in patients with a CD4 <100 cells/ml followed for two years and was 41.9% in patients followed for 27 months who had a CD4 count of <50 cells/ml.[110,111]

Even when the retinitis is peripheral, treatment appears to be beneficial.[112] Some authors recommend ophthalmic assessment in patients with CD4 <50 every 2–3 months.[113]

The retinitis is necrotizing, with multiple white dots coalescing into a leading edge, often associated with haemorrhage. There are usually only 1–3 sites of activity in the retina. Frosted branch retinopathy is an atypical form of CMV retinitis.

Treatment[114,115]

Standard therapy for CMV disease is with lifelong intravenously administered drugs, ganciclovir, foscarnet (see below) or cidofovir. Induction doses are followed after 2–3 weeks with half-dose maintenance. Ocular implant and direct intravitreal injections are other options but do not protect the other eye or the rest of the body.

Even on maintenance treatment, patients can show reactivation, necessitating recommencement of induction doses. If a patient does not respond after 3–5 weeks of treatment, other causes of retinitis should be considered.

One controversy over which drug to use stems from the results of a study suggesting the patients treated with foscarnet have a survival advantage over those treated with ganciclovir.

Oral ganciclovir[116]

This may lessen the incidence of neutropenia and avoid line sepsis. However, compared with IV ganciclovir the retinitis may progress faster and there is more frequent subsequent involvement of the uninfected eye in those presenting with uniocular disease. It is not recommended for treatment of CMV.

Intravitreal injection[117,118]
Induction:

- 200 mg intravitreal ganciclovir in 50 ml 2–3 times a week

Maintenance:
- 200 mg weekly indefinitely
- Foscarnet, cidofovir and antisense oligonucleotides have been used as alternative therapies.

Problems:
- endophthalmitis (0.6%)
- vitreous haemorrhage (3%)
- retinal detachment (8%) ?due to CMV disease
- optic atrophy (8%) ?due to CMV disease
- transient closure of central retinal artery (less risk with 50 ml volume)
- no protection for uninvolved eye
- no systemic protection

Use:
- consider if patient unresponsive to maximum systemic therapy

Intravitreal implant
- Release drug for 3–6 months
- Detachment occurs in 12% – may require vitrectomy and silicone oil.
- Infection endophthalmitis 1.2%
- Patients have blurring of vision for up to four weeks
- Does not protect other eye or systemically

Ganciclovir resistance[119]
Resistance to ganciclovir can occur in up to 10% of patients who have had continuous ganciclovir therapy for three months. Luckily these strains remain sensitive to foscarnet.

Clinical resistant disease is suspected after three weeks on treatment dosage of ganciclovir with no evident response on examination; refer to an ophthalmologist. A therapeutic trial of foscarnet may be recommended. If the patient has disease that is close to the optic nerve then combined treatment with foscarnet and ganciclovir may be advised.

Cidofovir
This nucleotide analogue has good tissue penetration to the eye.
- Long intracellular half-life
- Renal toxic
- Once-weekly dosing
- Need to take probenecid (to increase uric acid excretion and thus improve renal handling of cidofovir)
- May be crossreactive with ganciclovir

Preparations
Ampoule 375 mg/5 ml

Treatment dosage
5 mg/kg IV once a week for two weeks in 100 ml NaCl 0.9%

Maintenance
- A central venous line is not necessary
- Handle drug with caution
- Store in fridge

Administration
- The patient should be prehydrated with 1 litre N saline 1 hour predose.
- Drug given over one hour
- Probenecid should be given 2 g po 3 hours before, 1 g po 2 hours after, 1 g po 8 hours after.

OTHER CMV-RELATED DISEASES

CMV colitis[120]

The natural history of this condition is characterized by relapse and remission.

Symptoms
Fever, weight loss, anorexia, abdominal pain, diarrhoea are frequently present. Haemorrhage and perforation can be life-threatening.

Diagnosis
Radiographic appearance mimics other inflammatory bowel conditions. Colonoscopic or rectal biopsy identifies characteristic intranuclear and intracytoplasmic inclusions. *N.B. The finding of the above does not always imply that CMV is causing symptoms.*

Culture may not implicate CMV as the cause of symptoms.

Treatment
A 62-patient double-blind, placebo-controlled safety and efficacy study of 14 days of ganciclovir therapy for CMV colitis showed a reduction in urine and colonic cultures for CMV. Colonoscopy score improved and fewer ganciclovir-treated patients developed extracolonic disease after 14 days of therapy. Placebo patients also had a mean loss of 1.5 kg body weight while ganciclovir patients remained stable. Twenty-five percent of patients on placebo developed CMV retinitis in only 14 days.

The optimal treatment course and need for maintenance therapy remains of concern in this population and treatment of this condition is still a subject of debate. No data on cidofovir treatment of colitis are available.

Treatment failure
The efficacy of foscarnet treatment has been assessed in patients who have failed standard ganciclovir induction therapy for CMV GI disease. Overall, 67% of patients had a remission on foscarnet but median survival was only five months. Reversible nephrotoxicity occurred in 12%. A combination of the two agents has also been used and was well tolerated.

No data on cidofovir have been presented.

Maintenance
Time to relapse is about nine weeks, but relapse-free intervals of one year can occur. It is probably wise to give maintenance after one relapse.

CMV oesophagitis

Symptoms
Causes dysphagia.

Diagnosis
Extensive large shallow ulcers of the distal oesophagus are characteristic. Histological evidence is required; culture alone should not be relied upon to establish the diagnosis.

CMV polyradiculopathy[121]

Symptoms/signs
This neurological syndrome is characterized by subacute onset of leg weakness progressing to paraparesis/paraplegia. Patients usually have had a prior AIDS diagnosis. Bladder dysfunction and concomitant CMV retinitis are common.

CSF findings
CSF shows low glucose and increase in neutrophils; CMV can be grown or found by PCR in about 50%. Imaging shows no spinal lesion but may show leptomeningeal enhancement or crowding of nerve roots.

Other studies
EMG/nerve conduction supports the diagnosis of a radiculopathy.

Treatment
Lifelong treatment as for retinitis has been beneficial in some patients.

CNS infection[122]
Subacute encephalitis caused by CMV has been reported but the diagnosis can be confirmed by brain biopsy with evidence of periventricular necrosis, giant cells, intranuclear and giant cell inclusions and isolation or other identification of the virus by antigen or nucleic acid. The CSF usually shows a polymorphonucleocytosis and moderate to low glucose concentration. There are few data on the efficacy of treating the syndrome.

Pneumonia[123]
Isolation of CMV from pulmonary secretions is common. A true pathogenic role of the virus in the disease process is usually not apparent.

Diagnosis
Although CMV can cause fatal pneumonitis in transplant patients it is debatable how much it contributes to lung disease in HIV infection.

The diagnosis of CMV pneumonia can be considered:

- when there is a positive CMV culture from lung tissue or pulmonary secretions;
- in the presence of pathognomonic cells with intranuclear inclusion bodies;
- when there is CMV antigen or nucleic acid in the tissue;
- in the absence of other pathogenic organisms causing interstitial pneumonia.

In bone marrow transplant patients, high-dose intravenous immunoglobulin with ganciclovir is used but there are no data supporting this combination in patients with HIV.

Specific drug treatment

Ganciclovir
Preparations
Vial 500 mg

Induction therapy
5 mg/kg 12-hourly IV for 14–21 days

Maintenance
6 mg/kg IV as a single daily dose for five days per week *or* 5 mg/kg IV as a single daily dose for seven days per week. *N.B. Some units continue treatment dose as maintenance.*

Ganciclovir has to be given as an infusion (maximum concentration 10 mg/ml) in either sodium chloride 0.9% or 5% dextrose. Local practice is for each dose to be added to 100 ml of sodium chloride and infused over one hour. Caution should be exercised in the handling and disposal of ganciclovir as it shares some of the properties of antitumour agents. Polythene gloves and safety glasses should be worn during reconstitution: if mucosal contact occurs, the area should be washed thoroughly with soap and water. Infusions can be prepared by pharmacy.

Solutions of ganciclovir have a high pH (9–11) and may cause phlebitis and/or pain at the infusion site. It should therefore only be infused into veins with a good blood flow to allow rapid dilution and distribution. Patients in whom it is decided to continue with maintenance therapy will require long-term central venous access, e.g. a Hickman line or Port a cath.

What to monitor

	Treatment	Maintenance
Full blood count	Twice a week	Once a month
U&Es	Twice a week	Once a month
Liver function tests	Twice a week	Once a month
Fundoscopy	After two weeks	Once a month
Weight		Once a month

Side effects/toxicity
Decreased renal function
Ganciclovir should be used with caution in patients with impaired renal function: peak plasma levels and the plasma half-life may be increased.

Serum creatinine (mmol/l)	Dose (mg)	Dose interval (h)
125–225	2.5	12
226–398	2.5	24
>398	1.25	24

In patients who are taking ganciclovir for secondary prophylaxis of CMV disease, the dosage above that is appropriate for the patient's renal function should be halved.

Haematological toxicity
Neutropenia is reported to occur in 40% of patients receiving ganciclovir. It typically occurs during the first or second week of induction therapy. Counts may normalize within 2–5 days after either discontinuing treatment or decreasing the dose.

If neutrophils $<0.5 \times 10^9/l$, stop ganciclovir and start foscarnet *or* give a colony-stimulating factor (G-CSF or GM-CSF) and continue ganciclovir.

If platelets $<50 \times 10^9/l$, stop ganciclovir and start foscarnet.

Drug interactions
Use with caution:

- zidovudine (AZT): cumulative haematological toxicity may occur. Zidovudine should be stopped during the induction period. Patients on maintenance ganciclovir therapy may tolerate zidovudine. It should be reintroduced at a lower dose and titrated carefully
- myelosuppressive agents, e.g. pyrimethamine, sulphonamides, dapsone, cytotoxic agents: cumulative haematological toxicity may occur
- imipenem-cilastin: generalized seizures have been reported. N.B. Acyclovir for the treatment of herpes infections is not required during ganciclovir therapy.

Foscarnet

Preparations
Bottle 6 g/250 ml and 12 g/500 ml, i.e. 2 mg/ml

Induction therapy
90 mg/kg 12-hourly IV for 14–21 days

Maintenance
90 mg/kg once daily IV

Foscarnet is licensed to be given as a continuous IV infusion but local practice is to use twice-daily dosing which is more convenient for patients and staff. Each dose should be given over two hours with 1 l of 0.9% sodium chloride. If it is being given peripherally it is important that the saline runs concurrently with the foscarnet to reduce the incidence of thrombophlebitis. If the patient has a high sodium level, 5% dextrose can be used as the hydration fluid. Foscarnet contains 120 mmol sodium per litre.

There is some evidence that the renal toxicity of foscarnet can be minimized by prehydrating the patient, e.g. 1 l of 0.9% sodium chloride, before each dose.

What to monitor

	Treatment	Maintenance
Full blood count	Twice a week	Once a month
U&Es	Twice a week	Every two weeks
Liver function tests	Twice a week	Once a month
Calcium and magnesium	Twice a week	Once a month
24-h urine collection	Once a week	Once a month

Side effects/toxicity
Decreased renal function
As foscarnet is excreted entirely via the renal route the dose must be adjusted according to the patient's renal

function. A 24-h collection should be used for accurately calculating creatinine clearance (the dose can be started according to the creatinine clearance estimated from the patient's serum creatinine).

The following nomogram is suggested for the treatment of CMV retinitis with foscarnet according to renal function:

CrCl (ml/min/kg)	Foscarnet treatment dose in mg/kg given every 8 h	Foscarnet treatment dose in mg/kg given every 12 h. Prophylactic dose in mg/kg given every 24 h
1.6	60	90
1.5	57	85.5
1.4	53	79.5
1.3	49	73.5
1.2	46	69
1.1	42	58.5
1.0	39	52.5
0.8	32	48
0.7	28	42
0.6	25	37.5
0.5	21	31.5
0.4	18	27

Electrolyte disturbances
Low levels of calcium, magnesium and potassium can occur in patients on foscarnet. Initially IV supplements should be given and then levels should be maintained within the normal range using oral supplements. If levels of potassium and magnesium are difficult to maintain within the normal range, amiloride and/or spironolactone can be tried.

Anaemia
If haemoglobin <9 g/dl or the patient is symptomatic, transfuse as necessary.

Penile ulceration
This is thought to be due to a local effect caused by foscarnet concentrating in the urine. To avoid this the patient should be advised to wash the genital area with soap and water after urinating.

Drug interactions
Absolute contraindications:

- nephrotoxic drugs, e.g. treatment doses of pentamidine and amphotericin, high-dose frusemide, aminoglycoside antibiotics: additive renal toxicity
- prophylactic doses of nephrotoxic drugs
- high-dose acyclovir

N.B. Acyclovir for the treatment of herpes infections is *not required* during foscarnet therapy.

Complications/problems

Disease progression on drug maintenance
Consider:

- inadequate drug dosing;
- patient has become severely immunosuppressed;
- viral resistance to ganciclovir (occurs in up to 8% patients on ganciclovir for longer than three months);
- giving alternative treatment.

CMV prophylaxis

CMV viral load
A substudy of the Syntex/Roche 1654 oral GCV prophylaxis trial showed:

- 15% PCR CMV in blood negative;
- 45% PCR CMV in blood positive and will develop CMV end-organ disease;
- PCR positive more likely to develop end-organ disease if CMV viral load >50 000 copies per ml.[124]

One strategy for deciding who to put on prophylaxis is to measure CMV viral load in plasma in those who are CMV urine positive and if the plasma level is ≥50 000 copies per ml, to start prophylaxis.

Drugs used in CMV prophylaxis
Oral ganciclovir
Two randomized placebo-controlled trials have been completed.[125,126]

The two studies had conflicting results:

- 49% reduction in the Spector study;
- CPCRA 023 oral ganciclovir no better than placebo;
- different management strategies may explain conflicting results;
- little resistance (<1% after 8.3 months) seen in people on CMV prophylaxis who develop viruria. ?Relevance of this in long term.

Dosage and administration
Capsule   250 mg, 500 mg six times daily with food (fatty meals), 1000 mg three times daily with food (fatty meals)

Side effects/toxicity
- Granulocytopenia 18%
- Anaemia 12%
- Thrombocytopenia 6%
- Also GI nausea, vomiting, diarrhoea and rash

ACTG 204/valaciclovir
Increased mortality in valaciclovir group.[127] N.B. Can cause haemolytic uraemic syndrome.

HERPES SIMPLEX VIRUS (HSV) INFECTION (HSV-1 AND HSV-2)

HSV mucocutaneous lesions are less prone to spontaneous resolution in HIV patients. The tendency for chronicity may be related to the degree of immunosuppression. Data suggest that recurrences are more frequent when CD4 <50 cells/μl.

Clinical manifestations

Anogenital outbreaks are common and lead to:

- pain
- ± itching
- tenesmus
- discharge
- constipation

Associated features may include:

- fever
- inguinal lymph nodes
- sacral neuropathy

There are no data to suggest that keratitis, encephalitis and meningitis due to HSV are increased. N.B. All oral and perianal ulcers and fissures and all atypical skin blisters/ulcers should be cultured for HSV.

Other sites

Orolabial
- Large, multiple lesions may interfere with feeding, etc.
- Cutaneous lesions on limbs, etc.

Oesophageal
Symptoms are the same as for candida (see p.224) Endoscopic and culture evidence is needed for diagnosis.

Encephalitis
- Can be typical, i.e. acute with seizures, focal signs and obtundation or subacute with headache, personality change, etc.

- Ocular
- Pneumonitis

Diagnosis

- Virus culture leads to cytopathic effect in 24–72 h
- Virus type HSV-1 or -2 by monoclonal antibodies

Alternative methods of diagnosis
- Direct antigen detection
- Tzanck smear
- PCR, especially in CSF
- Antibodies useful epidemiologically, not as an individual tool

Treatment

Acyclovir is virustatic and doesn't eradicate latent virus.

Oral disease
Case reports suggest that early treatment of oral lesions with acyclovir may be beneficial but these lesions may be resistant to acyclovir when chronic.

Primary and recurrent genital HSV
- 200 mg five times daily orally or 5 mg/kg tds IV
- 800 mg twice daily may be effective
- Decreases virus shedding
- Decreases time to healing

Chronic suppression
All these regimens have been shown to be effective in immunocompetents and safe for up to three years:

- 200 mg tds
- 400 mg bd
- 800 mg bd

Resistance to acyclovir
Can be found *in vitro* even if the patient has never been treated with acyclovir. Clinically relevant resistance to acyclovir is associated with HIV infection. The patient usually has anogenital HSV-2 and CD4 <50 cells μ/l and is thymidine kinase deficient. Treat with foscarnet 40 mg/kg eight-hourly.

Other agents
Foscarnet
Use for acyclovir-resistant HSV. A foscarnet cream has been used.

Famciclovir
This is more bioavailable than acyclovir. Use twice-daily dosing because of the longer intracellular half-life.

Valaciclovir
This prodrug of acyclovir shows greater bioavailability.

Trifluridine
A topical solution used in eye disease, it can be used in mucocutaneous disease and in those with acyclovir resistance.

Specific drug treatment

Acyclovir – preparations
- Tablets: 200 mg
    400 mg
- Suspension: 200 mg/5 ml
    400 mg/5 ml
- Intravenous: 250 mg vial reconstituted in 10 ml of 0.9% NaCl infused over one hour
    500 mg vial

Dosage and administration
Mucocutaneous infection mild: 200 mg (–800 mg) od orally five times daily for five days
Mucocutaneous infection severe: 15 mg/kg/day IV for five days
Recurrent mucocutaneous infection: 200–400 mg orally tds
Systemic/visceral infection: 30 mg/kg/day IV

Side effects/toxicity
- Rapid increase in urea and creatinine can occur when given IV.
- Reversible neurological reactions, e.g. confusion, tremor, somnolence, convulsions.
- Rarely, elevation of liver enzymes, rashes and fever can occur.
- Anaemia, leucopenia and thrombocytopenia can occur.

Renal disease
- CrCl<10 ml/s/min – adjust dose of oral to 200 mg bd
- For IV preparation:
    CrCl 25–50 ml/min: 5–10 mg/kg 12-hourly
    CrCl 10–25 ml/min: 5–10 mg/kg 24-hourly
    CrCl 0–10 ml/min: 2.5–5 mg/kg 24-hourly

PROTOZOAL DISEASES

CRYPTOSPORIDIOSIS[128,129]

This intracellular protozoan parasite is ubiquitous, causing disease in many animal hosts. It is considered a major cause of diarrhoeal illness in the world. Transmission is by ingestion of highly resilient oocysts. Modes of transmission are:

- person to person;
- animal to person;
- environment (water, swimming pools, nosocomial).

It is not known how cryptosporidium causes disease:

- ?toxins
- ?villi damage
- ?bacterial overgrowth
- uncommon in patients on HAART.

Clinical features

- Diarrhoea can be scant, severe (up to 17 litres a day!), intermittent, continuous.
- Crampy abdominal pain can accompany diarrhoea.
- Spontaneous resolution can occur when CD4 >180 cells/µl.
- Low-grade fever.
- Malaise, nausea, vomiting, loss of appetite.

Associated features

- Cholecystitis, sclerosing cholangitis
- Hepatitis
- Pancreatitis
- Reactive arthritis
- Pulmonary disease

Laboratory features

- Leucocytosis *not* a feature.
- Abnormal absorption of fat, D-xylose and B_{12} have been noted.
- CT/XR non-specific.

Diagnosis

- Identification of oocyst in stool, duodenal aspirates, bile, etc.
- Identification of organism in small intestinal biopsies.
- Can concentrate faeces using flotation methods.
- ELISAs and PCR being developed.

Diagnostic techniques
- Phase contrast microscopy
- Stain with modified Ziehl–Neelsen stain
- Fluorescent stain
- Monoclonal antibodies
- PCR

Prognosis

A study has suggested that HIV-associated cryptosporidial diarrhoea does not always have a uniformly poor prognosis, with 11 out of 38 patients having a clinical remission of their diarrhoeal symptoms especially in patients with higher absolute lymphocyte counts (1100 compared with 550×10^6/l). Median survival was 66 weeks for the patients who remitted compared with 11.5 for those who did not.[128]

Recovery from infection and duration of symptoms depend on host immune status and usually occur when CD4 <200. Sclerosing cholangitis, hepatitis, pancreatitis, reactive arthritis and respiratory problems can occur.

Treatment[129,130]

The treatment of cryptosporidiosis remains problematic. At least 95 agents have been tried but there are no approved or reliably specific effective therapies available. In one small trial of seven patients who received a short course of paromomycin 500 mg $\times$ 4 daily, all patients experienced some type of initial beneficial response.[130] HAART appears to be of benefit.

By 1990, 95 agents had been tried with only anecdotal success. Most attention has been given to:

- octreotide acetate, a synthetic analogue of somatostatin which inhibits secretory diarrhoea but does not eliminate infection;
- diclazuril sodium, an anticoccidial agent. Poorly absorbed and ineffective. Letrazuril has better bioavailability;
- azithromycin, which is effective in the rat model;
- paromomycin, which can give symptomatic relief but is nephrotoxic.

Treatment with zidovudine has been associated with remission and cases of long-term remission in patients on HAART have been reported.[131] Immunomodulators such as bovine transfer factor or bovine dialysable extract and hyperimmune bovine colostrum have had anecdotal success. Supportive care with careful attention to nutrition and hydration with the use of antidiarrhoeal agents remains the backbone of therapy.

Prevention

- No effective disinfectant
- Good hygiene
- Boil water
- Drink only purified/processed water (documented to remove oocysts)

MICROSPORIDIOSIS

- Ubiquitous, small, intracellular, spore-forming protozoa. Two infecting HIV patients are *Enterocytozoon bieneusi* and *Septata intestinalis*.
- Found in up to 64% of AIDS patients with chronic unexplained diarrhoea.
- Are they pathogens?

Clinical features

- Chronic watery diarrhoea
- Weight loss, malaise
- Cholangitis, keratoconjunctivitis and visceral infections can occur

Diagnosis

- Electron microscopy
- Light microscopy and staining of stool or GI samples
- PCR

Treatment

- In uncontrolled studies albendazole 400 mg bd for 28 days has reduced symptoms. It does not eradicate *E. bieneusi* but does eradicate *S. intestinalis*.
- Thalidomide.

ISOSPORA BELLI

- Rare pathogenic parasite in normal host.
- Commoner in AIDS patients in developing world.
- Typically, infection is concentrated in the small bowel.
- Dissemination has been described in AIDS.

Clinical features

Same as cryptosporidium.

Diagnosis

Identify large oval oocysts in stool or GI samples with modified Kenyoun acid-fast stain.

Treatment

Co-trimoxazole 960 mg $\times$ 2 bd for 2–4 weeks.

Secondary prophylaxis

Co-trimoxazole (see PCP, p. 214) or pyrimethamine-sulfadoxine.

BACTERIAL PNEUMONIA[134-137]

About 10% of pneumonias in AIDS patients in one study were due to community-acquired bacteria. The majority are due to the *pneumococcus* or *H. influenzae*.

Bacteraemia is commoner in HIV-positive patients and recurrence is common. Other bacterial species such as *Staph. aureus, Legionella mycoplasma*, etc. have been isolated. In one study there was a 46% prevalence of bacteraemia in HIV-seropositive patients with pneumococcal pneumonia.

Symptoms

The presentation of bacterial pneumonia in AIDS patients is similar to that seen in non-immune suppressed hosts, with an abrupt onset, fever, productive cough, dyspnoea and pleuritic chest pain. Local signs – crackles, bronchial breathing, etc. – are common.

Chest X-ray

Lobar or segmental consolidation is more common than diffuse disease.

Sputum

Gram staining of sputum can reveal neutrophils and organisms. Sputum culture should be performed.

Bloods

Blood cultures are positive in 40–80% of cases and a relative leucocytosis is often found.

Treatment

Empiric treatment should include coverage for most common pathogens, usually with a second-generation cephalosporin, ampicillin-clavinulate or semisynthetic penicillin and an aminoglycoside. Amoxycillin plus erythromycin is the BTS recommendation in severe community-acquired pneumonia.

In patients for whom a definitive diagnosis is not made and for those who fail to respond to therapy, diagnostic evaluation for pneumocystosis should be performed. Up to 10% of HIV-infected patients with pyogenic pneumonia may have concomitant PCP.

SALMONELLOSIS[138,139]

In San Francisco AIDS patients, the incidence of salmonellosis was 20-fold greater than in controls. Bacteraemia occurred in 45% versus 9% controls. One-third of salmonella infections occurred before an AIDS diagnosis.

Symptoms

- Diarrhoea with cramping, bloating and nausea and fever.
- Febrile illness *without* diarrhoea in up to 45%.

Diagnosis

Stool and blood culture.

Treatment

- Ciprofloxacin 750 mg bd or tds
- Ampicillin, dependent on sensitivity
- Chloramphenicol, dependent on sensitivity
- Co-trimoxazole, dependent on sensitivity

Bacteraemia should be treated for 14 days. Enteric disease alone should be treated for 7–14 days.

Recurrence

Use ciprofloxacin long term 750 mg bd.

REFERENCES

1. Wall, P.G., Porter, K., Noone, A. *et al.* (1993) Changing incidence of *Pneumocystis carinii* pneumonia as initial AIDS defining disease in the United Kingdom. *AIDS*, **7**, 1523–1525.
2. Bacellar, H., Monoz, A., Hoover, Dr. *et al.* (1994) Incidence of clinical AIDS conditions in a cohort of homosexual men with CD4+ count <100 mm/mm^3. *J. Infect. Dis.*, **170**, 1284–1287.
3. Murray, J.F. and Mills, J. (1990) Pulmonary infectious complications of human immunodeficiency virus infection. *Am. Rev. Respir. Dis.*, **141**, 1582–1598.
4. Richie, T.L., Yamaguchi, E., Virani, N.A. *et al.* (1989) Extrapulmonary *Pneumocystis* infection. *Ann. Intern. Med.*, **111**, 339.
5. Ng, V.L., Yajko, D.M. and Hadley, W.K. (1997) Extrapulmonary pneumocystosis. *Clin. Micro. Rev.*, **10**, 401–418.
6. Goodman, P.C. and Gamsu, G. (1987) Pulmonary radiographic findings in the acquired immunodeficiency syndrome. *Postgrad. Radiol.*, **7**, 3–15.
7. Levine, S.J. and White, D.A. (1988) *Pneumocystis carinii* infections in patients with AIDS. *Clin. Chest Med.*, **9**, 395–423.

8. Smith, D.E., McLuckie, A., Wyatt, J. et al. (1989) Severe exercise hypoxaemia with normal or near normal x-rays: a feature of *Pneumocystis carinii* infection. *Lancet*, **ii**, 1049–1051.
9. Orenstein, M., Weber, C.A., Cash, M. et al. (1986) Value of broncho-alveolar lavage in the diagnosis of pulmonary infection in acquired immune deficiency syndrome. *Thorax*, **41**, 345–349.
10. Shaw, R.J., Roussak, C., Forster, S.M. et al. (1988) Lung function abnormalities in patients infected with the human immunodeficiency virus with and without overt pneumonitis. *Thorax*, **43**, 436–440.
11. O'Doherty, M.J., Page, C.J., Bradbeer, C.S. et al. (1987) Alveolar permeability in HIV antibody positive patients with *Pneumocystis carinii* pneumonia. *Genitourinary Med.*, **63**, 268–270.
12. Miller, R.F. and Mitchell, D.M. (1995) *Pneumocystis carinii* pneumonia. *Thorax*, **50**, 191–200.
13. Hopewell, P.C. (1988) *Pneumocystis carinii* pneumonia: diagnosis. *J. Infect. Dis.*, **157**, 629–632.
14. Zaman, M.K. and White, D.A. (1988) Serum lactate dehydrogenase levels and *Pneumocystis carinii* pneumonia. *Am. Rev. Respir. Dis.*, **137**, 796–800.
15. Luce, J.M. and Clement, M.J. (1989) Pulmonary diagnostic evaluation in patients suspected of having an HIV-related disease. *Semin. Respir. Infect.*, **4**, 93–101.
16. Hadley, W.K. and Ng, V.L. (1989) Organization of microbiology laboratory services for the diagnosis of pulmonary infections in patients with human immunodeficiency virus infection. *Semin. Respir. Infect.*, **4**, 85–92.
17. Ng, V.L., Garner, I., Weymouth, L.A. et al. (1989) The use of mucolysed induced sputum for the identification of pulmonary pathogens associated with human immunodeficiency virus infection. *Arch. Pathol. Lab. Med.*, **113**, 488–493.
18. Walzer, P.D. (1993) Editorial review. *AIDS*, **7**, 1293–1305.
19. Tu, J.U., Biem, J.H. and Detsky, A.S. (1993) Bronchoscopy versus empirical therapy in HIV infected patients with presumptive PCP: a decision analysis. *Am. Rev. Respir. Dis.*, **148**, 370–377.
20. Bozzette, S.A., Sattler, F.R., Chiu, J. et al. (1990) A controlled trial of early adjunctive treatment with corticosteroids for *Pneumocystis carinii* pneumonia in the acquired immunodeficiency syndrome. *N. Engl. J. Med.*, **323**, 1451–1457.
21. Montaner, J.S.G., Lawson, L.M., Levitt, J. et al. (1990) Corticosteroids prevent early deterioration in patients with moderately severe *Pneumocystis carinii* pneumonia and the acquired immunodeficiency syndrome (AIDS). *Ann. Intern. Med.*, **113**, 14–20.
22. Gagnon, S., Boota, A.M., Fischl, M.A. et al. (1990) Corticosteroids as adjunctive therapy for severe *Pneumocystis carinii* pneumonia in the acquired immunodeficiency syndrome – a double-blind, placebo-controlled trial. *N. Engl. J. Med.*, **323**, 1444–1450.
23. The National Institutes of Health, University of California Expert Panel for Corticosteroids as Adjunctive Therapy for Pneumocystis Pneumonia (1990) Consensus statement on the use of corticosteroids as adjunctive therapy for pneumocystis pneumonia in the acquired immunodeficiency syndrome. *N. Engl. J. Med.*, **323**, 1500–1504.
24. Gachot, B., Clair, B., Wolff, M. et al. (1992) Continuous positive airway pressure by face mask or mechanical ventilation in patients with HIV and severe PCP. *Intens. Care Med.*, **18**, 155–159.
25. Wachter, R.M., Rossi, M.B., Bloch, A. et al. (1991) *Pneumocystis carinii* pneumonia and respiratory failure in AIDS: improved outcomes and increased use of intensive care units. *Am. Rev. Respir. Dis.*, **143**, 251–256.
26. Bennett, R.L., Gilman, S.C., George, L. et al. (1993) Improved outcomes in intensive care units for AIDS-related *Pneumocystis carinii* pneumonia: 1987–1991. *J. Acq. Immune Defic. Syndr.*, **6**, 1319–1321.
27. Wharton, J.M., Coleman, D.L., Wofsy, C.D. et al. (1986) Trimethoprim-sulfamethoxazole or pentamidine for *Pneumocystis carinii* pneumonia in the acquired immunodeficiency syndrome: a prospective randomized trial. *Ann. Intern. Med.*, **105**, 37–44.
28. Sattler, F.R., Cowar, N., Neilson, D.M. et al. (1988) Trimethoprim-sulfamethoxazole compared with pentamidine for treatment of *Pneumocystis carinii* pneumonia in the acquired immunodeficiency syndrome. *Ann. Intern. Med.*, **109**, 280–287.
29. Sattler, F.R., Cowar, N., Neilson, D.M. et al. (1988) Trimethoprim-sulfamethoxazole compared with pentamidine for treatment of *Pneumocystis carinii* pneumonia in the acquired immunodeficiency syndrome. *Ann. Intern. Med.*, **109**, 280–287.
30. Wharton, J.M., Coleman, D.L., Wofsy, C.B. et al. (1986) Trimethoprim-sulfamethoxazole or pentamidine for *Pneumocystis carinii* pneumonia in the acquired immunodeficiency syndrome. A prospective randomised trial. *Ann. Intern. Med.*, **105**, 37–44.

31. Klein, N.C., Duncanson, F.P., Lenox, T.H. et al. (1992) Trimethaprim-sulfamethoxazole for *Pneumocystis carinii* in AIDS patients: results of a large prospective randomised treatment trial. *AIDS*, **6**, 301–305.
32. Black, J.R., Feinberg, J., Murphy, R.L.L. et al. (1994) Clindamycin and primaquone therapy for mild-to-moderate episodes of *Pneumocystis carinii* pneumonia in patients with AIDS. ACTG 044. *Clin. Infect. Dis.*, **18**, 905–913.
33. Medina, I., Mills, J., Leoung, G. et al. (1990) Oral therapy for PCP in the acquired immunodeficiency syndrome. *N. Engl. J. Med.* **323**, 771–782.
34. Sarffrin, S., Finkelstein, D.M., Feinberg, J. et al. (1996) Comparison of three regimens for the treatment of mild to moderate PCP in patients with AIDS. *Ann. Intern. Med.* **124**, 792–802.
35. US Public Health Service Task Force on Antipneumocystis Prophylaxis in Patients with Human Immunodeficiency Virus Infection (1993) Recommendations for prophylaxis against *Pneumocystis carinii* pneumonia for persons infected with human immunodeficiency virus. *J. Acq. Immune Defic. Syndr.*, **6**, 46–55.
36. Hardy, W.D., Feinberk, J., Finkelstein, D.M. et al. (1992) A controlled trial of trimethoprim-sulfamethoxazole or aerosolized pentamidine for secondary prophylaxis of *Pneumocystis carinii* pneumonia in patients with the acquired immunodeficiency syndrome. *N. Engl. J. Med.*, **327**, 1842–1848.
37. Schneider, M.M.E., Hoepelman, A.I.M., Jan Karel M. et al. (1992) A controlled trial of aerosolized pentamidine or trimethoprim-sulfamethoxazole as primary prophylaxis against *Pneumocystis carinii* pneumonia in patients with human immunodeficiency virus infection. *N. Engl. J. Med.*, **327**, 1836–1841.
38. Masur, H. (1992) Prevention and treatment of *Pneumocystis pneumonia*. *N. Engl. J. Med.*, **327**, 1853–1860.
39. Bozzette, S.A., Findelstein, D.M., Spector, S.A., et al. (1995) A randomised trial of three antipneumocystis drugs in patients with advanced HIV infection. *N. Engl. J. Med.*, **332**, 693–699.
40. USPHS/IDSA (1997) Guidelines for the prevention of opportunistic infections in persons infected with human immunodeficiency virus. *Morbid. Mortal. Wkly Rep.*, 461–478.
41. Toma, E., Poisson, M., Phaneuf, D. et al. (1989) Clindamycin with primaquine for *Pneumocystis carinii* pneumonia. *Lancet*, **i**, 1046–1048.
42. Ruf, B. and Pohle, H.D. (1989) Clindamycin/primaquine for *Pneumocystis carinii* pneumonia. *Lancet*, **ii**, 626–627.
43. Leoung, G.S., Mills, J., Hopewell, P.C. et al. (1986) Dapsone-trimethoprim for *Pneumocystis carinii* pneumonia in the acquired immunodeficiency syndrome. *Ann. Intern. Med.*, **105**, 45–48.
44. Falloon, J., Kovacs, J., Hughes, W. et al. (1991) A preliminary evaluation of 566C80 for the treatment of *Pneumocystis* pneumonia in patients with the acquired immunodeficiency syndrome. *N. Engl. J. Med.*, **325**, 1534–1538.
45. El Sadr, W., Murphy, R., Lustein-Hawk, R. et al. (1997) Atovaquone versus dapsone in the prevention of PCP in patients intolerant to trimethoprim and or sulfamethoxole. CPCRA 034/ACTG 769. IDSA 35th Meeting, San Francisco.
46. Ruskin, J. and LaRiviere, M. (1991) Low-dose co-trimoxazole for prevention of *Pneumocystis carinii* pneumonia in human immunodeficiency virus disease. *Lancet*, **337**, 468–471.
47. Kemper, C.A., Tucker, R.M., Lang, O.S. et al. (1990) Low-dose dapsone prophylaxis of *Pneumocystis carinii* pneumonia in AIDS and AIDS-related complex. *AIDS*, **4**, 1145–1148.
48. Clotet, B., Sirera, G., Romeu, J.M. et al. (1991) Twice-weekly dapsone-pyrimethamine for preventing PCP and cerebral toxoplasmosis. *AIDS*, **5**, 601–602.
49. Oksenhendler, E., Charreau, I. and Tournerie, C.I. (1994) *Toxoplasma gondii* infection in advanced HIV infection. *AIDS*, **8**, 483–487.
50. Luft, B.J., Hafner, R., Dorzun, A.H. et al. (1993) Toxoplasmic encephalitis in patients with the acquired immunodeficiency syndrome. *N. Engl. J. Med.*, **329**, 995–1000.
51. Grant, I.H., Gold, J.W.M., Rosenblum, M. et al. (1990) *Toxoplasma gondii* serology in HIV-infected patients: the development of central nervous system toxoplasmosis. *AIDS*, **4**, 519–521.
52. Post, M.J., Kursunoglu, S.J., Hensley, G.T. et al. (1985) Cranial CT in acquired immunodeficiency syndrome: spectrum of diseases and optimal contrast enhancement technique. *Am. J. Radiol.*, **145**, 929–940.
53. Levy, R.M., Mills, C.M., Posin, J.P. et al. (1990) The efficacy and clinical impact of brain imaging in neurologically symptomatic AIDS patients: a prospective CT/MRI study. *J. Acq. Immune Defic. Syndr.*, **3**, 461–471.
54. Grant, I.H., Gold, J.W.M., Rosenblum, M. et al. (1990) *Toxoplasma gondii* serology in HIV-infected patients: the development of central nervous system toxoplasmosis. *AIDS*, **4**, 519–521.
55. Danneman, B., McCutchan, J.A., Israelski, D. et

al. (1992) Treatment of toxoplasmic encephalitis in patients with AIDS. *Ann. Intern. Med.*, **116**, 33–43.
56. Selik, R.M., Starcher, E.T. and Curran, J.W. (1987) Opportunistic diseases reported in AIDS patients: frequencies, associations and trends. *AIDS*, **1**, 175–182.
57. Carr, A., Tindall, X.X., Brew, B.J. *et al.* (1992) Low-dose trimethoprim-sulfamethoxozole prophylaxis for toxoplasma encephalitis in patients with AIDS. *Ann. Intern. Med.*, **117**, 106–111.
58. Klein, R.S., Harris, C.A., Small, C. *et al.* (1984) Oral candidiasis in high risk patients as the initial manifestation of the acquired immunodeficiency syndrome. *N. Engl. J. Med.*, **311**, 354–358.
59. Pons, V., Greenspan, D., Debruin, M. and the Multicenter Study Group (1993) Therapy for oropharyngeal candidiasis in HIV infected patients: a randomized prospective multi-center study of oral fluconazole vs clotrimazole troches. *J. Acq. Immune Defic. Syndr.*, **6**, 1311–1316.
60. Glatt, A.E. (1993) Editorial – therapy for oropharyngeal candidiasis in HIV-infected patients. *J. Acq. Immune Defic. Syndr.*, **6**, 1315–1318.
61 Schuman, P., Capps, L., Peng, G. *et al.* (1997) Weekly fluconazole for the prevention of mucosal candidiasis in women with HIV infection. A randomized, double-blind, placebo-controlled trial. Terry Beirn Community Programs for Clinical Research on AIDS. *Ann. Intern. Med.*, **126**(9), 689–696.
62. Wood, J.J. (1994) Oral azole drugs as systemic antifungal therapy. *N. Engl. J. Med.*, **330**, 263–272.
63. Wheat, J.L. (1993) Diagnosis and management of fungal infections in AIDS. *Curr. Opin. Infect. Dis.*, **6**, 617–627.
64. British Society for Antimocrobial Chemotherapy Working Party (1992) Antifungal chemotherapy in patients with acquired immunodeficiency syndrome. *Lancet*, **340**, 648–651.
65. Moskovitz, B.L., Wilcox, C.M., Daroviche, R. *et al.* (1996) Itraconazole solution compared with fluconazole for the treatment of oesophageal candidiasis. Abstract MO B116. XI International Conference on AIDS, Vancouver.
66. Galgiani, J.N. (1987) Antifungal susceptibility tests. *Antimicrob. Agents Chemother.*, **31**, 1867–1870.
67. Cameron, M.L., Schell, W.A., Bruck, S. *et al.* (1993) Correlation of *in vitro* fluconazole resistance of Candida isolates in relation to therapy and symptoms of individuals seropositive for human immunodeficiency virus type 1. *Antimicrob. Agents Chemother.*, **37**, 2449–2453.
68. Review (1993) Ambisome – liposomal amphotericin B. *Drugs Therapeut. Bull. Rev.*, **31**, 93–96.
69. Dismukes, W.E. (1988) Cryptococcal meningitis in patients with AIDS. *J. Infect. Dis.*, **157**, 624–628.
70. Powderly, W.G., Saag, M.S., Cloud, G.A. *et al.* (1992) A controlled trial of fluconazole or amphotericin B to prevent relapse of cryptococcal meningitis in patients with the acquired immunodeficiency syndrome. *N. Engl. J. Med.*, **326**, 793–798.
71. Van der Horst, C., Saag, M.S., Cloud, G.A. *et al.* (1997) Treatment of cryptococcal meningitis associated with the acquired immune deficiency syndrome. *N. Engl. J. Med.*, **337**, 15–21.
72. Wheat, L.J., Connolly-Stringfield, P., Baker, R.L. *et al.* (1993) Disseminated histoplasmosis in the acquired immunodeficiency syndrome: clinical findings, diagnosis and treatment, and review of the literature. *Medicine*, **69**, 361–374.
73. Conces, D.J., Stockberger, S.M., Tarver, R.D. *et al.* (1993) Disseminated histoplasmosis in AIDS: findings on chest radiographs. *Am. J. Roentgenol.*, **160**, 15–19.
74. Wheat, L.J., Hafner, R.E., Ritchie, M. *et al.* (1992) Itraconazole is effective treatment for histoplasmosis in AIDS: prospective multicenter non-comparative trial [abstract]. *Prog. Abstr. Intersci. Conf. Antimicrob. Agents Chemother.*, 1206.
75. Houston, S. (1994) Histoplasmosis and pulmonary involvement in the tropics. *Thorax*, **49**, 598–601.
76. Wheat, J., Hafner, R. and Korzun, A.H. (1995) Itraconazole treatment of disseminated histoplasmosis in patients with the acquired immunodeficiency syndrome. *Am. J. Med.*, **98**, 336–342.
77. Ampel, N.M., Dols, C.L., Galgiani, J.N. *et al.* (1993) Coccidioidomycosis during human immunodeficiency virus infection: results of a prospective study in a coccidioidal endemic area. *Am. J. Med.*, **94**, 235–240.
78. Fish, D.G., Ampel, N.M., Galgiani, J.N. *et al.* (1993) Coccidioidomycosis during human immunodeficiency virus infection: a review of 77 patients. *Medicine*, **69**, 384–391.
79. Catanzaro, A., Galgiani, J.N., Levine, B.E. *et al.* (1995) Fluconazole in the treatment of chronic pulmonary and nonmeningeal disseminaged coccidiomycosis. NIAD Mycoses Study Group. *Am. J. Med.*, **98**, 249–256.
80. Denning, D.W., Follansbee, S.E., Scolaro, M. *et al.* (1991) Pulmonary aspergillosis in the acquired immunodeficiency syndrome. *N. Engl. J. Med.*, **324**, 654–662.

81. Minamoto, G.Y., Barlam, T.F. and Vander Els, N.J. (1992) Invasive aspergillosis in patients with AIDS. *Clin. Infect. Dis.*, **14**, 66–74.
82. Pappas, P.G., Pottage, J.C., Powderly, W.G. *et al.* (1992) Blastomycosis in patients with the acquired immunodeficiency syndrome. *Ann. Intern. Med.*, **116**, 847–853.
83. Bradsher R.W. (1996) Histoplasmosis and blastomycosis. *Clin. Infect. Dis.*, **22**(suppl 2), S102–111.
84. Horsburgh, C.R. (1991) *Mycobacterium avium* complex infection in the acquired immunodeficiency syndrome. *N. Engl. J. Med.*, **324**, 1332–1338.
85. Nightingale, S.D., Bird, L.T., Southern, P.M. *et al.* (1992) Incidence of *Mycobacterium avium intracellulare* complex bacteraemia in HIV positive patients. *J. Infect. Dis.*, **165**, 1082–1085.
86. Inderlied, C.B., Kemper, C.A. and Bermudez, L.E.M. (1993) The *Mycobacterium avium* complex. *Clin. Microbiol. Rev.*, **6**, 266–310.
87. Horsburgh, C.R., Havelik, J., Ellis, D.A. *et al.* (1991) Survival of patients with acquired immunodeficiency syndrome and disseminated *Mycobacterium avium* complex infection with and without antimycobacterial chemotherapy. *Am. Rev. Respir. Dis.*, **144**, 557–559.
88. Chaisson, R.E., Moore, R.D., Richmond, D.D. *et al.* (1992) Zidovudine Epidemiology Study Group. Incidence and natural history of *Mycobacterium avium* complex infection in patients with advanced human immunodeficiency virus disease treated with zidovudine. *Am. Rev. Respir. Dis.*, **146**, 285–289.
89. Dautzenberg, B., Saint Marc, T., Meyohas, C. *et al.* (1993) Clarithromycin and other antimicrobial agents in the treatment of disseminated *Mycobacterium avium* infections in patients with acquired immune deficiency syndrome. *Arch. Intern. Med.*, **153**, 368–372.
90. (1993) Recommendations on prophylaxis and therapy for disseminated *Mycobacterium avium* complex for adults and adolescents infected with human immunodeficiency virus. *Morbid. Mortal. Wkly Rep.*, **42**.
91. Shafran, S.D., Singer, J. and Zarowny, D.P. (1996) Comparison of two regimens for Mycobacterium avium complex bacteraemia in AIDS. *N. Engl. J. Med.*, **335**, 377–383.
92. Nightingale, S.D., Cameron, D.W., Gordin, F.M. *et al.* (1993) Two controlled trials of rifabutin prophylaxis against *Mycobacterium avium* complex infection in AIDS. *N. Engl. J. Med.*, **329**, 828–833.
93. Havlir, D.V., Dube, M.P. and Sattler, F.R. (1996) Prophylaxis against disseminated Mycobacterium avium complex with weekly azithromycin, daily rifabutin or both. *N. Engl. J. Med.*, **335**, 392–398.
94. Pierce, M., Crampton, S., Henry, D. *et al.* (1996) A randomized trial of clarithromycin as prophylaxis against disseminated mycobacterium avium in patients with acquired immune deficiency syndrome. *N. Engl. J. Med.*, **335**, 384–391.
95. Horsburgh, C.R. and Selik, R.M. (1989) The epidemiology of disseminated nontuberculosis mycobacterial infection in the acquired immunodeficiency syndrome (AIDS). *Am. Rev. Respir. Dis.*, **139**, 4–7.
96. Research Committee, British Thoracic Society. (1994) *Mycobacterium kansasii* pulmonary infection: a prospective study of the results of nine months of treatment with rifampicin and ethambutol. *Thorax*, **49**, 42–445.
97. Barnes, P.F., Boch, A.B., Davidson, P.T. *et al.* (1991) Tuberculosis in patients with human immunodeficiency virus infection. *N. Engl. J. Med.*, **324**, 1644–1650.
98. Centers for Disease control year testing and preventive therapy for HIV infected persons. Revised recommendations. *Morbid. Mortal. Wkly Rep.*, **46**, RR1–15.
99. L.P. Ormerod, for a Subcommittee of the Joint Tuberculosis Committee (1990) Chemotherapy and management of tuberculosis in the United Kingdom: recommendations of the Joint Tuberculosis Committee of the British Thoracic Society. *Thorax*, **45**, 403–408.
100. Subcommittee of the Joint Tuberculosis Committee of the British Thoracic Society (1992) Guidelines on the management of tuberculosis and HIV infection in the United Kingdom. *Br. Med. J.*, **304**, 1231–1233.
101. Centers for Disease Control and Prevention (1993) Initial therapy for tuberculosis in the era of multidrug resistance. Recommendations of the Advisory Council for the Elimination of Tuberculosis. *Morbid. Mortal. Wkly Rep.*, **42**.
102. Wood, A.J.J. (1993) Treatment of multidrug-resistant tuberculosis. *N. Engl. J. Med.*, **329**, 784–791.
103. Wilkinson, D.H. (1994) High-compliance tuberculosis treatment programme in a rural community. *Lancet*, **343** 647–648.
104. Weis, S.E., Slocum, P.C., Blais, F.X. *et al.* (1994) The effect of directly observed therapy on the rates of drug resistance and relapse in tuberculosis. *N. Engl. J. Med.*, **330**, 1179–1184.
105. Pape, J.W., Jean, S.S., Ho, J.L. *et al.* (1993) Effect of isoniazid prophylaxis on incidence of

active tuberculosis and progression of HIV infection. *Lancet*, **342**, 268–272.
106. Centers for Disease Control and Prevention (1990) Guidelines for preventing the transmission of tuberculosis in health-care settings, with special focus on HIV-related issues. *Morbid. Mortal. Wkly Rep.*, **39**.
107. PHLS Communicable Disease Surveillance Centre (1994) Joint statements on preventing tuberculosis. *Commun. Dis. Wkly Rep.*, **4** (8).
108. Jabs, D.A., Enger, C. and Bartlett, J.G. (1989) Cytomegalovirus retinitis and acquired immunodeficiency syndrome. *Arch. Ophthalmol.*, **107**, 75–80.
109. Desmet, M.D and Nussenbatt, R.B. (1991) Ocular manifestations of AIDS. *J. Am. Med. Assoc.*, **266**, 3019–3022.
110. Crowe, S.M., Carlin, J.B., Stewart, K.I et al. (1991) Predictive value of CD4 lymphocyte numbers for the development of opportunistic infections and malignancies in HIV-infected persons. *J. Acq. Immune. Defic. Syndr.*, **4**, 770–776.
111. Pertel, P., Hirschtick, R., Phair, J. et al. (1992) Risk of developing cytomegalovirus retinitis in persons infected with the human immunodeficiency virus. *J. Acq. Immune. Defic. Syndr.*, **5**, 1069–1074.
112. Spector, S.A., Weingeist, T., Pollard, R.B. et al. (1993) A randomized, controlled study of intravenous ganciclovir therapy for cytomegalovirus peripheral retinitis in patients with AIDS. *J. Infect. Dis.*, **168**, 557–563.
113. Luckie, A. and Everett, A. (1993) Diagnosis and management of cytomegalovirus retinitis in AIDS. *Curr. Opin. Ophthalmol.*, **4**, 81–89.
114. Studies of Ocular Complications of AIDS Research Group, in Collaboration with the AIDS Clinical Trials Group (1992) Mortality in patients with the acquired immunodeficiency syndrome treated with either foscarnet or ganciclovir for cytomegalovirus retinitis. *N. Engl. J. Med.*, **326**, 213–220.
115. Jacobson, M.A., Causey, D., Polsky, B. et al. (1993) A dose-ranging study of daily maintenance intravenous foscarnet therapy for cytomegalovirus retinitis in AIDS. *J. Infect. Dis.*, **168**, 444–448.
116. Balfour, H.H., Chase, B.A., Stapleton, J.T. et al. (1989) A randomized, placebo-controlled trial of oral acyclovir for the prevention of cytomegalovirus disease in recipients of renal allografts. *N. Engl. J. Med.*, **320**, 1381–1387.
117. Cochereau-Massin, I., Lehoang, P., Lautier-Frau, M. et al. (1987) Efficacy and tolerance of intravitreal ganciclovir (dihydroxy propoxymethyl guanine) for cytomegalovirus retinitis in a patient with AIDS. *Am. J. Ophthalmol.*, **103**, 17–23.
118. Smith, T.J., Pearson, P.A., Blandford, D.L. et al. (1992) Intravitreal sustained-release ganciclovir. *Arch. Ophthalmol.*, **110**, 255–258.
119. Drew, W.L., Miner, R.C., Busch, D.F. et al. (1991) Prevalence of resistance in patients receiving ganciclovir for serious cytomegalovirus infection. *J. Infect. Dis.*, **163**, 716–719.
120. Dieterich, D.T., Kotler, D.P., Busch, D.F. et al. (1993) Ganciclovir treatment of cytomegalovirus colitis in AIDS: a randomized double-blind, placebo controlled multicenter study. *J. Infect. Dis.*, **167**, 278–282.
121. Young, S.K. and Hollander, H. (1993) Polyradiculopathy due to cytomegalovirus: report of two cases in which improvement occurred after prolonged therapy and review of the literature. *Clin. Infect. Dis.*, **17**, 32–37.
122. Morgello, S., Cho, E.S., Nielsen, S. et al. (1987) Cytomegalovirus encephalitis in patients with acquired immunodeficiency syndrome. An autopsy study of thirty cases and a review of the literature. *Hum. Pathol.*, **18**, 289–297.
123. Millar, A.B., Patou, G., Miller, R.F. et al. (1990) Cytomegalovirus in the lungs of patients with AIDS. Respiratory pathogen or passenger? *Am. Rev. Respir. Dis.*, **141**, 1474–1477.
124. Spector, S.A., Pilcher, M., Lamy, P. et al. (1996) PCR of plasma for CMV DNA identifies HIV-infected persons most likely to benefit from oral ganciclovir prophylaxis. Abstract ThB302. XI International Conference on AIDS, Vancouver.
125. Spector, S.A., McKinley, G.F., Lalezari, J.P. et al. (1996) Oral ganciclovir for the prevention of CMV disease in persons with AIDS. *N. Engl. J. Med.* **334**, 1491–1497.
126. Brosgart, C.L., Craig, C., Hillman, D. et al. (1996) CPCRA 023: final results from a randomised placebo-controlled trial of the safety and efficacy of oral ganciclovir for the prophylaxis of CMV retinal and gastrointestinal muscosal disease. Abstract ThB301. XI International Conference on AIDS, Vancouver.
127. Feinberg, J., Cooper, D., Hurwitz, S. et al. (1995) Phase III study of valaciclovir for CMV prophylaxis in patients with advanced HIV disease. Abstract 112. 35th ICAAC, San Francisco.
128. McGowan, I., Hawkins, A.S. and Weller, I.V.D. (1993) The natural history of cryptosporidial diarrhoea in HIV-infected patients. *AIDS*, **7**, 349–354.
129. Current, W.L. and Garcia, L.S. (1993)

Cryptosporidiosis. *Clin. Microbiol. Rev.*, **4**, 325–358.
130. Fichtenbaum, C.J., Richards, D.J. and Powderly, W.G. (1993) Use of paromomycin for treatment of crytosporidiosis in patients with AIDS. *J. Infect. Dis.*, **16**, 298–300.
131. Mileno, M.D., Tashima, K., Farrar, D. *et al.* (1997) Resolution of AIDS-related opportunistic infections with addition of protease inhibitor treatment. 4th Conf. Retro. and Opportun. Infect. (Jan 22–26): 129 (Abstract no. 355).
132. Soave, R. and Johnson, W.D. Jr (1988) Cryptosporidium and *Isospora belli* infections. *J. Infect. Dis.*, **157**, 225–229.
133. Pape, J.W. and Johnson, W.D. Jr (1991) *Isospora belli* infections. *Prog. Clin. Parasitol.*, **2**, 119–127.
134. Polsky, B., Gold, J.W.M., Whimbe, E. *et al.* (1986) Bacterial pneumonia in patients with the acquired immunodeficiency syndrome. *Ann. Intern. Med.*, **104**, 38–41.
135. Selwyn, P.A., Feingold, A.R., Harel, D. *et al.* (1988) Increased risk of bacterial pneumonia in HIV-infected drug users without AIDS. *AIDS*, **2**, 267–272.
136. Mundy, L., Autwater, P., Butron, A. *et al.* (1991) Etiology of community acquired pneumonia (CAP): HIV+ vs HIV- patients. Abstracts Programs and Abstracts of Thirty-first Interscience Conference on Antimicrobial Agents and Chemotherapy.
137. Yamaguche, E., Charache, P. and Chaisson, R.E. (1991) Increasing incidence of pneumococcal infections (PI) associated with HIV infection in an inner-city hospital, 1985–1989. Abstract 619. 1990 World Conference on Lung Health, Boston, Massachusetts.
138. Celem, C.L., Chaisson, R.E., Rutherford G.W. *et al.* (1987) Incidence of salmonellosis in patients with AIDS. *J. Infect. Dis.*, **156**, 998–1001.
139. Levine, W.C., Buehler, J.W., Bean, N.H. *et al.* (1991) Epidemiology of nontyphoidal *Salmonella* bacteremia during the human immunodeficiency virus epidemic. *J. Infect. Dis.*, **164**, 81–87.

17 The diagnosis and management of tumours in patients with AIDS

F.C. Boag and E.M. Carlin

INTRODUCTION

It is thought that immune surveillance plays an important role in the prevention of malignancies and therefore it might be expected that immunosuppression would give rise to an increased incidence of malignant disease. This has certainly been seen in the case of iatrogenic immunosuppression, in congenital immunodeficiency syndromes and it is also the case in AIDS. Some malignancies occur with greater frequency than others, particularly Kaposi's sarcoma and lymphomas, but cervical cancer, lung cancer, squamous carcinoma of the anus, testicular tumours, glioma and squamous cell tumours of the head and neck are also seen. This may be because they have an infective aetiology (for instance, the Epstein–Barr virus has been implicated in the pathogenesis of AIDS-related non-Hodgkin's lymphoma). Or it could be that some malignancies have a longer natural history than others and we are only seeing the ones that develop quickly. In the survivors of Hiroshima and Nagasaki, there was an increased incidence first of leukaemia and thyroid tumours, then lymphoma, then breast cancer and later other solid tumours. If this is the case, then perhaps the picture may change as increasing awareness of how to manage the opportunistic infections that occur in AIDS enables the patients to live longer with profound immunosuppression.

In an HIV-positive individual, Kaposi's sarcoma, non-Hodgkin's lymphoma and cervical cancer are AIDS-defining illnesses.

Standard recommendations are that all chemotherapy and radiotherapy must be supervised and directed by an oncologist or radiotherapist. We endorse and adhere to these guidelines.

KAPOSI'S SARCOMA

Kaposi's sarcoma (KS) was first described by Moritz Kaposi in 1872. It was originally found in Ashkenazi Jews and elderly men of Mediterranean origin, in whom it runs an indolent course (classic KS).

Endemic KS, first described in the 1950s, comprises a spectrum of KS ranging from indolent skin lesions to aggressive disease of the skin, lymph nodes and multiple organs. Endemic KS occurs in equatorial Africa, mainly in young black males and occasionally in children.

KS also occurs in people who are iatrogenically immunosuppressed, for instance for renal transplantation.

KS associated with HIV infection is known as epidemic KS. Increased notification to the CDC of KS and of pneumocystis pneumonia occurring in young males in the early 1980s led to the discovery of AIDS.

KS occurs most frequently in homosexual males, but also relatively frequently in Africans. It is by far the most commonly occurring tumour in AIDS. In 1981 in the USA, KS occurred in approximately 48% of homosexual men with AIDS.[1] It is now less frequently the presenting illness but occurs later in the disease (Table 17.1).

The biggest risk group for KS is homosexual men, especially those from California or having partners from New York. There is also a reported link with oroanal sex and KS has been known to occur in HIV-negative gay men. In this group KS is benign and the patients have no evidence of immunosuppression. Caucasian women who develop KS are most likely to have acquired HIV from a bisexual male. This led to the suggestion that a sexually transmitted agent is a cofactor in the development of this tumour (Table 17.2). The cofactor is human herpes virus type 8 (HHV 8); its mode of transmission is not clear.

Table 17.1 Variants of KS

Type	Patient group	Clinical course
Classic	Older males from the Eastern Mediterranean	Indolent
Endemic	Africa – young adult males	Indolent or progressive
	Africa – children	Usually indolent
Iatrogenic	Iatrogenic immunosuppression	Usually indolent
Epidemic	HIV+ individuals	Depends upon degree of immunosuppression. Rapidly progressive in advanced immunosuppression or pregnancy

Table 17.2 Risk groups for KS in AIDS

Homosexual males	Highest risk group, particularly men living in New York or California or with partners from New York. Also a link with oroanal sex
Africans	
Heterosexuals	Seen mainly in women who have acquired HIV from a bisexual male
Intravenous drug users	
Transfusion recipients	

Pathology

Immunosuppression plays an important role in the development of KS, as evidenced by the relatively high incidence in iatrogenic immunosuppression and the fact that it regresses in about 30% of cases where the immunosuppressive drug is withdrawn. There is some evidence that various growth factors and cytokines can promote the development of KS, such as interleukins 1 and 6, tumour necrosis factor, granulocyte-macrophage colony-stimulating factor, basic fibroblast growth factor and others, many of which are released by the lesions themselves. HIV itself may also be a cofactor. KS is thought to originate from vascular or possibly lymphatic endothelium. Histological examination reveals the presence of abnormal, dilated vascular channels and spindle-shaped stromal cells, as well as interstitial infiltration by inflammatory cells and extravasated erythrocytes. HHV 8 has been found in KS in every risk group and the presence of HHV 8 in HIV-positive persons is a predictive factor for the development of KS.

Staging

The two staging systems commonly in use are those suggested by Mitsuyasu and Groopman[2] (Table 17.3) and by the AIDS clinical trials group[3] (Table 17.4).

Clinical features

KS can occur at any stage of HIV infection, irrespective of the CD4 count or prior AIDS-defining illnesses. There is a tendency for the tumour to be stable when the CD4 count is high, becoming more widespread and involving the viscera as the count falls or as opportunistic infections appear. One of the most striking features of KS is its psychosocial impact. It is often disfiguring and causes distress as it advertises to the

Table 17.3 KS staging (Mitsuyasu and Groopman)

Stage I	Cutaneous disease, less than 10 lesions, confined to one anatomical area
Stage II	Cutaneous disease, more than 10 lesions and/or involvement of more than one area
Stage III	Visceral disease without cutaneous involvement
Stage IV	Visceral and cutaneous disease

Further subdivided into A or B depending on the absence or presence of 'B'-symptoms (unexplained fevers, weight loss of >10%, night sweats) or of previous opportunist infection

Table 17.4 KS staging (ACTG)

	Good risk (all of the following)	Poor risk (any of the following)
Tumour	Confined to skin and/or lymph nodes and/or minimal oral disease*	Tumour-associated oedema or ulceration. Extensive oral KS. Gastrointestinal KS. KS in other non-nodal viscera
Immune system	CD4 count >200 × 10⁶/l	CD4 count <200 × 10⁶/l
Systemic illness	No history of opportunistic infections or candida. No 'B'-symptoms.** Karnofsky performance status at least 70%	History of opportunistic infections and/or candida. 'B'-symptoms present. Performance status less than 70%. Other HIV-related disease (e.g. neurological disease or lymphoma).

* Minimal oral disease is non-nodular KS confined to the palate.
** 'B'-symptoms are unexplained fever, night sweats, >10% involuntary weight loss or diarrhoea persisting for >2 weeks

Table 17.5 Skin biopsy

- Choose a lesion which appears most typical of KS.
- Choose a lesion which the patient wants to be rid of.
- Avoid oedematous/necrotic lesions.
- Infiltrate with local anaesthetic.
- Remove the lesion in an ellipse of surrounding skin.
- Suture.

world that the patient has AIDS. KS is a multifocal disease and can affect the skin, oropharyngeal mucosa, the conjunctiva, the gastrointestinal tract, the respiratory tract, the lymph nodes and other organs (but not the central nervous system and very rarely the skeleton).

Mucocutaneous KS

The usual manifestation is on the skin or mucous membranes of the mouth. The typical appearance is of small pink or purple macules, which enlarge into nodules and eventually into plaques. The lesions are often surrounded by areas of bruising and may be associated with oedema, due to involvement of the lymphatics. Advanced lesions may ulcerate. KS is usually painless; pain arises when the tumour is necrotic or ulcerated or compressing other structures. There is also a tendency for it to occur on the extremities (ear, nose, limbs) and it sometimes exhibits the Koebner phenomenon, e.g. after shingles.

There is a wide differential diagnosis, including bruises, malignant melanoma, secondary syphilis, haemangioma, basal cell carcinoma and bacillary angiomatosis. A skin biopsy is recommended; KS does not bleed more than any other skin biopsy and provided oedematous or necrotic areas are avoided, the scar heals well (Table 17.5).

Lymph nodes

KS presents in lymph nodes as firm lymphadenopathy, with or without pain. If pain is present this is usually due to rapid growth, pressure on adjacent structures or infection. It may or may not be associated with cutaneous disease. The areas drained by involved nodes become oedematous, a frequent occurrence on the feet, penis and in periorbital tissue. If the face and neck are involved patients can present with rapid onset of facial swelling, tracheal compression or the inability to open the eyes or mouth. Involvement of the abdominal lymph nodes can present with abdominal pain and obstruction due to pressure on the gastrointestinal tract.

Gastrointestinal system

At postmortem up to 50% of patients with cutaneous KS have been found to have gastrointestinal (GI) involvement, in the majority of cases without previous symptoms. Notably, GI disease can occur in the absence of cutaneous KS. It can present with bleeding (haematemesis, melaena, rectal bleeding or anaemia), complete or partial obstruction due to pedunculated lesions, perforation or protein-losing enteropathy due to involvement of the GI lymphatics (Table 17.6).

The diagnosis is usually made at upper GI endoscopy, sigmoidoscopy or colonoscopy. If these are negative, CT scanning, barium studies or laparoscopy proceeding to laparotomy may be required to identify the cause of the GI symptoms.

The endoscopic appearance of KS is of vascular nodules, papules or telangiectasia and ulcers can occur. KS is initially submucosal, later involving the mucosa

Table 17.6 Symptoms of gastrointestinal KS

Asymptomatic
Bleeding
Bowel obstruction
Partial/intermittent obstruction
Perforation
Protein-losing enteropathy

itself and then invading deeply. Biopsy can usually be performed without excessive bleeding, but care must be taken when the lesions are already haemorrhagic.

Pulmonary KS

Postmortem studies have shown pulmonary involvement in 20–50% of patients with KS. Again, pulmonary KS is usually seen in conjunction with cutaneous disease but it can be the only site. Pulmonary KS presents with breathlessness, often with a period of moderate symptoms, then a sudden onset of dyspnoea, or with chest pain (which may be pleuritic or dull and boring in nature), fever, haemoptysis and/or haemopneumothorax. The patient is frequently hypoxic and the chest X-ray may show discrete nodules, effusions and reticulonodular shadowing due to interstitial infiltration, or it may be normal. A bronchoscopy is often diagnostic, with erythema or purple plaques visible in the main bronchi or bronchial tree. The diagnosis may be confirmed by histological examination of a transbronchial biopsy if required (as the lesions bleed freely, biopsy is deferred if the clinical diagnosis is certain). Occasionally, open lung biopsy is required. Although KS in other sites is rarely life-threatening, advanced pulmonary disease has a poor prognosis, usually about three months (Table 17.7), prior to the use of highly active antiretroviral therapy (HAART).

Other organs

Visceral KS is often asymptomatic and the diagnosis is made when a space-occupying lesion is found on radiographic examination, e.g. of the liver or kidney. It may present with pain due to compression (e.g. of the mediastinum) or with organ failure (e.g. of the adrenal glands or pancreas) or as a palpable mass (e.g. in the testicle). Diagnosis is by biopsy.

'B'-symptoms

KS is sometimes diagnosed during investigation for 'B'-symptoms (fever, night sweats, malaise and weight loss), for example revealed by chest X-ray, endoscopy, liver ultrasound, abdominal CT, etc. Occasionally no other cause is found and the symptoms are attributed to the KS.

Table 17.7 Pulmonary KS – investigations

Blood gases
Chest X-ray
Bronchoscopy
Transbronchial biopsy if necessary
Open lung biopsy if necessary

Treatment

KS is not usually life-threatening (except for pulmonary disease), but it can be painful, psychologically disturbing and the oedema can be troublesome.

KS is not curable, the underlying immunosuppression remains. It is important to bear this in mind and not to make the situation worse with the side effects of treatment. There is a variety of treatment options which in most cases offer good palliation.

Early disease, if HAART is not indicated, is best treated by camouflage or by local therapies such as excision, radiotherapy or intralesional injection of dilute chemotherapy drugs. HAART is emerging as an effective measure in controlling moderate KS, probably due to immune reconstitution and/or control of levels of HHV 8. Widespread, advanced or rapidly progressing disease may respond to HAART, but systemic therapy, often with cytotoxic drugs, may be needed in addition if the patient has presented late.

Local treatment is used for cosmetically troublesome or painful lesions. Excision, cryotherapy, radiotherapy, intralesional chemotherapy and various experimental regimens can be offered. Care must be taken as radiotherapy to KS can result in hyperpigmented patches in dark- or olive-skinned people which can look worse than the original disease. In addition, local treatments are used in patients with limited KS in the full knowledge that it will not prevent new lesions from developing. Very sick patients can get symptomatic relief even if they are too ill to have systemic treatment.

Systemic therapy is required for widespread, visceral and rapidly progressive KS. There is considerable overlap between the treatment options; radiotherapy may be given concurrently with systemic treatment to give rapid relief of a local problem.

Camouflage

Advice on cosmetic camouflage with make-up and restyling of hair, facial hair and clothes is vital to reduce the psychological impact of KS whilst awaiting response from specific therapies.

Infection

If there is infection present, swabs and blood cultures should be taken, then antibiotics and local dressings commenced.

Diuretics and steroids

Oedema should be treated with elevation, diuretics (e.g. frusemide) and, for more severe cases, steroids (e.g. prednisolone 30 mg daily; this should be given as a short course and with suitable prophylactic cover against *Mycobacterium tuberculosis* such as isoniazid 300 mg daily).

Antiretroviral drugs

It is not uncommon to see a response after starting antiretroviral drugs. There is therefore an argument for delaying other treatments and considering antiretroviral therapies as first-line treatment, provided the KS is not too severe.

Cryotherapy

Cryotherapy is useful for stable KS and treatment of cosmetically troublesome lesions. Two freeze/thaw cycles lasting 11–60 per cycle, repeated every three weeks for three treatments per lesion gave a greater than 50% 'cosmetic improvement' in 20 patients.[4] The side effects seen were blistering and local pain and hypopigmentation in dark-skinned subjects. Patients without prior opportunistic infections tended to do best. In our own centre, infection has been noted when treating patients with very low CD4 counts and so we now exclude this patient group.

Laser therapy

The argon laser can be used to photocoagulate cutaneous KS. However, it is extremely time consuming and is therefore not often used. The carbon dioxide laser vaporizes the KS but is also time consuming.

All-trans retinoic acid

Topical application of All-trans retinoic acid 1% daily for three months resulted in fading of colour and some flattening of lesions in eight patients. The side effects included dryness of the skin and inflammation.[5] Like all retinoids, this medication should be avoided by women if they are likely to become pregnant as it is teratogenic.

α-Interferon

IFNα-2a and IFNα-2b can be used both locally and systemically. They have been shown to produce prolonged remissions in some subjects with good-prognosis KS (40% with high dose, 10% with low dose) and to have a synergistic effect with antiretroviral drugs. However, there is a high incidence of significant morbidity with severe flu-like symptoms and significant neutropenia. This toxicity and relatively low response rates preclude its use in many centres. The dose for systemic therapy is 10–20 million units daily.

Interferons plus zidovudine

IFNα 10–18 million units daily administered by IM or SC injection and zidovudine 600 mg daily have been used for KS. Response rates are about 40% in patients with CD4 >200 × 10^6/l. Neutropenia, occasional anaemia and flu-like symptoms are the main side effects and many patients cannot tolerate such regimens.

Intralesional chemotherapy

Intralesional chemotherapy is a safe and effective treatment for early lesions (less than 1 cm across) and oral disease. A response rate of 88% was achieved in one study.[6] The most widely used preparation is vinblastine 0.2 mg in 1 ml made up in an insulin syringe; 0.3–1.0 ml is injected superficially into the lesion until blanching occurs.

The most common side effect is local pain at time of injection (often described as being like a bee sting), which may be severe; local anaesthesia brings little relief. This is followed by an ache for a few days afterwards and occasionally ulceration occurs. There are sometimes flu-like symptoms over the next 24 h. Care must be taken where the injection site overlies a nerve, as neuropathies have been reported.

Radiotherapy

For small skin lesions, superficial X-ray treatment may be given. A 50–300 kV X-ray machine gives its maximum dose to the skin with very little penetration into underlying tissues. The maximum field size is limited and at least a 10 mm margin is allowed around the greatest extent of the lesion. A wider field may be needed to cover the disease if the lesion has a poorly defined margin with surrounding bruising. The most commonly used doses are 16 Gy (gray) in four daily fractions, 8 Gy in a single fraction or 20 Gy in 10 daily fractions. If necessary, the treatment can be repeated up to a maximum of three times in a given area.

The main side effects are transient erythema, rather like sunburn, lasting about two weeks. There may be a degree of residual pigmentation after the KS has responded – the KS itself very often leaves behind a permanent haemosiderin deposit and the radiotherapy can cause a faint pigmented patch in the irradiated area (this may be more pronounced in individuals with darker skin). Many people find the most troublesome aspect of radiotherapy to be the time it takes, sometimes requiring daily visits to a distant department for several weeks.

The skin of HIV-positive individuals appears to be excessively sensitive to radiation, particularly on the foot, where ulceration can occur. This can be minimized by longer treatment periods with smaller fractions and by only treating one foot at a time to allow the patient a degree of continued mobility.

Despite this, it is important to treat these lesions as extensive plaques of KS in the foot often break down and become infected.

Where a larger radiotherapy field is required, a better alternative is to use megavoltage electrons generated by a linear accelerator at energies of 5–20 MeV, depending on available equipment and depth of penetration required. The dose is 16 Gy in four daily fractions or 20–30 Gy in 10–15 daily fractions. Total and subtotal skin electron therapy has been used successfully.

Occasionally it is necessary to treat a block of tissue, for instance in gastrointestinal, oropharyngeal or pulmonary disease. This is best achieved with megavoltage photons from a linear accelerator or a cobalt source, using parallel opposed fields. The energy is deposited deeply, sparing the skin. If an adequate skin dose is also required, for instance to treat a limb covered with extensive plaques of KS, layers of wax can be put over the skin to bring the dose up to the surface. The dose is 20–30 Gy in 10–15 daily fractions. Irradiation of the abdomen commonly causes severe nausea and vomiting, necessitating adequate antiemetics.

The increased sensitivity of HIV-positive patients to radiation is particularly marked in the oropharynx and severe mucositis can result from even modest doses. Thus irradiation of the oropharynx is best avoided if at all possible and other modes of treatment are preferable. However, if there is no alternative, the treatment can be given as weekly fractions of 4 Gy to a total dose of 16 Gy, allowing the reaction to be monitored and therapy delayed if it becomes severe.

There are some sites which are best treated by brachytherapy (the local application or implantation of a radioactive source). This allows a high dose to be given to a small area, with sparing of the surrounding tissues. Examples are the use of strontium moulds for conjunctival lesions, implantation of oral disease and treating the penis with a jig to hold linear iridium sources in a ring around the shaft. The dose depends upon the activity of available sources and equipment.

Chemotherapy
Cytotoxic chemotherapy can be an effective treatment for KS; even where existing lesions do not respond, it very often arrests the development of new ones. It must be remembered, however, that all chemotherapy has the potential for serious adverse effects and treatment, once started, is difficult to stop as the KS returns. Therefore it is reserved for severe or rapidly progressing disease. The side effects common to most regimens in use for KS are transient fatigue (usually lasting 24 h or so) and varying degrees of myelosuppression. The effect of cytotoxic drugs on the immune system in AIDS has not been carefully studied but most oncologists, concerned about accelerating immunosuppression, err on the side of caution in patients who are relatively immunocompetent. Because of the combined myelosuppressive effects of HIV itself, many of the treatments prescribed and chemotherapy, it is necessary before each dose to ensure the availability of a recent full blood count. Vincristine, vinblastine, bleomycin, doxorubicin, etoposide and epirubicin have all been used as single-agent chemotherapeutic agents for KS. Regimes using combination chemotherapy or the new liposomal preparations are to be preferred as higher rates and duration of response can be achieved. The regimens most commonly in use are detailed below.

Bleomycin and vincristine (BV)
This is the combination of choice in the UK. It is generally very well tolerated, with little nausea or myelosuppression, and gives a response in up to 83% of subjects. The usual dose is vincristine 1 mg and bleomycin 15 units, given fortnightly as an intravenous bolus injection. Hydrocortisone 100 mg IV is also given to prevent the fever and flu-like symptoms otherwise commonly associated with bleomycin; this can be increased to 200 mg if necessary. Variants of this regimen include weekly alternating vincristine and bleomycin and higher doses given as a three-weekly cycle (Table 17.8).

ABV
This is BV with the addition of doxorubicin (hydroxydaunorubicin or adriamycin). This regimen is generally not used in the UK as it is considered rather toxic. However, it is in common use in Europe and the USA and response rates of up to 88% have been reported. The dose is doxorubicin 20 mg/m^2, bleomycin 10 mg/m^2 and vincristine 2 mg, given by intravenous injection every 14 days (Table 17.8).

Etoposide (VP16)
Etoposide is not usually given as first-line treatment in the UK, but can be used if there is no response to other drugs or if they are not tolerated. Response rates of up to 70% have been reported, although they are likely to be lower if the KS is refractory to other agents. The drug can be given intravenously or orally. The intravenous dose is 100 mg/m^2 daily for three days, repeated every 21 days. The oral dose is 50 mg daily for 14 days out of 28 (or seven days out of 28 if the bone marrow function is poor) (Table 17.8).

Epirubicin
Epirubicin is an alternative for second-line therapy. It is usually well tolerated when given weekly at a dose of 20 mg by intravenous injection, with a response rate of 35%. A greater response is seen with higher doses (up to 90 mg/m^2 every 14 days), but this is very toxic.

Liposomal daunorubicin and doxorubicin
These are useful agents, available as monotherapy, but are reserved for patients failing on bleomycin and vincristine in the UK, as they are relatively expensive.

Table 17.8 Side effects of cytotoxic drugs used in KS

Anthracycline antibiotics	Daunorubicin and doxorubicin (adriamycin)	Nausea and vomiting Alopecia Myelosuppression Cardiotoxicity (related to cumulative dose, sometimes fatal) Pain and tissue necrosis if extravasated
	Epirubicin	Similar to daunorubicin and doxorubicin but tend to be milder
Epipodophyllotoxins	Etoposide	Myelosuppression Alopecia Nausea and vomiting
Vinca alkaloids	Vincristine	Neurotoxicity (generally peripheral neuropathy, related to cumulative dose) Pain and tissue necrosis if extravasated
	Vinblastine	Myelosuppression Less neurotoxic than vincristine
Non-anthracycline antibiotics	Bleomycin	Flu-like symptoms Rashes (characteristically flagellate dermatitis) Pigmentation Pulmonary fibrosis (related to cumulative dose)

Liposomes are particles consisting of a phospholipid bilayer enclosing an aqueous core. Drugs can be enclosed within the liposome: fat-soluble drugs in the phospholipid membrane and water-soluble drugs in the aqueous phase. They can be manufactured with various physical properties (size, membrane fluidity, etc.) which enable them to be targeted, for instance to a tumour or a site of infection. They are carried in the bloodstream, enter the target cell (probably by endocytosis) and are then broken down, releasing their contents. The benefits are a high concentration at the site of action, with sparing of other sites (e.g. the bone marrow) and a lower total dose can be administered.

The anthracyclines daunorubicin and doxorubicin have been used as the preparations DaunoXome and Doxil respectively. These have shown activity against solid tumours including KS and lower toxicities are seen than with the free drug.

Free daunorubicin is very active against certain tumours, but its use is limited by severe toxicity (see Table 17.8). It causes extensive tissue damage if extravasated. It is currently not licensed for use in the UK, but still forms part of many leukaemia regimens.

Our centre has participated in a multicentre study of DaunoXome in the treatment of advanced KS.[7] Ten patients were evaluated: nine had extensive KS, eight prior opportunistic infections and the median CD4 count was $25 \times 10^6/l$. DaunoXome 40 mg/m^2 was administered by intravenous infusion every 14 days, with dose reduction if significant toxicity occurred, and a delay of therapy in the presence of intercurrent infection. After two cycles all patients showed a response and so treatment continued. Clinical status, cardiac, biochemical and haematological parameters were evaluated. A partial remission was achieved in four patients, stabilization of disease in the remaining six and the drug was well tolerated. No cardiotoxicity was seen, the main problems being anaemia (requiring transfusion) and neutropenia (requiring granulocyte colony-stimulating factor in one patient). Overall, the side effects were mild compared to those seen with the free drug.

A group in California[8] reported on the results of their patients recruited to the same study. They achieved a higher response rate, with a complete remission in 8.3% and a partial response in 54.2%. They also noted an improvement in the quality of life. Again, the toxicities were mainly haematological. In both centres, a response was observed in patients who had failed on other chemotherapy regimens.

The safety, efficacy and improvement in quality of life demonstrated by DaunoXome suggest that this is a useful agent for treating KS. The results of another study comparing DaunoXome with BV/ABV are awaited.

Our centre is currently evaluating the effects of DaunoXome against observation in patients with 'early' KS (fewer than 20 cutaneous lesions). Patients are initially randomized to treatment or observation for three months and the two groups then cross over. The aim of the study is to establish the optimum time for commencing systemic chemotherapy and the required duration of treatment if started early.

LYMPHOMA

Non-Hodgkin's lymphoma (NHL) is a frequent problem in individuals with AIDS, affecting between 4% and 10% of all cases. It is usually a high-grade B-cell lymphoma although lymphoblastic and Burkitt's-type lymphomas also occur. Extranodal disease is very common, particularly of the central nervous system, gastrointestinal tract and bone marrow. Hodgkin's lymphoma (HL) is being closely monitored but a causal association with HIV infection has not been proven and it is not an AIDS-defining diagnosis.

The Epstein–Barr virus (EBV) is thought to play an important role in many of these tumours. Evidence of infection with EBV is frequently found and EBV DNA has now been detected in the cerebrospinal fluid (CSF) of patients with cerebral lymphoma.

NHL and primary cerebral lymphoma in an HIV-positive individual are AIDS-defining illnesses. Both usually occur in the presence of profound immunodeficiency although NHL has been described in patients with high T4 counts.

The clinical presentation differs from that seen in cohorts of untested or HIV-seronegative individuals as AIDS-related lymphomas tend to be particularly aggressive, advanced stage IV disease is usually found at presentation, the response to treatment is not as good and if a response is achieved early, relapse frequently occurs. To date, HAART has not altered the incidence or prognosis of lymphomas in the presence of HIV.

Non-Hodgkin's lymphoma

In the USA NHL occurs with a 60-fold increased frequency in HIV-positive individuals compared with that expected in the general population. The majority of tumours are high-grade B-cell lymphomas, with lymphoblastic (immunoblastic) or Burkitt's lymphomas also occurring.

In developed countries there is about a 5% chance of developing NHL irrespective of risk factor for HIV acquisition. However, fewer cases are reported in Africa and the Caribbean (approximately 1%).[9] Possibly this may be due to the propensity for NHL to develop in the severely immunosuppressed. Hence it is seen more often in those with access to treatment and prophylaxis of opportunistic infections and antiretroviral therapy which may prolong survival with profound immunodeficiency.

Clinical presentation

NHL develops both in the lymph nodes and in many extranodal sites such as the bone marrow, tonsils, pharynx, gastrointestinal tract, liver, lungs, heart, central nervous system (mainly the meninges) and the skin (Table 17.9).

About 75% of patients present with 'B'-symptoms of fever, drenching night sweats, weight loss of over 10% and fatigue.

NHL of the lymph nodes presents as a rapidly enlarging gland. As it swells it may become painful and symptoms of obstruction are common, for example, superior vena caval obstruction and tracheal compression. The glands continue to grow extremely quickly with lymphomas such as Burkitt's having a doubling time of 24 h. When NHL is within the abdominal nodes the presentation is usually with abdominal pain and intermittent or total bowel obstruction. Thoracic gland involvement may present with pain or respiratory symptoms but may be asymptomatic and detected on a routine chest X-ray. Pharyngeal and tonsillar NHL presents with pain, dysphagia and/or a visible abscess.

Anorectal lymphomas cause pain, bleeding and mucoid discharge and/or an abscess. As they may occur adjacent to fistulae and fissures, to which homosexual men are prone, a delay in diagnosis often occurs as the symptoms are attributed to the coexisting condition.

Meningeal lymphoma presents with headache, cranial nerve lesions, a numb chin or a stiff neck, often in the presence of lymphoma elsewhere.

Bone marrow involvement may result in pancytopenia.

In HIV disease the manifestations of secondary disorders are often non-specific. Fever, night sweats, weight loss, fatigue, generalized lymphadenopathy,

Table 17.9 Common presenting symptoms of NHL

'B'-symptoms – fever, night sweats, weight loss >10%, fatigue
Rapidly enlarging lymph nodes
Pharyngeal or tonsillar abscess
Abdominal pain, intermittent or total bowel obstruction
Perianal pain, bleeding or discharge
Headache
Focal central nervous system signs
Pancytopenia
Mass on chest X-ray

rashes, pruritus or gastrointestinal upset are common presenting features of many opportunistic infections and often several different diagnoses coexist. Therefore, the diagnosis of lymphoma requires constant vigilance on the part of the clinician. Any suspicious mass or ulcer should be biopsied and examined histologically whilst samples of aspirated fluid or CSF should be examined cytologically.

Staging

Once a tissue diagnosis of lymphoma is made the decision on whether to treat or not needs to be fully discussed with the patient. If treatment is decided on, then the patient should be fully staged with a full haematological and biochemical screen, bone marrow examination, chest X-ray and whole-body tomography (either with CT or MRI scan) which is also important for treatment planning. Lymphoma, both NHL and HL, is staged according to the Ann Arbor scheme (Table 17.10).[10]

Treatment

Rapid histological confirmation of NHL is essential so that therapy to relieve pain and possible obstruction can commence. In some cases symptomatic treatment with analgesia, local radiotherapy and steroids to reduce oedema is commenced on the basis of a clinical diagnosis as the patients are too sick to wait for histological confirmation. This is frequently the case in the presence of superior vena caval or tracheal obstruction.

Survival is determined by many factors including the tumour histology and staging, the CD4+ lymphocyte count, the presence or absence of 'B'-symptoms, the Karnofsky index or WHO performance status and prior AIDS illnesses (Table 17.11). The overall prognosis is very poor with a median survival time of six months; 20–70% of patients die from opportunistic infection, 35–55% from the lymphoma. Studies indicate that a subgroup of patients aged under 60 years with a CD4+ lymphocyte count over $100 \times 10^6/l$, no 'B'-symptoms, good performance status (Karnofsky score over 70%, WHO performance status less than 2), no previous AIDS-defining illnesses and absence of bone marrow involvement do better. Those with stage IV disease do badly. Histological subtype seems to be a less clearly defined prognostic indicator although immunoblastic lymphoma does less well than Burkitt's lymphoma.

Table 17.11 Poor prognostic factors for NHL

Tumour stage IV
Primary CNS lymphoma
Age greater than 60 years
CD4+ lymphocyte count less than $100 \times 10^6/l$
'B'-symptoms
Karnofsky score less than 70%
Prior AIDS-defining illness

Treatment regimens developed for NHL in the non-AIDS setting have given complete response (CR) rates of 86% with long-term survival in 65% of cases.[11] These regimens have much lower CR rates in AIDS patients (17–56%) with overall survival rates of 4–6 months. In addition, the myelosuppression caused by therapy is much more severe than in the non-AIDS situation.

An intensive chemotherapeutic regimen with intent to cure was given to 141 patients with AIDS-related NHL, none of whom had primary CNS lymphoma. A CR rate of 68% was achieved but despite this, the median survival was only nine months, similar to the 5–7 month survival seen with less intensive regimens not aiming for cure. Within the study group 66 patients without poor prognostic factors (histology other than immunoblastic lymphoma, CD4+ lymphocyte count over $100 \times 10^6/l$, WHO performance status over 1, no prior AIDS diagnosis) had a 50% probability of survival at two years.[12]

Optimal therapy for AIDS lymphomas is as yet undefined. A pan-European study run by an EORTC group has commenced, studying the outcome of different treatment options in the presence of the various features discussed above, which should help to define the best treatment options. Meanwhile, an individual decision between oncologist and patient for

Table 17.10 Ann Arbor staging of lymphoma

Stage*	Features
Stage I	Single lymph node area involved or single extralymphatic site (Ie)
Stage II	Two nodal areas involved on the same side of the diaphragm or involvement of limited contiguous extralymphatic organ or tissue (IIe)
Stage III	Involvement on both sides of the diaphragm; may include spleen (IIIs) or limited extralymphatic site (IIIe, IIIes)
Stage IV	Multiple or disseminated involvement of one or more extralymphatic organs or tissues, with or without lymphatic involvement

*Stages further subdivided into A and B according to the absence or presence of 'B'-symptoms (fever, night sweats, weight loss of more than 10%)

treatments ranging from aggressive 'intent to cure' regimens, lower dose chemotherapy, local treatment or no treatment is made based on the above criteria and on patient preference.

'Intent to cure' therapy usually requires chemotherapy, which is usually very aggressive, although much milder regimens may be used for palliation. The gold standard of NHL chemotherapy is CHOP (cyclophosphamide, doxorubicin, vincristine and prednisolone), although doses may need to be modified in HIV disease. Very intensive regimens such as PACE-BOM (prednisolone, doxorubicin, cyclophosphamide, etoposide, bleomycin, vincristine and methotrexate) have also been used. Bleomycin and vincristine are well tolerated and may give good palliation where bone marrow function is poor. Radiotherapy can be used in localized disease and is also important in symptom control. Usually, low-dose palliative radiotherapy is employed which has a low morbidity, although some oncologists advocate high-dose radiotherapy on the grounds that if a remission is achieved prolonged survival may occasionally be possible. Surgery may be offered to remove or 'debulk' the tumour mass where necessary if this is achievable, particularly if the lymphoma is causing symptoms due to obstruction.

Unfortunately, there is no consensus at present as to the best treatment and the results of multicentre trials are eagerly awaited. As a broad guide, patients with good prognostic features should be offered 'intent to cure' chemotherapeutic regimens, preferably within the context of a prospective study; patients with poor prognostic signs should be offered palliative therapy with either less toxic chemotherapeutic regimens, radiotherapy and/or steroids. In patients with poor prognostic signs electing for palliative therapy, it is important to explain that high-intensity regimens will not prolong survival and that symptomatic relief can be achieved using less toxic regimens with fewer side effects.

During induction chemotherapy antiretroviral drugs and any drugs which may cause myelosuppression should be discontinued if possible. Prophylaxis against opportunistic infection should be offered. It is generally agreed that *Pneumocystis carinii* pneumonia (PCP) prophylaxis should be recommended for all. However, recommendations for prophylaxis against candida, herpes and *Mycobacterium avium* complex (MAC, MAI) vary between units. After induction chemotherapy antiretroviral therapy should recommence.

Primary CNS lymphoma

Primary CNS lymphoma accounts for 20% of cases of AIDS-related lymphoma, usually occurring in the profoundly immunosuppressed, most frequently in those with a previous AIDS diagnosis. In a USA study 72% had a previous AIDS diagnosis and the mean CD4+ lymphocyte count was $64 \times 10^6/1$ (range $0-279 \times 10^6/1$).[13]

Clinical presentation

The presentation is usually as a space-occupying lesion with headaches, confusion, lethargy, convulsions or focal neurological symptoms and signs such as cranial nerve palsies or hemiparesis. Signs may be more obtuse with clouding of consciousness, memory loss, change in personality or mental state with the development of mania or apathy. The differential diagnosis includes cerebral toxoplasmosis and progressive multifocal leucoencephalopathy (PML).

Primary CNS lymphoma occurs at any site within the brain but does not spread beyond it. It is diagnosed as a mass on a CT or MRI scan. Most lesions are iso- or hypodense prior to contrast media and oedema and a mass effect are variably present. The lesions are usually single (unlike cerebral toxoplasmosis where the lesions are usually multiple) and larger than 3 cm when detected.[14] Initially, therapy is often given for presumed cerebral toxoplasmosis and a failure to respond is strong evidence for a missed diagnosis of lymphoma. Many cases remain unconfirmed in life as, at the present time, brain biopsy is not performed routinely in most UK centres.

Mean survival is extremely poor, of the order of three months. It has been shown that in an untreated group of 26 patients with primary cerebral lymphoma the median survival was one month, whilst in a smaller group of 10 patients subjected to irradiation, chemotherapy and partial tumour resection, despite achieving a CR in six, the median survival was only 5.5 months.[11]

Treatment

The most commonly used therapy is whole-brain irradiation. In one study patients were given a total dose of 2200–5000 cGy over 2–4 weeks. However, as not all patients had histological confirmation of primary cerebral lymphoma the outcome data may not be reliable.

Prospective studies are needed to define the optimum therapy for this tumour.

Hodgkin's disease

The incidence of Hodgkin's disease (HD) is higher in homosexuals, injecting drug users and haemophiliacs infected with HIV compared to population-based rates. When it occurs, it is usually of the mixed cellularity or lymphocyte-depleted type rather than the more favourable nodular-sclerosing type.

Clinical presentation

In HIV-positive patients HD usually presents with 'B'-symptoms such as fever, drenching night sweats and weight loss. Disseminated disease is usually found at presentation, frequently involving the bone marrow, but mediastinal lymphadenopathy is rare. The diagnosis is often made during investigation of the 'B'-symptoms, usually by bone marrow or liver biopsy.

Staging of the HD is, like NHL, according to the Ann Arbor scheme (see Table 17.10).[10]

Treatment

Using standard treatment regimens, the relapse rate and morbidity are high. Protocols to evaluate appropriate treatment regimens are required in order to provide more effective therapy.

CERVICAL DYSPLASIA/NEOPLASIA

Women who are immunosuppressed have an increased risk of cervical dysplasia or intraepithelial neoplasia (CIN) and cancer. In addition, there are considerable data linking human papillomavirus (HPV), particularly types 16, 18 and 31, and cervical carcinoma and both cervical dysplasia and HPV have been found frequently in HIV-positive women. One study found CIN in 52% of women with HIV disease with concurrent HPV infection compared with 18% of women with either virus alone and demonstrated a strong association between HPV and CIN in symptomatic HIV-positive women which was most marked in younger women, aged under 35 years, and those of black or Hispanic origin.[15] It is difficult to assess whether the higher prevalence of cervical HPV (particularly HPV 16 and 18) amongst HIV-positive women is caused by more frequent reactivation, the establishment of a chronic active HPV infection, replication of the HPV virus to a higher level than in immunocompetent women or whether the acquisition of both viruses is merely associated with similar risk behaviour. Neither is it known whether cervical HPV infection may facilitate HIV transmission by disrupting the mucosal integrity or altering the local immunity. Whatever the explanation, the combined presence of the two viruses appears to increase the risk of developing cervical dysplasia.

Dysplasia, suggested by dyskaryosis seen on cytological examination of a Papanicolaou smear (Table 17.12), is confirmed only by histological examination of a cervical biopsy and may be divided into three stages (Table 17.13).

CIN is more common and more severe in HIV-infected women with an odds ratio of 4.9 in meta-analysis, particularly for those who also have evidence of immunosuppression.[16] In a study of 111 HIV-seropositive women, 76 HIV-seronegative women and 526 HIV-untested gynaecology outpatient attendees, cervical dysplasia/neoplasia was found in 41% of the HIV-positive women, five of whom had invasive cervical cancer; 9% of the HIV-negative group; 4% of the outpatient group, two of whom had invasive cervical cancer. In addition, cytological changes consistent with HPV occurred four times more frequently in the HIV-tested groups compared with the untested outpatient group. The authors also found that both the frequency and severity of dysplasia appeared to increase in those HIV-positive women with lower CD4+ lymphocyte counts.[17] The effect of immunosuppression was confirmed by another study in which HIV-positive and -negative women were matched for age, age at first coitus, lifetime number of partners and smoking habit. They found that only in the presence of immunosuppression was the prevalence of dysplasia greater in the HIV-positive compared to HIV-negative women.[18]

When cervical cancer develops, the majority, like those occurring in young women in the general population, are squamous cell in origin arising in the transformation zone. However, they tend to be higher grade lesions, involve a greater proportion of the cervix and may be multifocal. Less commonly, adenocarcinoma, mixed cell, lymphoma, sarcoma or melanoma may occur. Cervical cancer progresses in stages from severe dysplasia through carcinoma *in situ* (where the

Table 17.12 Cervical cytology

Normal
Inflammatory
Borderline
Mild dyskaryosis
Moderate dyskaryosis
Severe dyskaryosis
Invasive carcinoma suspected

Table 17.13 Cervical Intraepithelial Neoplasia (CIN)

Stage	Description
CIN 1	Mild dysplasia – less than 1/3 of the thickness of the epithelium is involved
CIN 2	Moderate dyskaryosis – up to 2/3 of the thickness of the epithelium is involved
CIN 3	Severe dysplasia – involving 2/3 to full thickness of the epithelium (includes carcinoma *in situ*)

Table 17.14 FIGO (International Federation of Gynaecology and Obstetrics) staging of cervical carcinoma

Stage		Features
0	0	CIN3 or carcinoma *in situ*
I – confined to cervix	Ia1	Minimally microinvasive carcinoma
	Ia2	Microinvasive carcinoma <5 mm deep and <7 mm wide
	Ib	Invasive carcinoma confined to the uterus only (including the body of the uterus)
II – extending beyond cervix	IIa	Invasive carcinoma extending beyond the cervix into the upper 2/3 of the vagina
	IIb	Invasive carcinoma extending into the parametrium but not reaching the pelvic side wall
III – extending on to pelvic wall or to lower third of vagina	IIIa	Carcinoma extending into the lower third of the vagina
	IIIb	Carcinoma extending into the pelvic side wall or causing ureteric obstruction
IV – extending beyond pelvis, invading adjacent organs or distant metastases	IVa	Carcinoma extending beyond the pelvis or involving the mucosa of the bladder or rectum
	IVb	Distant metastases

basement membrane is not breached) to invasive carcinoma (Table 17.14).

Cervical cancer in an HIV-positive woman became an AIDS-defining illness in early 1993. In the UK, invasive cervical carcinoma and HIV has not been reported, but in the USA cases have been. It is aggressive with a poor response to therapy.

Clinical presentation

Cervical dysplasia is almost always asymptomatic and is detected by routine cervical cytology or histological examination of a cervical biopsy. All immunosuppressed HIV-positive women should be offered annual cervical screening. By adopting regular cytology and colposcopy screening for any degree of dysplasia or inflammatory changes, dysplasia can be detected and treated at an early stage before invasion occurs.[19] Recently, relapse has been described after accepted therapy for dysplasia, so close prolonged follow-up is essential.

Invasive cervical cancer presents with abnormal vaginal bleeding, pain or discharge. The tumour is usually palpable on vaginal examination as ulceration, induration or an exophytic growth. Spread is locally down the vaginal vault, into the parametrium (where it may involve the ureters) and regionally to the pelvic lymph nodes. Bloodborne metastases to the liver, lungs and bones and lungs occur late.

In HIV-positive women in the USA, cervical carcinoma presents with high-grade, advanced stage tumours and death may occur before any other features of AIDS have developed. Oncologists there advise young women presenting with advanced squamous cell carcinoma of the cervix to consider having an HIV antibody test. In one study 16 (19%) out of a cohort of 84 women with cervical cancer were HIV positive.[20] Those women who were HIV positive generally presented with advanced disease, only one had early-stage disease compared with 40% of the HIV-negative women and despite 13 of the 16 undergoing therapy, 82% had persistent or recurrent disease. Women with a CD4+ lymphocyte count greater than $500 \times 10^5/l$ tended to have a better outcome than those with lower CD4+ lymphocyte counts. Most of the women (nine) died of cervical cancer, two of AIDS-related illnesses and five were still alive at the end of the study.

Investigation and treatment

CIN can usually be treated under colposcopic control by local excision, most commonly by loop excision although sometimes a general anaesthetic (GA) with a knife cone biopsy is necessary. Other destructive methods (such as diathermy, laser or cryotherapy) are not recommended as it is difficult to ensure that the tissue edge is free of disease.

Staging investigations include an examination under anaesthetic with endometrial biopsy and abdominal and pelvic ultrasound, CT or MRI scan.

Optimal therapy for invasive carcinoma in HIV-positive women has not been defined. The regimens used at present, dependent on staging, are debulking surgery with or without local radiotherapy and/or combination chemotherapy.

SQUAMOUS CELL CARCINOMA OF THE ANUS

The incidence of squamous cell carcinoma of the anus (SCCA) in young single men has been rising in the USA over the past 10 years. It is not yet clear whether this can be directly attributed to HIV infection but the two are certainly seen together. Known risk factors in HIV-negative patients include male homosexuality (although SCCA is commoner overall in women than in men), anal condylomata acuminata, chronic anal irritation and smoking. HPV has been implicated in the pathogenesis of SCCA and there appears to be some similarity with cervical cancer in women, in that there is a preinvasive dysplastic stage (anal intraepithelial neoplasia – AIN) analogous to cervical intraepithelial neoplasia (CIN).

Palefsky has studied homosexual men with CDC IV HIV disease in the San Francisco area.[21] He found a high incidence of HPV infection, with viral DNA detected in 52 of 97 subjects (54%). Abnormal cytology on anal smears was found in 38%, with AIN in 15%. The proportion with AIN rose in those subjects who were followed up. He concluded that these men were at high risk of developing anal carcinoma and that if they survived long enough from their HIV disease (which he thought likely), then this would develop.

Staging

SCCA spreads locally into the anal sphincter and via the lymphatics into the inguinal, pelvic and abdominal lymph nodes. Distant haematogenous spread occurs into the liver, lungs and skin (Table 17.15).[22]

Presentation

SCCA usually presents as anal bleeding, pain, discomfort, discharge or a palpable lump. These symptoms may be attributed to perianal disease such as haemorrhoids, fistula, warts or abscesses which frequently coexist. Thus patients with HIV disease and anal symptoms require early investigation. The tumour is easily palpable by digital examination of the rectum. Biopsy is essential.

Table 17.15 TNM staging of SCCA

Primary tumour	T0	No primary tumour
	Tis	Carcinoma *in situ*
	T1	<2 cm greatest dimension
	T2	– 5 cm greatest dimension
	T3	>5 cm greatest dimension
	T4	Invasion of adjacent organs
Regional lymph nodes	N0	No lymph node metastases
	N1	Perirectal lymph nodes involved
	N2	Unilateral internal iliac and/or inguinal lymph nodes involved
	N3	Bilateral internal iliac and/or inguinal lymph nodes involved
Distant metastases	M0	Absent
	M1	Present

Treatment

The outcome of treatment of SCCA in HIV-positive patients is often poor. The tumours are usually very aggressive and therapy is usually aimed at palliation of symptoms. Wide local excision may be used for small tumours confined to the perianal skin. Larger tumours are treated with radiotherapy, often in combination with chemotherapy.

TESTICULAR TUMOURS

Testicular tumours are rare, occurring in approximately 1 in 50 000 males in the UK. However, an increased incidence of testicular tumours in HIV-positive men of 57-fold[23] and 68-fold[24] compared to the general population rate has been reported and they often run a more aggressive course. Mostly the tumours are seminomas although germ cell tumours have also been seen.

Clinical presentation

Testicular tumours may develop in both asymptomatic and CDC stage IV patients. The usual presentation is with a hard, craggy mass in the testicle. Hence, routine testicular examination by clinical staff or the patient is essential to detect these tumours early.

Treatment

Treatment within a specialist unit should be offered if the patient opts to have therapy. Only when these units have had sufficient experience with testicular tumours in HIV-infected men will a specific management protocol emerge. Early tumour progression after therapy has been observed in HIV-positive men. It appears that those without an AIDS diagnosis tend to tolerate chemotherapy best and are more likely to obtain a tumour response.

LUNG CANCER

HIV-positive patients developing lung cancer differ from general population controls, as demonstrated in one study which showed that they were younger (mean age 38 versus 60.5 years), were more likely to have metastatic disease at presentation and had a shorter survival (four weeks versus 25.5 weeks).[25] All the patients reported were smokers and the predominant risk factor for HIV acquisition was injecting drug use. As expected in a younger group of patients, most of the tumours are adenocarcinomas.

Clinical presentation

The main presenting symptoms are cough, haemoptysis, chest pain, dyspnoea, fever, weight loss and anorexia. The diagnosis is confirmed by chest X-ray and bronchoscopy.

Treatment

There are no specific therapies for HIV-positive patients with lung cancer. Standard active or palliative therapy is offered depending on the clinical state and wishes of the patient.

REFERENCES

1. Buchbinder, A. and Friedman-Kier, A.E. (1991) Clinical aspects of epidemic Kaposi's sarcoma. *Cancer Surv.*, **10**, 39–42.
2. Mitsuyasu, R.T. and Groopman, J.E. (1984) Biology and therapy of Kaposi's sarcoma. *Semin. Oncol.*, **11**, 53–59.
3. Krown, S.E., Metroka, C. and Wernz, J.C. (1989) Kaposi's sarcoma in the acquired immune deficiency syndrome: a proposal for uniform evaluation, response and staging criteria. *J. Clin. Oncol.*, **7**, 1201–1207.
4. Tappero, W., Berger, G., Kaplan, D. et al. (1991) Cryotherapy for cutaneous Kaposi's sarcoma (KS) associated with acquired immunodeficiency syndrome (AIDS): a phase II trial. *J. Acquired Immunodeficiency Syndrome*, **4**, 839–846.
5. Bonhomme, L., Fredj, G., Averous, S. et al. (1991) Topical treatment of epidemic Kaposi's sarcoma with all-trans-retinoic acid. *Ann. Oncol.*, **2**, 234–235.
6. Boudreaux, A.A., Smith, L.L., Cosby, C.D. et al. (1993) Intralesional vinblastine for cutaneous Kaposi's sarcoma associated with acquired immunodeficiency syndrome. A clinical trial to evaluate efficacy and discomfort associated with infection. *J. Am. Acad. Derm.*, **28**(1), 61–65.
7. Money-Kyrle, J.F., Bates, F., Ready, J. et al. (1993) Liposomal daunorubicin in advanced Kaposi's sarcoma: a phase II study. *Clin. Oncol.*, **5**, 367–371.
8. Presant, C.A., Scolaro, M., Kennedy, P. et al. (1993) Liposomal daunorubicin treatment of HIV-associated Kaposi's sarcoma. *Lancet*, **341**, 1242–1243.
9. Obrams, G. and Grufferman, S. (1991) Epidemiology of HIV-associated non Hodgkin lymphoma. *Cancer Surv.*, **10**, 91–102.
10. Carbone, P.T., Kaplan, H.S., Musshoff, K. et al. (1971) Symposium (Ann Arbor). Staging in Hodgkin's disease. *Cancer Res.*, **31**, 1860–1861.
11. Northfelt, D. and Kaplan, L. (1991) Clinical manifestations and treatment of HIV-related non Hodgkin lymphoma. *Cancer Surv.*, **10**, 121–133.
12. Gisselbrecht, C., Oksenhendler, E., Tirelli, U. et al. (1993) Human immunodeficiency virus related lymphoma treatment with intensive combination chemotherapy. *Am. J. Med.*, **95**, 188–196.
13. Levine, A.M., Sullivan-Halley, J., Pike, M.C. et al. (1991) HIV-related lymphoma: prognostic factors predictive for survival. *Cancer*, **68**, 2466–2472.
14. Gill, P.S., Levin, A.M., Meyer, P.R. et al. (1985) Primary central nervous system lymphoma in homosexual men. Clinical, immunologic and pathologic features. *Am. J. Med.*, **78**, 742–748.
15. Vermund, S.H., Kelley, K.F., Klein, R.S. et al. (1991) High risk of human papillomavirus infection and cervical squamous intraepithelial lesions among women with symptomatic human immunodeficiency virus infection. *Am. J. Obstet. Gynecol.*, **165**, 392–400.
16. Mandelblatt, J.S., Fahs, M. and Garibaldi, K. (1992) Association between HIV infection and cervical neoplasia: implications for clinical care of women at risk for both conditions. *AIDS*, **6**, 173–178.
17. Schäfer, A., Friedmann, W., Mielke, M. et al. (1991) The increased frequency of cervical dysplasia-neoplasia in women with the immunodeficiency virus is related to the degree of immunosuppression. *Am. J. Obstet. Gynecol.*, **164**, 593–599.
18. Smith, J.R., Kitchen, V.S., Botcherby, V. et al. (1993) Is HIV infection associated with an increase in the prevalence of cervical neoplasia? *Br. J. Obstet. Gynaecol.*, **100**, 149–153.
19. Maiman, M., Tarricone, N., Vieira, J. et al. (1991) Colposcopic evaluation of human immunodeficiency virus-seropositive women. *Obstet. Gynaecol.*, **78**, 84–88.
20. Maiman, M., Fruchter, R., Guy, L. et al. (1992) Human immunodeficiency virus infection and cervical cancer. *Cancer*, **71**, 402–406.
21. Palefsky, J.M., Gonzales, J., Greenblatt, R.M. et al. (1990) Anal intraepithelial neoplasia and anal papillomavirus infection among homosexual males with group IV HIV disease. *J. Am. Med. Assoc.*, **263**(21), 2911–2916.
22. International Union Against Cancer (1989) *TNM Atlas*, 3rd edn, Springer, Berlin, pp. 90–97.
23. Wilson, W.T., Frenkel, E., Vuitch, F. et al. (1992) Testicular tumours in men with human immunodeficiency virus. *J. Urol.*, **147**, 1038–1040.
24. Moyle, G., Hawkins, D.A. and Gazzard, B.G. (1991) Seminoma and HIV infection. *Int.J.STD and AIDS*, **2**, 293-294.
25. Karp, J., Profeta, G., Marantz, P. and Karpel, J. (1991) Lung cancer in patients with immunodeficiency syndrome. *Chest*, **103**, 410-413.

Index

Page numbers in **bold** indicate a figure; *italic* page numbers refer to tables.

abacavir 196
acanthosis nigrans 81
acneiform rash 190
acquired immunodeficiency syndrome, *see* AIDS
acyclovir 107, 193, 216
adriamycin 260
agglutination assays 38–9
AIDS 19, 37, 49, 50, 61, 79, 185, 187, 191, 194
   education in schools 3
   morbidity and mortality associated with 1
   opportunistic infections in 211–53
   tumours in 255–68
   see also HIV infection
alfuzocia 155
amikacin 233–4
aminoglycosides 216, 228
aminopeptidases 23
aminoquinolones 154
amoebiasis 78
amphotericin 225, 226, 229
amphotericin B 192, 216–20
ampicillin 116, 128, 247
amprenavir 196
anal examination in child abuse 16
angiokeratoma 78
anogenital carcinoma 79
anogenitorectal syndrome 117
antacids 239
antibiotics 77
antifungal cream 88
aphthous ulceration 192
aspergillosis 230
atovaquone 217–18, 223
azithromycin 232, 234, 246

bacterial infection 76–7
bacterial pneumonia 247

bacterial vaginosis (BV) 83, 88–92
   and pelvic infammatory disease (PID) 162–3
   clinical features 90
   complications 91–2
   diagnosis 90–1
   diagnostic methods 31–3
   epidemiology 89–90
   grading of smears *32*
   microbiology 89
   screening 92
   symptoms 89
   terms used **89**
   treatment 91
balanitis 77
Bartholin's gland abscess 144
basal cell epithelioma 79
BCG vaccination 237
Behçet's disease 73, 78
benign familial pemphigus 80
benign mucous membrane pemphigoid 73, 80
benign orgasmic cephalalgia 181
benign transient lymphagiectasis 80
bimanual pelvic examination 8
bleomycin 260
blood tests, homosexual men 12
Bowenoid papulosis (BP) 142–3
Bowen's disease 79
bullous eruptions 80
Buschke–Lowenstein tumour 143

Calymmatobacterium granulomatis 13, 26, 118
Candida 33, 123, 224–7
   disseminated 225
   drug resistance 225
   oesophageal 224–5
   oral 225
   oral/oesophageal 225

Candida albicans 17, 33, 87–8, 192
Candida glabrata 192
candidiasis 77, 83, 87–8
  clinical features 87
  complications 88
  diagnostic procedures 33–4
  differential diagnosis 87
  epidemiology 87
  immunology 87
  laboratory diagnosis 88
  microbiology 87
  oral and oropharyngeal 224
  pathology 87
  pseudomembranous 224
  recurrent 88
  secondary prophylaxis 224
  treatment 88, 224
  vaginal 224
carbuncles 144
case report forms (CRFs) 52, 56
ceftriaxone 128
cephalosporins 116, 128
cerebrospinal fluid (CSF), examination 13–14
Cernilton 154
cervical carcinoma 108, 191
  staging *266*
cervical cytology 7, *265*
cervical dysplasia/neoplasia 265–6
cervical intraepithelial neoplasia (CIN) 141, 265, *265*
cervix
  cleaning 7
  examination 6
  sampling 7, 28
chancre 12, 112
chancroid 13, 115–16
  clinical presentation 115
  diagnostic procedures 24–6
  epidemiology 115
  follow-up 116
  investigations 116
  management 116
  partner notification 116
  transmission 115
chaperone 6
chickenpox 77
child sexual abuse 16–17
  anal signs *17*
  sexually transmitted diseases in *17*
  vulvovaginal signs *16*
Chlamydia 27, 98
Chlamydia trachomatis 5, 7, 8, 10, 11, 13, 17, 27–31, 116, 123, 124, 126–8, 130, 131, 151, 154, 164–6
chlamydiae 27
  detection 28–9
  serological tests 29–30
chloramphenicol 118, 247
chorioamnionitis 92
chronic inflammatory bowel disease 80
cidofovir 240
ciprofloxacin 116, 129, 233–4, 247
circinate balanitis 98
clarithromycin 232, 233
clindamycin 91, 92, 214, 216, 222–3
clinical examination 5–18
  homosexual men 11–14
  male 9
  women 5–9
clinical service, *see* genitourinary medicine clinic service
clinical trials 49–58
  audit 56
  data monitoring committee 56
  ethical considerations 54–5
  ethics committee submission 52
  good clinical practice 54–6
  informed consent 55
  outset 50–4
  principal investigator 52–3
  publication 56–7
  recruitment 53–4
  statistical analysis 51
  study design 50
  subject eligibility 50
clomipramine 173
clotrimazole 88, 224
CMV colitis 240–1
CMV oesophagitis 241
CMV polyradiculopathy 241
CMV prophylaxis 243–4
CMV retinitis 239
  treatment 243
co-amoxiclav 91
coccidiodocomycosis 229–30
colposcopy **64–5**
Committee on Safety of Medicines (CSM) 154
condylomata acuminata 139–43
  management 141–2, *142*
  treatment 141–2, *142*
condylomata lata 145
conjunctival specimens 14, 28
contact dermatitis **75–6**
contraception 2, 162
corticosteroids 221, 228, 239
Corynebacterium minutissimum 76
Corynebacterium vaginalis 89
co-trimoxazole 118, 154, 213–15, 218, 246, 247
cryptoccocal meningitis 227–9
cryptococcal meningitis, maintenance 228–9

Cryptococcus neoformans 151
cryptosporidiosis 245–6
Cusco speculum removal 8
CVA, sexual adjustment 181
cystine tryptic agar (CTA) 23
cystitis, honeymoon 157
cytokines 202
cytology 98
cytolytic vaginitis 92

dapsone 216–17, 219, 239
Darier's disease 81
dark-field microscopy 21
dark-ground examination 99
daunorubicin 260–1
DaunoXome 261
deciduitis 92
declazuril sodium 246
delavirdine 194, 196
dermatitis artefacta 76
dermatological conditions
    common presentations *74*
    diagnosis 73–81
    examination 73
    history 73
    investigations 73
developing countries, genitourinary medicine clinic service 69–71
diabetes mellitus 5
diazepam 155
didanosine 194, 196, 215, 216
dipstick analysis *9*
direct antigen tests 40
direct fluorescent antibody (DFA) test 19, 27–9
DNA
    amplification assays 29
    fingerprinting 15
    probes 31
Donovan body detection 27
donovanosis 13, 101, 118
    clinical manifestations 118
    diagnostic procedures 26–7
    epidemiology 118
    follow-up 118
    investigations 118
    management 118
    partner notification 118
    serological tests 27
    transmission 118
dot-blot assays 39
doxorubicin 260–1
doxycycline 118, 129
drug eruptions 75–6
dynamic infusion cavernosometry 174

dyspareunia 177–8
dysplasia 75

eczema 75–6
Efavirenz 196
elementary bodies (EBs) 27
ELISA 38–40
endocervical infection *41*
enzyme immunoassays (EIAs) 19, 27, 29, 33, 35
eosinophilic granuloma 81
Epidermophyton 77
epirubicin 260
Epstein–Barr virus (EBV) 192
Epstein–Barr virus (EBV) mononucleosis 186, *186*
erectile dysfunction 173–5, *173*, *174*
erysipelas 77
erythema multiforme 76
erythematous dermatosis 75
erythrocyte sedimentation rate (ESR) 163
erythromycin 77, 116, 130, 132, 133
erythromycin minocycline 118
erythromycin stearate 129
erythroplasia of Queyrat 79
ethambutol 232, 233, 236–9
etoposide 260
external genitalia, infectious lumps 139–47, *139*
extramammary Paget's disease 79

famciclovir 244
fibroma 78
filariasis 78
fixed drug eruptions 76, 98
flucloxacillin 77
fluconazole 88, 151, 192, 224–30, 239
flucytosine 228
fluorescent treponemal antibody (FTA) test 20
5-fluorouracil 79
folinic acid 221–3
follicle-stimulating hormone (FSH) 179
folliculitis 144, 190
Fordyce spots 75
foscarnet 216, 242, 244
Fox–Fordyce disease 80
fungal infection 14, 77
fungal skin conditions 190–1
furunculosis 144

ganciclovir 239, 241
Gardnerella vaginalis 31, 89
genital examination
    male 9–11
    women 6

genital herpes 104
  atypical 105
  classification according to serological response **105**
  in HIV-infected individuals 110
  presentation 105
  risk of acquisition *108, 109*
  risk to foetus and neonate 109
  transmission *109*
  treatment 101
genital ulceration 12
genital warts 139–43
  aetiology 140
  clinical features 141
  diagnosis 141
  epidemiology 140
  pathogenesis 140–1
  pathology 140–1
genitourinary medicine clinic service 59–68
  advertising 66–7
  appointments 64
  arrangements with other departments 66–8
  colposcopy 64–5
  design 63
  developing countries 70
  equipment 65–6
  family planning services 64
  funding 60–1
  furnishings 63–4
  need for service 59–60
  patients' notes 66
  problems cases 64
  services to be offered 64–5
  staffing 61–3
  support services 61
gentamicin 216
gingivitis 192
gonorrhoea 123–4
  antibiotic susceptibility 24
  diagnostic procedures 22–4
Gram-stained smear 32
granuloma inguinale, *see* donovanosis

haemangioma 78
Haemophilus ducreyi 13, 24–6, 99, 115–16
Haemophilus influenzae 164
Haemophilus vaginalis 89
Hailey–Hailey disease 80
headaches, sexually related 180–2
hepatitis B 15, 16
hepatitis C, risk factors *12*
herpes cervicitis 105
herpes proctitis 106

herpes simplex virus (HSV) 7–8, 12–13, 99, 102–10, 191, 244–5
  and cervical cancer 108
  and pregnancy 108–9
  antiviral therapy 107
  as risk factor for acquisition of HIV 110
  clinical disease 102–3, *102*
  clinical presentation 104–5
  complications 106
  diagnostic procedures 34–6, 106, *106*
  epidemiology 102
  follow-up 110
  infectivity 108
  neonatal infection 108–9
  partner notification 110
  psychological support 107–8
  recurrence rate 108
  seroepidemiology 103, *103*
  serological tests 35
  transmission 103–4
  vaccine development 110
herpes zoster virus (VZV) 77, 191
hidradenitis suppurativa 144
histoplasmosis
  diagnosis 229
  symptoms and signs 229
  treatment 229
history taking 2–3
HIV infection 15, 16, 19, 37–41, 49, 61, 63, 64, 69–70, 152, 157, 185–209, 265, 266
  and herpes simplex virus (HSV) 110
  and urethritis 134
  antibody response 37–8
  antibody screening tests 38
  antiretroviral therapy 194–6
  asymptomatic infection 187–9
  clinical presentation 186–9
  clinical trials 53
  compartment penetration 199
  constitutional symptoms 189
  current CDC classifications **188**
  drug interactions 198–9, **198**
  drug reactions 190, *190*
  drug resistance and cross resistance 199–200
  early symptomatic infection 189
  education in schools 3
  immunomodulatory therapies 201–2
  initial therapy 196–7
  opportunistic infections 211
  oral conditions 192
  primary infection 194–5
  prognostic indicators 38–41
  risk factors *12*
  skin conditions 190–2
  surrogate markers 193–4

tests 12, *12*
transfer of humoral immunity 200–2
see also AIDS
Hodgkin's disease 264–5
homosexual men
  blood tests 12
  clinical examination 11–14
  investigations 11–12
  NGU in 126
honeymoon cystitis 157
human herpes virus (HHV) infection 189
human immunodeficiency virus, *see* HIV
human papillomavirus (HPV) 141, 265
  and anogenital neoplasia 141
  diagnostic procedures 36–7
  DNA detection **37**
  molecular procedures 36
  serological tests 36–7
  subtypes 191
hydradenitis suppurativa 80
hydroxydaunorubicin 260
hydroxyurea 196
hymen, child abuse 16
Hypertension and sexual dysfunction 182, *182*
hyperuricaemia 238

ICP (intracorporeal injection of papaverine) 174
imipenem-cilastin 242
immune deficiency 191
immune responses 201
implantable penile prosthetic splint 174
indinavir 194, 196
indirect immunofluorescence assays (IFA) 39
infestation 77–8
in situ hybridization 39–40
interferons 142
intracranial space-occupying lesions 181
intrauterine contraceptive device (IUCD) 162, 164, 166
isoniazid 236–9
isospora belli 246
itraconazole 88, 224–6, 228, 229, 239

Kaposi's sarcoma (KS) 79, 189, 192, 255–62
  'B'-symptoms 258
  clinical features 256–8
  gastrointestinal 257–8, *257*
  mucocutaneous 257
  pathology 256
  pulmonary 258, *258*
  risk groups *256*
  skin biopsy *257*
  staging 256, *256*, *257*
  treatment 258–62
  variants *256*
ketoconazole 192, 224–6, 239
KOH test 7, 32

laboratory facilities 5
Lactobacillus acidophilus 88
lamivudine 194, 196
laparoscopy 163–4
Lentivirinae 37
Leptothrix vaginalis 92
Leptotrichia buccalis 77
lichen planus 73, 75
lichen sclerosus et atrophicus 79
lichen simplex chronicus 76
ligase chain reaction (LCR) 27, 29
line immunoassays (LIA) 39
lipoma 78
liposomes 261
low back pain 181–2
lower genital tract infection *41*
Loxosceles reclusa 78
lumbar pain 181–2
lung cancer 267–8
luteinizing hormone (LH) 179
lymphocele 80
Lymphogranuloma venereum (LGV) 13, 27, 100–1, 116
  clinical manifestations 117
  epidemiology 116
  follow-up 117
  investigations 117
  management 117
  partner notification 117
  pathogenesis 116–17
  secondary stage 117
  tertiary stage 117
  transmission 116
lymphoma 262–5

male
  clinical examination 9
  victims of rape 15–16
  see also homosexual men
malignant tumours 79
melanoma 79
metronidazole 78, 91, 133
miconazole oral gel 225
microimmunofluorescence (MIF) test 29
microsporidiosis 246
minimum inhibitory concentration (MIC) 24
minocycline 129

molluscum contagiosum virus (MCV) 77, 143–4, 191
  aetiology 143
  clinical features 143–4
  diagnosis 143–4
  epidemiology 143
  management 144
  pathogenesis 143
  pathology 143
  treatment 144
Monks Report 59, 60
monoclonal antibodies 202
MTB *232*, 235
Mycobacterium avium complex (MAC) 230–4
  diagnosis 231
  presenting features 231
  prevention *234*
  primary prophylaxis 232
  prophylaxis 234
  screening 231
  treatment 231–2, *231*
Mycobacterium kansasii 234
Mycobacterium tuberculosis 151, 157
Mycoplasma fermentans 30
Mycoplasma genitalium 30, 31, 123–6, 128, 131
Mycoplasma hominis 30
Mycoplasma penetrans 30
Mycoplasma pirum 30
Mycoplasma pneumoniae 30, 31
Mycoplasma primatum 30
Mycoplasma spermatophilum 30
mycoplasmal infections 30
  serological tests 31
myocardial infarction and sexual dysfunction 182

natural killer (NK) cells 140
necrolytic migratory erythema 81
Neisseria gonorrhoeae 7–10, 17, 22–4, 123, 126, 127, 129, 151, 154, 155, 164, 165
  identification 23–4
  see also Gonorrhoeae
Neisseria lactamica 23
Neisseria meningitidis 23, 123
nelfinavir 194, 196, 232
neurological examination 11
neutralizing antibodies 201–2
nevirapine 194, 196
non-gonococcal urethritis (NGU) 123–6
  clinical features 132
  complications 132
  microbiology 130
  persistent and recurrent 129–34
  treatment 130
  treatment of partner 130

non-Hodgkin's lymphoma (NHL) 262–4, *262*
  clinical presentation 262–3
  prognostic factors *263*
  staging 263, *263*
  treatment 263
non-treponemal tests 20
norfloxacin 118, 129
nucleic acid probes 27
Nugent scoring system for vaginal smears **90**
nystatin 225

octreotide acetate 246
oesophageal candida 224–5
opportunistic infections
  AIDS 211–53
  HIV infection 211
oral contraception 162
oral hairy leucoplakia (OHL) 192–3
oral herpes 104
orgasmic dysfunction 176–7
orogenital herpes simplex infection 73
oxytetracycline 129

papaverine 174
papular dermatosis 75
paraphimosis 75
parasitic infection 77–8
paromomycin 86, 246
particle adherence test 39
pelvic examination, bimanual 8
pelvic inflammatory disease (PID) 91, 161–9
  and bacterial vaginosis 162–3
  antimicrobial chemotherapy 165
  clinical diagnosis 163
  definition 161–2
  diagnosis 163
  epidemiology 162
  laparoscopic diagnosis 163–4
  management protocol 166–7
  microbial aetiology 164
  microbiological diagnosis 163
  non-antibiotic therapy 165
  pathogenesis 164–5
  prevention 166
  risk factors 162–3
  treatment 165
  treatment of contacts 165–6
pelvic pain 177–8
pemphigus 73
pemphigus vulgaris 80
penicillin 24, 113, 128, 130
penile papillae 75
penile plethysmography 180

pentamidine 213, 214, 216–19
pentoxifylline 202
perianal region 10
pharyngeal sampling 9
phentolamine 174
phenytoin 228, 239
phimosis 75
pigmented naevus 79
PIPE (pharmacologically induced penile erection) 174
plasma cell balanitis 78
plasmids 116
Pneumocystis carinii pneumonia (PCP) 193, 211, 217
  common presenting symptoms 211
  drug treatments 214–18
  examination 211
  investigations 211–13
  primary and secondary prophylaxis 218
  prophylaxis 214
  treatment 213–18
pneumonia 241
podophyllin 142
podophyllotoxin 142
polymerase chain reaction (PCR) 19–20, 27, 29, 31, 35, 40, 186, 213, 235
polymorphonuclear leucocytes (PMNLs) 123
polyuria 5
postcoital migraine 181
postexertional migraine 181
postgonococcal urethritis 129
  follow-up 132–4, *133*
  management 132–4, *133*
  treatment 132–4, *133*
posttraumatic stress disorder 181
prazocin 154
pregnancy
  and herpes simplex virus (HSV) 108–9
  syphilis in 114
  test 9
premalignant conditions 79
premature ejaculation 172–3
preterm birth 92
primaquine 216
primary CNS lymphoma 264
primary cutaneous diphtheria 77
primary vaginismus 175
prostaglandin E1 174
prostate gland **149**
prostatic symptoms 13
prostatitis 132, 149–59
  acute bacterial (ABP) 150, 152
  aetiology 151–2
  bacterial 150
  chronic 152–4

chronic bacterial (CBP) 150
chronic non-bacterial 150–1
classification 149
diagnosis 149
management 154–5
organisms implicated *150*
symptoms *150*
treatment 154, *154*
prostatodynia 151
  management 154–5
protease inhibitors 239
protozoal diseases 245–6
psoriasis 75, 190
psychosexual problems 171–84, *172*
  associated with STDs 182
  counselling 178
  in men 172
  in women 175–8
psychosexual services 65
pubic lice 14
pyoderma gangrenosum 80
pyogenic granuloma 78
pyrazinamide 236, 237, 239
pyrimethamine 219, 221–3
pyrimethamine sulfadoxine 246

rape, male victims 15–16
rapid plasma reagin (RPR) card test 20
reflex anal dilatation 17
Reiter's disease 80, 98
retarded ejaculation 175
retinoic acid gel 193
reverse transcriptase (RT) assays 39
rifabutin 232, 233
rifampicin 228, 236–8
rimactazid 237
ritonavir 194, 196
rosacea 190

Schistosoma haematobium 78
salmonellosis 247
sampling
  cervix 7, 28
  HIV 38
  pharyngeal 9
  rectum 8
  urethra 8
  urine 28, 29
  vagina 7
saquinavir 194, 196
scabies 77–8
  identification 14
schistosomiasis 78
scrotum 9–10

sebaceous cysts 78
seborrhoeic eczema 190
seborrhoeic wart 78
seminal fluid, detection 15
sensate focus 172, *173*
septrin 214
serological tests 20–1
   chlamydiae 29–30
   effect of treatment 20–1
   granuloma inguinale 27
   HSV 35
   human papillomavirus (HPV) 36–7
   mycoplasmal infections 31
   syphilis (STS) 12, *21*
sexual assault 14–15
sexual dysfunction
   and ageing 181
   assessment 178–80
   history taking 178–9, *178*
   initial tests 179
   neurological investigations 180
   physical examination 179
   radiological investigations 180
sexual health care 1, 2
sexually related headaches 180–2
sexually transmitted diseases (STDs) 69–70, 101–2, 139, 162, 166
   control 1–3, *1*
   identification and significance 17
   in child sexual abuse *17*
   psychosexual problems associated with 182
sexually transmitted genital tract infections in women, recommendations for investigating *41*
shingles 77
skin biopsy 74–5
skin conditions, HIV infection 190–2
skin rashes, drug-related 190
skin reactive disorders 75–6
specimen collection 5, 10–11
spectinomycin 116, 128
spider bite 77–8
squamous cell carcinoma of the anus (SCCA) 266–7, *267*
squamous dermatosis 75
Staphylococcus aureus 144, 151
stavudine 194, 196
Stevens–Johnson syndrome 73, 76
sulfonamides 116
sulphadiazine 215, 216, 221, 223
sulphamethoxazole 214
sulphonamides 116, 154
syphilis 12, 16, 111–15
   clinical presentation 112, *112*
   diagnostic procedures 19–22
   epidemiology 111, *111*
   follow-up 115
   in pregnancy 114
   management 113–14
   partner notification 114–15
   pathogenesis 111
   primary 112–13
   secondary 73, 113
   serological tests (STS) 12, *21*, 99
   transmission 111–12
   treatment 101, *114*

testicular tumours 267
tetracycline 116, 118, 128–30
thalidomide 202, 246
tinea cruris 77
tinidazole 91
topical steroid 73, 74
toxic shock syndrome 92
toxoplasma encephalitis (TE) 219
   clinical presentation 219–20
   differential diagnosis 220
   focal signs 220
   investigations 220–1
   treatment 221
Toxoplasma gondii 219, 224
toxoplasmosis 219–24
   drug treatment 221–3
   primary prophylaxis 223–4
   secondary prophylaxis 223
trauma, physical and sexual 76
Treponema carateum 19
Treponema pallidum 111
Treponema pallidum haemagglutination (TPHA) test 20
treponemal tests 20
trichloroacetic acid 142
Trichomonas foetus 86
Trichomonas vaginalis 10, 17, 83–6, 125
   diagnostic procedures 33
trichomoniasis 83, 85–6
   complications 86
   diagnostic tests 86
   pathology 85–6
   symptoms and signs 86
   treatment 86
Trichophyton 77
trifluridine 245
trimethoprim 116, 214, 216–17
trimoxazole 214
tuberculin skin tests 235
tuberculosis (TB) 235–9
   clinical presentation 235
   diagnosis 235
   drug resistance 236

genital 77
  multiresistant 236–9
  prevention strategies 237
  secondary prophylaxis 236
  treatment 235, 237–9
tumour necrosis factor α (TNFα) 202
tumours in AIDS 255–68

ulceration 97–119
  differential diagnosis *101*
  differential diagnosis algorithm **100**
  examination 99
  history 97
  investigations 99
  management 97, 101
  oral cavity 99
  past medical history 98
  sexual history 98, *98*
  treatment 101
uncomplicated gonococcal infection (UGI) 22
unheated serum reagin (USR) test 20
upper genital tract infection *41*
Ureaplasma urealyticum 30, 89, 123–5, 128, 129, 131, 151
urethra, sampling 8
urethral culture 10
urethral discharge in males 123–38
  aetiology 123–6
  epidemiology 123
urethral smear 10
urethral specimens 28
urethral tests 10
urethritis 123
  aetiology 124, *126*
  and HIV 134
  complications 126
  diagnosis 126–8
  non-gonococcal (NGU) 123–6
  psychological aspects 134
  treatment 128–9
  treatment of partners 129
urinary tract infection 155–7, *156*
urine culture 9
urine samples 28, 29
urine tests 8–11

vaccines 201
vagina
  examination 6
  sampling 7
vaginal candidiasis 224
vaginal discharge 83–95, **83**
  management *84*
  miscellaneous causes 92
vaginal infection *41*
vaginal lactobacillosis 92
vaginal physiology 84–5, *84*
vaginismus 175–6, *176*
  primary 175
  secondary 175
valaciclovir 245
varicella-zoster virus (VZV) 77, 191
vasoactive intestinal peptide (VIP) 174
Veneral Disease Research Laboratory (VDRL) test 20
vestibulitis 65
vinblastine 260
Vincent's fusiform organisms 77
vincristine 260
viral diseases 239–45
viral infection 77
viral transport medium (VTM) 34
viral warts 191
virus culture 39
vitiligo 75
vulva, child abuse 16
vulval clinic 65
vulvitis 77
vulvovaginitis 5

warfarin 228, 239
western blot (WB) assay 39
women
  clinical examination 5–9
  genital examination 6

yohimbine 175

zalcitabine 194, 196, 215, 216
zidovudine 190, 194–7, 199–201, 242
Zoon's balanitis 78